THE GREENING OF CONSERVATIVE AMERICA

THE GREENING OF CONSERVATIVE AMERICA

John R. E. Bliese

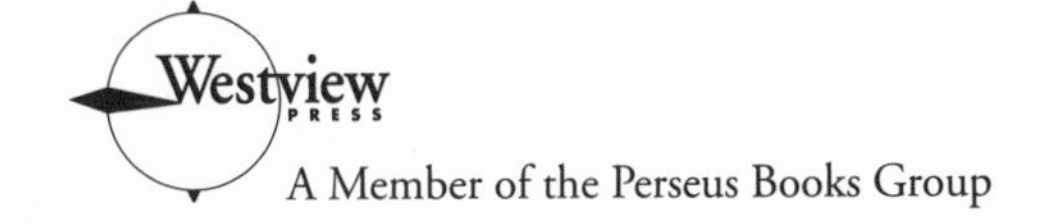

A Member of the Perseus Books Group

Published in 2001 in the United States of America by Westview Press, 5500 Central Avenue, Boulder, Colorado 80301-2877, and in the United Kingdom by Westview Press, 12 Hid's Copse Road, Cumnor Hill, Oxford OX2 9JJ

Find us on the World Wide Web at www.westviewpress.com

Library of Congress Cataloging-in-Publication Data
Bliese, John R.
The greening of conservative America / John R. Bliese.
p. cm.
Includes bibliographical references and index.
ISBN 0-8133-3802-6
1. Environmental policy—United States. 2. Conservatism—United States. I. Title.

GE180.B57 2001
363.7'0525'0973—dc21

00-065430

The paper used in this publication meets the requirements of the American National Standard for Permanence of Paper for Printed Library Materials Z39.48-1984.

10 9 8 7 6 5 4 3 2 1

CONTENTS

PREFACE

Conservatives should be conservationists—that seems logical enough, or so I always thought. As environmental consciousness came to the forefront in the late 1960s and early 1970s, the Republican Party seemed to be generally on the right track, building on the heritage of Theodore Roosevelt. Most of the landmark environmental laws were passed with broad bipartisan support, and many were signed into law during the Nixon and Ford administrations.

Just a few years later, however, the newly elected Ronald Reagan made several terrible appointments to key environmental positions. That was not what I had in mind. And that is when I, and many other Republicans, began joining the Sierra Club and other leading environmental organizations. I could at least show support for the environment by joining and making an occasional donation.

Then came the near disaster of 1995, which occurred after the Republicans gained control of Congress. The situation was most dire in the House. Even though the Contract with America did not mention the environment, the House leadership launched an unprecedented attack on virtually every environmental law on the books. There was no attempt at all to improve these laws, only to weaken them.

This was clearly *not* what they had been elected to do—poll after poll showed that Republican voters *opposed* their party on environmental policy and overwhelmingly favored *stronger* environmental laws.

Voters were clearly right and the Republican leaders were clearly wrong. Although they claimed to be conservatives, their attacks on the environment were, in fact, the very opposite of conservatism. They were violating some of the most important principles of the conservative political philosophy.

There was—and still is—a huge discrepancy between the anti-environmental agenda of many Republican politicians and the opinions of Republican voters. And there is an equally huge discrepancy between the conservative political philosophy and the anti-environmental antics of these politicians, despite the fact that they call themselves conservatives.

But what, then, *should* be the conservative approach to environmental policies? What conservative principles are important here? What kinds of policies do those principles suggest as effective approaches to protecting our environment, conserving our resources, and preserving our natural heritage? When I searched,

I found very little in the literature that even considers those questions. So, I decided to try to develop an answer. I wrote a series of articles on a conservative environmental agenda, and this book is the final result.

The public debate on the environment includes several attacks on environmentalists that are now standard fare: that environmentalists are opposed to our capitalistic system, that they are nothing more than prophets of gloom and doom, and that protecting the environment is harmful to the economy. The only problem with these charges is that they are dead wrong. A fundamental assumption about conservatism is that it is supposed to back up whatever the business community wants. But the principles of the conservative political philosophy show clearly that it is *not* a cover for greed and irresponsibility and it does *not* justify victimizing the public (e.q., with polluted air and water) to maximize profits.

The works of the great scholars and thinkers who developed the conservative political philosophy for our time include several fundamental principles that are relevant to environmental policy issues—and they all support environmental protection. They often suggest policies that are rather different from the ones the liberals and bureaucrats have given us. But the policies they support would be *effective* in protecting the environment, in conserving our natural resources, and in preserving nature.

In this book you will find a refutation of the fallacies and misconceptions that play a major role in the anti-environmental rhetoric. You will find a discussion of several principles of conservatism that are important for environmental policy. And you will find an application of those principles to several major issues, to see what kinds of policies the conservative philosophy might support.

The policies presented here are meant to suggest illustrations of the kinds of thinking conservatives should be doing on environmental issues. I am not, in other words, making any attempt to solve all of our environmental problems. I am only attempting to demonstrate that there *can* be and *should* be a *positive* conservative environmental policy agenda. As Russel Kirk, one of our "founding fathers," once wrote, "Nothing is more conservative than conservation."

John R. E. Bliese

Introduction

"CONSERVATIVE ENVIRONMENTALIST" is not an oxymoron. That it is now widely believed to be a contradiction in terms is the reason for this book.

Conservatism is a venerable political philosophy that was developed for our time by such sterling minds and principled personalities as Russell Kirk, Richard Weaver, Frank Meyer, Friedrich Hayek, Leo Strauss, and Willmoore Kendall, to name just a few of the earlier authors. These scholars were not themselves "environmentalists" in today's sense, for most of their works were produced in the years before our environmental problems became acute. But if you read their works, you will find that the principles of the conservative political philosophy that they developed clearly support environmental protection. That is, of course, what you would expect. Conservatives ought to be in favor of conserving things: our cultural heritage, our civilization, our basic political and social institutions that established our freedom—as well as our natural heritage, our natural resources, and our planet.

Unfortunately, over the past thirty years and more, conservatives have contributed very little that is positive to the analysis of our environmental problems or to the formation of environmental policies. Conservative scholars have taken virtually no part in the public debate over these issues. A search through the contents page of, for example, *Modern Age* or *Intercollegiate Review* will yield virtually nothing on conservatism and the environment. (Both journals recently published articles of mine, so there is apparently no reluctance on the part of the editors to cover the environment. But conservative scholars have produced almost nothing on the subject.) The one exception at the scholarly level is a small group of libertarian economists who advocate "free market environmentalism." Much of this work has been done by two organizations headquartered in Bozeman, Montana: the Political Economy Research Center (PERC) and the Foundation for Research on Economics and the Environment (FREE).

At a more journalistic level, the Rockford Institute supports environmental protection in general, but it ranks very low among the institute's priorities. Its journal, *Chronicles*, on rare occasions carries an article on the environment, but it also gives space to anti-environmentalists. *National Review* comments on practically every subject under the sun, but on the environment you will find virtually nothing beyond a few shallow pieces that merely attack some fringe group of "eco-freaks" or give a forum to some maverick who denies the general scientific consensus on an environmental problem. The one exception at this level is the Heritage Foundation. Although environmental issues are a low priority among its concerns, over the years *Policy Review* has carried a number of interesting and positive articles.

If conservative scholars have largely ignored the environment, so, for a long time, did conservative politicians. That attitude has left a general impression that conservatism is at best completely unconcerned about environmental problems. Thus, from the beginning, conservatives have basically abdicated on environmental policy, leaving the field to liberals and bureaucrats.

Although conservative scholars have continued to ignore the environment in recent years, many politicians who call themselves conservatives most certainly have not. Through the late 1980s and early 1990s, many of these politicians became increasingly hostile to any and all attempts to protect our environment or to conserve our natural resources. When the Republicans took over the 104th Congress (in 1995), they launched an unprecedented broadside attack on virtually every environmental protection law on the books—laws that had originally been adopted with broad bipartisan support.[1] On the basis of conservative principles, many of these laws should be reformed, but that is not what the Republican leaders did. They proposed virtually nothing to make environmental protection better; they wanted only to make it weaker. Our natural resources also came under ever greater assault. The resource committees in both the House and the Senate became dominated by politicians who treated the United States like a banana republic—plundering our land to benefit their campaign contributors.

The attack on the environment was so virulent that one Republican voter wrote in a letter to the editor of the *Seattle Times*: "Please give me back the Democrats! They only wanted to take my money to give to their favored constituents. These Republicans want to take my health, my quality of life, and the natural heritage my children and I might enjoy."[2] So now the impression given to the public is that conservatism is opposed to environmental protection. But there is, as we will see, *nothing* conservative about this.

The worst excesses of these newly powerful Republicans in the 104th Congress were eventually blocked by a combination of several factors: reenergized environmental groups (that delivered a petition to Congress bearing over a million signatures); a group of moderate Republicans, especially in the House, who were increasingly willing to break with their leadership; a Democratic administration that eventually regained its morale and its willingness to confront congressional excesses with its veto power; and, presumably, a series of polls that showed a majority of Republicans opposed their own party on the environment.

In the 105th and 106th Congresses, the conservatives for the most part turned their attention elsewhere. Their attitude toward environmental protection remained basically unreformed, but some of the worst anti-environmentalists had been defeated and some who were reelected had their margins of victory severely reduced. They generally tried only stealth attacks on the environment in the form of riders on appropriation bills, almost all of which were eventually dropped.

If many Republican politicians are guilty of what Theodore Roosevelt IV has called "a betrayal of the very definition of conservative," the pundits and radio entertainers are no better.[3] They have disastrously muddied the waters of public discourse by distorting conservatism, and often the facts as well, when dealing with environmental issues.

Republican voters, on the other hand, who generally think of themselves as conservatives, have always been strongly in favor of environmental protection. Polls have consistently shown that Republicans think the environment is very important and that they do not trust their own party to protect it.[4] But party officials have not even attempted to follow their lead and offer a positive policy agenda. They still give the erroneous impression that conservatives are anti-environmentalists. The Republican voters are on the right track and their politicians are not.

In sum, there is an enormous discrepancy today between the ideas and principles of the conservative political philosophy, on the one hand, and the antics and opportunism of many conservative politicians and pundits, on the other. The near disaster of the 104th Congress should never have happened. *Principled* conservative politicians would have proposed conservative *solutions* to environmental problems, not the assault on environmental protection that we in fact got.

Too many conservative politicians have made a bad choice—to serve the industries that contribute to their campaigns rather than to offer conservative policies to protect the environment. This is part of a more general development, which Russell Kirk saw happening as early as 1989: "The fundamental division among conservative groups today, it seems to me, is the gulf fixed between (on the one side) all those conservative men and women who, taking long views, argue that intellectual activity and rousing of the imagination are required urgently; and (on the other side of the canyon) all those professedly 'pragmatic' persons who think of a conservative government as one that holds office through placating or serving certain powerful interests." Kirk warns that such "time-servers" or "placemen" cannot govern well. He calls on conservatives to steer clear of "opportunism" of just the sort we are dealing with today.[5] This opportunism is actually an abdication of important principles.

A few conservatives have warned that what Kirk was referring to is now happening on environmental issues. According to John C. Vinson Jr., "much that passes for conservatism today is more concerned with cash than character, possessions more than posterity. Its only close attachment to the land is on the fairway and the putting green at the country club. Environmental activists rightly despise this type of conservatism, but the reason it is despicable is that it really isn't conservative."[6]

Likewise, Paul Weyrich is concerned about the "widespread public perception . . . that conservatives are out to rape the environment." Unfortunately, "there are a good many people who call themselves conservatives who are guilty of holding [this] attitude. Some hold [it] because they benefit personally from [it], because they make money by destroying the environment. . . . Until we make it clear that conservatism is not just a convenient fig leaf to plaster over greed, we will fairly get saddled with bad reputations in [this] area. . . . We, the defenders of traditional Western values, are made to look like violators of those values and therefore as hypocrites because of these perceptions."[7]

In times like these we need to return to the wisdom of Richard Weaver, who advises conservatives to argue and act from principle. If they abandon principles that establish their proper mode of thinking and act as opportunists, they will govern no better than the liberals.[8]

The negative approach to these issues taken by leading conservative politicians hides the fact that there *can* be, and *should* be, a positive environmental agenda for conservatives. What, we must ask, would a principled conservative do to solve our environmental problems? What would a conservative environmental policy agenda look like? This book is an attempt to answer those questions or at least suggest some possible answers. I will analyze several relevant principles of the conservative political philosophy and then look at several major environmental issues and consider what kinds of policies those principles would suggest.[9]

First, however, I will have to dispose of some widely disseminated misconceptions and myths. There are several erroneous ideas about environmentalism and environmentalists that must be cleared up. There is also a major misconception about conservatism that stands in the way and can block serious analysis of policy. I address these issues in Chapters 1 and 2.

"Conservatism" is perhaps the single most abused word in our political vocabulary so I will offer a short explanation of what I mean by conservatism as the source of the principles that I will consider in Chapter 3. (The principles themselves will fill in some of the details.)

Conservatism, as it developed in America, includes two primary schools of thought: free market libertarianism, which is based on the heritage of Adam Smith, and traditionalism, whose intellectual roots go back to Edmund Burke.[10]

Libertarians stress freedom above all else, at least in the political and economic realms. They generally want a government with little but police and military power to protect the citizens and judicial power to try criminals, enforce contracts, and judge tort cases. They believe that a free market is the best way to maximize human creative and productive potential. Generally, they are little concerned—at least in their political or economic theory—with the values and goals of individuals or society.[11] Prominent names in this school include Milton Friedman, Murray Rothbard, and Tibor Machan.

Traditionalists, on the other hand, are concerned above all with the health of our culture and the values of our society. They show little interest in economics. Their overriding concern in our time has been with the restoration of Western civilization and the reestablishment of community, which they believe have suffered severe blows in our materialistic mass society.[12] Among the leading traditionalists are Russell Kirk, Richard Weaver, and John Gray.

This libertarian-traditionalist split is widely recognized in conservative scholarship and has generated much debate over the years.[13] The tension between these two schools has provided much of the dynamism in conservative thought.

Early in this debate, Frank Meyer contended that the two sides were not arguing mutually exclusive positions but rather were emphasizing two different aspects of a shared philosophical heritage. He argued that these two trends of thought belong together—in rather complicated ways, to be sure, and in ways that will never result in perfect agreement, but belong together they do. His argument is contained in a marvelous article, "Freedom, Tradition, Conservatism," that originally appeared in 1960 in *Modern Age*. It is still the best short description of conservatism I know.

Meyer believes that both libertarians and traditionalists draw from the same heritage of Western civilization.

> Their opposition . . . is essentially a division between those who abstract from the corpus of Western belief its stress upon freedom and upon the innate importance of the individual person (what we may call the libertarian position) and those who—drawing upon the same source—stress value and virtue and order (what we may call the traditionalist position). . . . In fact the two positions which confront each other today in American conservative discourse both implicitly accept, to a large degree, the ends of the other. . . . On neither side is there a purposeful, philosophically founded rejection of the ends the other side proclaims. Rather, each side emphasizes so strongly the aspect of the great tradition of the West which it sees as decisive, that distortion sets in . . . and the complementary interdependence of freedom and virtue, of the individual person and political order, is forgotten. Their opposition is not irreconcilable, precisely because they do in fact jointly possess that very heritage. Extremists on one side may be undisturbed by the danger of the recrudescence of authoritarian status society if only it would enforce the doctrines in which they believe. Extremists on the other side may care little what becomes of ultimate values if only political and economic individualism prevails. But both extremes are self-defeating: truth withers when freedom dies, however righteous the authority that kills it; and free individualism uninformed by moral value rots at its core and soon brings about conditions that pave the way for surrender to tyranny. Such extremes,

> however, are not the necessary outcome of a dialectic between doctrines which emphasize opposite sides of the same truth.[14]

Meyer's article, as Edwin Feulner says, "achieved an intellectual *tour de force* which redefined the course of modern conservatism."[15] His position became known by the unlovely term "fusionism" and it rather quickly won the day. Although Meyer's essay was attacked by both sides when it first appeared, according to George Nash, "as the dust settled, many conservatives began to make a common discovery: that Meyer's fusionism had won."[16] It has now, Feulner claims, "become widely accepted in the conservative movement."[17]

There are, in fact, relatively few pure libertarians or pure traditionalists. Most conservative intellectuals combine a great deal from both schools in their own thinking, after the manner of William F. Buckley Jr., who, Meyer himself claims, "personified fusionism."[18]

This is not the place to try to extend the debate, which continues to this day, nor even to summarize the issues. It should suffice for our purposes at this point to add that Adam Smith and Edmund Burke were friends in their day and found themselves in general agreement.[19] For the most part, the same is true of their heirs today. Libertarians, traditionalists, and fusionists have long made common cause, at least at the level of policy analysis, even if they do not completely agree on finer points of political philosophy.

Conservatism thus is not a monolithic ideology. (Indeed, conservatives insist that it is not an ideology at all.) Rather, it is a spectrum of political and social thought that extends from the libertarian to the traditionalist, with most conservative intellectuals clustered around the "fusionist" middle of the range.

This range of thought encompasses at least nine principles (discussed in Chapter 3) that are relevant for confronting environmental issues:

1) Conservatism is not materialistic.
2) Freedom of the individual is extremely important.
3) But it has an ever present corollary, which is responsibility.

For people to exercise their freedom responsibly, conservatives support two important social institutions:

4) Private property
5) The free market

And in the responsible exercise of our freedom, three principles should govern our thinking and our actions and our policies:

6) Piety, especially piety toward nature.
7) The realization that society is intergenerational.

8) The realization that prudence is the most important political virtue.
Finally,
9) Conservatives are not ideologues; conservatism is not an ideology.

This is not, of course, a complete list of conservative principles—it is not my purpose to give a complete description of conservatism. I have selected these principles because they have important implications for environmental policy and because I believe that they are accepted, with greater or lesser emphasis, by conservatives across the spectrum.

These principles often imply environmental policies that are different from the ones liberals and bureaucrats have given us. They should form the basis or foundation for a *conservative* environmental policy agenda. When we look at specific environmental problems and issues later in this book, I will in each case try to indicate the lines of analysis implied or suggested by the relevant principles.

I make no attempt to present solutions to every environmental and resource problem. I could not do that even if I wanted to. The problems and policies considered here are limited to the United States. I do not deal with the environmental problems of the third world, not even with the ones that get considerable publicity, such as destruction of the tropical rain forests. Nor do I consider the environmental issues raised by international trade, for example, the North American Free Trade Agreement (NAFTA) and the World Trade Organization (WTO). Also, there are many U.S. issues that I leave to others, such as the environmental impact of immigration and increasing population, environmental justice, nuclear waste storage, indoor pollution, and preserving the great forests of the Northeast, to list just a few.

That will still leave a rather full plate: air and water pollution, hazardous waste disposal, resources on public lands, global climate change, loss of biodiversity, perverse subsidies, and some questions relating to sustainability. A consideration of these problems should be quite sufficient to indicate the kind of thinking principled conservatives should be doing on other environmental and resource issues as well.

In other words, I am not trying to present a complete conservative solution to every single environmental and resource problem we face. My purpose, rather, is to present a conservative environmental agenda that will be *positive*, for a change, and to suggest the directions in which conservatives should think in developing policies designed to solve the many real environmental problems we face. Regarding conservatives who are not closely involved with environmental issues but may be uncomfortable with the antics of some Republican politicians, I hope to show them that they are on the right track and the politicians are not. And I hope that this book will let environmentalists see that their conservative opponents are often violating the very principles they should be upholding. This should give environmentalists some ammunition to use for publicly humiliating certain politicians and pundits who richly deserve it.

Once before, during the Reagan administration, conservatives had an opportunity to take the lead and move environmental policy in better and more effective directions, but they did not even try. The Heritage Foundation summarizes their sorry record:

> The Reagan Administration . . . had a rare opportunity to offer an alternative policy model for environmental controls: to place greater emphasis on developing property rights and on other solutions using market incentives. It has failed to do so and has provided absolutely no vision for a tough environmental policy based on market mechanisms. . . . Although the Reagan Administration extolled the virtues of a free market, it has not offered a positive, market-oriented approach to environmental protection as an alternative to the command and control approach that it inherited. Instead of providing a different vision, it has appeared simply to offer lower budgets, federal inaction, and fewer restrictions on polluting industries.[20]

As we enter the twenty-first century, with the general shift of American politics to the right, conservatives have yet another chance. So far, they have let it go to waste and allowed the opportunists to take over. This must change. The task of conservatives in the new century is to reaffirm their principles and apply them in solving environmental problems and in preserving our natural heritage as careful stewards of the earth and as trustees for future generations.

When the environmental movement was first beginning, Russell Kirk heartily and rightly approved: "Nothing is more conservative than conservation."[21] Today, Gordon Durnil exhorts his conservative colleagues: "We should be leading this parade!"[22] Yes, we should.

And with that, let us try to clear up some of the myths that are muddying the waters of public discourse.

The literature on conservatism and the environmental issues addressed in this book is vast, and I have made no attempt to give anything close to a comprehensive list of sources for any given point. I cite only representative sources or the best ones known to me. At the same time, I cite supporting material for all points in order to do what I can to raise the evidentiary standards in a debate where they have become deplorably low.

About conservatism and about environmental problems there is, unfortunately, little to be learned from politicians or the press, and thus I have largely avoided both. For conservatism, you must turn to the scholars and thinkers who developed the principles of this political philosophy for our time. For environmental problems, you must turn to the scientists who study them and to the legal scholars, policy analysts, and economists who attempt to devise ways to deal with them. And that is where I have concentrated my research.

1

Misconceptions About Environmentalism and Conservatism

ANY CONSIDERATION OF CONSERVATIVE environmental policies must begin, unfortunately, with a survey of several widespread misconceptions about environmentalism and environmentalists. There seems to be a general belief that conservatives are the natural enemies of environmentalists and that conservatism is opposed to all but the most insignificant measures to preserve nature or to protect the environment. All too frequently charges are made (1) that environmentalists are leftists who are out to destroy the American capitalist system, (2) that environmentalists are tree-hugging pagan nature worshipers and that environmentalism is contrary to the Christian religion, (3) that environmentalists are mere prophets of gloom and doom who do not deserve to be taken seriously. Another fallacy that is of crucial importance is actually a misconception about conservatism itself: (4) that conservatives are, above all else, supposed to be "for business" and should support anything the business community wants. Finally, also of crucial importance, is the persistent but erroneous belief (5) that there is a trade-off between the environment and a healthy economy—that protecting the environment hurts the economy.

These fallacious ideas are expressed frequently everywhere, from articles in conservative magazines to comments by politicians and pundits, from conversations with people who are highly educated but not necessarily highly political to journalists' interviews with the "average person on the street." Some of these misconceptions seem to have reached a level of basic assumption among much of the general public. The persistence of these misguided notions is bad for both the environment and conservatism.

In dealing with myths and misconceptions about environmentalism, I assume that the reader is aware of these charges because they are so widespread. I do not refer to specific erroneous claims about environmentalism made by particular individuals for two reasons. First, to cite specific statements would involve a nearly endless task of refutation. To do justice to each and every statement and to be fair to their authors would require tracking down their sources, considering the context in which they made each claim, and refuting in detail all errors of fact as well as all misguided or fallacious reasoning. That would certainly be worth doing, and it would make an important contribution to the public debate. But to do it properly would require a small army of research assistants and would have to be a separate book, perhaps several volumes. Worthy attempts have been made, and I can strongly recommend two books in particular: Paul and Anne Ehrlich's *Betrayal of Science and Reason* and Edward Flattau's *Tracking the Charlatans.*[1] But that is not my task here.

Second, my primary purpose here is to be *positive*, to the greatest extent possible. I want to outline what an *affirmative* conservative environmental agenda would look like, an agenda that would attempt to *solve* major environmental problems rather than merely deny their existence or deflect public attention from them.

"ENVIRONMENTALISTS ARE ANTICAPITALISTS AND LEFTISTS"

Anti-environmentalists have spread far and wide the charge that environmentalists are leftists, socialists, or worse, who are out to destroy the capitalist system. This is simply nonsense. To the extent that the accusation is a misconception and not simply demagoguery, it probably arose because much of our pollution and environmental degradation is caused by businesses and industries. Fearing that their profit margins would be reduced if they had to act responsibly, those businesses have adamantly opposed even the mildest of measures to clean up the air and water and land. Many individual companies and business organizations, therefore, became the enemies of environmentalists.

But the environmental movement in the United States, in general, has never opposed the system of capitalism. The environmental movement in the United States, unlike some Green parties in Europe, has never had a broadly socialistic agenda. Environmentalists in this country are simply people who are concerned more or less actively about some or all aspects of the degradation of our planet. Insofar as they are united at all, they are united only on environmental issues; the movement as such has no wider agenda.[2]

Moreover, there is nothing about environmental concerns that is inherently leftist or anticapitalist. As Frances Cairncross of the *Economist* observes, "The environment is an issue without any obvious political home."[3] People of *any* political stance can be environmentalists. Russell Kirk, one of the "founding fathers" of contemporary conservatism, notes that "the issue of environmental quality is one which transcends traditional political boundaries. It is a cause which can attract, and very sincerely, liberals, conservatives, radicals, reactionaries, freaks, and middle-class straights."[4]

Indeed, some have claimed that protecting the environment actually has its natural home in the conservative political camp. British political philosopher John Gray contends: "Far from having a natural home on the Left, concern for the integrity of the common environment, human as well as ecological, is most in harmony with the outlook of traditional conservatism of the British and European varieties."[5] And Russell Kirk claims that "nothing is more conservative than conservation."[6]

Clearly, in terms of political doctrine there is nothing inherently left-wing about environmentalism. When we look at environmentalists in America, we also find empirically that this myth is simply false: environmentalists are *not* a left-wing group. Individuals who identify themselves as environmentalists come from all over the political spectrum. A Roper poll that ranked Americans on their environmental awareness and commitment found that the most active environmentalists contained the highest share of conservatives as well as of liberals.[7] Gordon Durnil observes, from his experience speaking with many local groups, that

> most of the people worried about an environmental problem are not motivated by political philosophy. They are motivated by a specific environmental problem in their community and by the desire to correct it. . . . They are not organized by the large environmental groups; in fact, they are hardly organized at all. . . . The environmental movement I observed is made up mostly of modest groups of ten or twenty people, seeking information from some level of government and getting none, or seeking redress from the local plant and being ignored, or being told by some authority that they are off base and just don't know what they are talking about.[8]

Moreover, there are now a number of libertarians who call themselves "free market environmentalists," and they are most certainly not opposed to capitalism.[9]

Examining the policies of the major environmental organizations in the United States reveals that they are *not* socialistic or anticapitalistic or radical leftist of any sort. The major environmental groups are clearly in the mainstream of the American political tradition. After all, they are made up, notoriously, almost exclusively of white, upper-middle-class professionals—hardly a group that seriously opposes our economic system!

It is true, of course, that environmentalists and the major environmental organizations tend to support Democrats—but for that the GOP has only itself to blame. It is reported that on the first Earth Day in 1970, Richard Nixon looked out a White House window at the thousands of well-dressed environmentalists and said, "Those are Republicans."[10] If Republicans had joined the environmental organizations from the first, they could have directed their policies in conservative, and more effective, directions.

Instead, the Republicans far too often have chosen to defend to the hilt those who are polluting our air and water and land and destroying our forests and streams. And this tendency has become much worse since the late 1980s. (This decision was probably based on the huge misconception that conservatives are supposed to be "for business.") This forced the environmentalists to turn to the Democrats, who gave them support.

Environmentalists have often looked to the federal government for action because some problems are national in scope, but also because they were blocked at lower levels of government by entrenched special interest politics. Ed Marston, an environmentalist and publisher of *High Country News*, explains what happened in an exchange with Chilton Williamson, who is one of those who accuse environmentalists of being big-government, anticapitalist leftists.

> Those who say that environmentalists use resource issues simply to impose progressive [i.e., big government] values have it backwards. Environmentalists care first and last about the land. But the only tools they know to protect the land with are so-called progressive tools: regulation through a strong federal government, lots of paper-driven process, lawsuits in federal courts, and so on. As environmentalism has matured, we have begun to see the possibilities of market-based solutions, local control, and the like. Our fear in abandoning big government doesn't come out of a love of big government. It comes out of the fear that without big government, we will be at the mercy of those who would destroy the West. . . . The Wise Users are engaging in Red Baiting against environmentalists in order to avoid the issue that created environmentalism: the degraded state of the West as a natural place.[11]

What Marston is saying applies equally well to environmentalists and their opponents in the rest of the country.

Of course, there are some fringe groups from the far left that have used pollution as a weapon for bashing capitalism, but as conservative Republican environmentalist Gordon Durnil says, "So what?"[12] To focus on them and their absurdities as a reason to reject environmentalism is exactly like looking only at white supremacists and paranoid militias and concluding that all of conservatism is ridiculous and should be rejected. Conservatives actually should sympathize with environmentalists on this score, since they have long been embarrassed by the lunacies of the radical right.[13]

Besides, with the collapse of communism in eastern Europe and the former Soviet Union everyone now can see that pollution is not merely the result of capitalism. The socialists have been far worse on all counts, and environmentalists were quick to criticize them too. It is time to put aside this nonsensical generalization that environmentalists are anticapitalists and get on with the important work of solving the very real problems our modern industries and lifestyles have created for all of us.

"ENVIRONMENTALIST ARE PAGAN NATURE WORSHIPERS; CHRISTIANS CANNOT BE ENVIRONMENTALISTS"

A second misconception, spread by some politicians and radio talk shows, is that environmentalists are tree-hugging pagan nature worshipers or New Age crystal gazers and that environmentalism is contrary to the Christian religion. This is even more absurd than the previous error, and it simply shows the accusers' profound ignorance of both environmentalism and Christianity. But the accusation is sufficiently widespread that we need to deal with it here.

Some fringe groups are pretty strange. But the membership of all of the large, national environmental organizations notoriously consists of, to repeat, white, upper-middle-class professionals—hardly a social group that tends to exhibit spiritual nonconformity. And in fact our Judeo-Christian heritage provides a solid foundation for protecting the environment and conserving natural resources. We can dispose of this charge at two levels—empirical and doctrinal. Finally, I want to examine a historical question.

In the first place, there are, in fact, many Christians and Jews who *are* environmentalists. And many churches and church organizations have actively addressed environmental issues. In 1993 an umbrella group was formed—the National Religious Partnership for the Environment. Its member organizations are the U.S. Catholic Conference, the Evangelical Environmental Network, the National Council of Churches of Christ, and the Coalition on the Environment and Jewish Life. This environmental organization serves faith groups with a combined membership of some 100 million souls.[14] It represents most of the major Christian denominations, across the spectrum from evangelical to mainline Protestant to Roman Catholic. Its mission is to "act in faith to cherish and protect God's creation. Our goal is to integrate commitment to global sustainability and environmental justice permanently into all aspects of religious life."[15] In 1994 its member organizations developed their own educational materials and sent them to some 140,000 congregations nationwide.[16]

The National Council of Churches has taken a very active role recently on the problem of global warming, calling it a "religious issue." The NCC is coordinating an "interfaith global warming strategy" to try to persuade the Senate to ratify the Kyoto treaty.[17] A coalition of evangelical Christians in 1996 organized an intensive lobbying campaign that played a key role in blocking attempts in Congress to weaken the Endangered Species Act.[18] The Catholic Church adopted a strong environmental stand in the U.S. Bishops' Statement, *Renewing the Earth: An Invitation to Reflection and Action on the Environment in Light of Catholic Social Teaching*. Following up on the Bishops' Statement, the U.S. Catholic Conference has distributed environmental action materials to its parishes.[19] The Central Conference of American Rabbis has proclaimed that protecting the Headwaters Forest in California is "part of the covenant with the Creator" and has asked Congress to take strong action to protect endangered species.[20] Clearly, the leadership of many Christian denominations in the United States not only believes that it is *possible* for Christians to be environmentalists but that Christians have a *religious obligation* to be environmentalists. Likewise the Jewish community in the United States.

In terms of doctrine there is, of course, plenty of material in Christianity and Judaism to support protection of the planet from degradation. Over the last twenty years or so, theologians have devoted an increasing amount of attention to developing a theology of ecology.[21] No attempt can be made here to present such a theology or even summarize the major doctrines. But it is important to note a couple of points.

The Judeo-Christian approach to ecology differs from some others. It is not anthropocentric, as are some of the pragmatic or instrumental approaches to valuing environmental protection. For example, "we should preserve biodiversity because in the future we may discover potential uses for these currently unknown creatures and plants." This argument is, of course, entirely acceptable from the Judeo-Christian perspective, but the religious doctrine goes much deeper than this. Nor is it biocentric, as in Deep Ecology. "All creatures have equal intrinsic worth." From a Judeo-Christian perspective, this claim is not acceptable without serious qualifications.

A Judeo-Christian approach to ecology, rather, is entirely *theocentric*.[22] "The earth is the Lord's and the fulness thereof, the world and those who dwell therein."[23] It is a central teaching of Christianity "that all creation, including humankind, is created by God for God's use" and is therefore valuable.[24] Similarly in Judaism. "The fact that God is Creator endows all of creation with an intrinsic significance and importance." The Talmud observes: "Blessed be He created in His world, He created nothing in vain." Nothing in creation is useless or expendable; everything manifests some divine purpose. It follows, therefore, that there is a divine interest in maintaining the natural order of the universe.[25] Consequently, as Wendell Berry, a Christian and an environmentalist, says, "our destruction of nature is not just bad stewardship, or stupid economics, or a betrayal of family responsibility; it is the most horrid blasphemy. It is flinging God's gifts into His face, as if they were of no worth beyond that assigned to them by our destruction of them."[26]

The world does not belong to us; it belongs to God. Nor should we assume that the world was created on our account and simply for our purposes. C. S. Lewis believed that we come to know the greatness of God because of the "hint furnished by the greatness of the material universe." But he adds, "I hope you do not think I am suggesting that God made the spiral nebulae solely or chiefly in order to give me the experience of awe and bewilderment. I have not the faintest idea why He made them; on the whole, I think it would be rather surprising if I had. As far as I understand the matter, Christianity is not wedded to an anthropocentric view of the universe as a whole. . . . It is, of course, the essence of Christianity that God loves man and for his sake became man and died. But that does not prove that man is the sole end of nature."[27]

As far as environmental policy is concerned, the key concept from our Judeo-Christian tradition is, of course, that of stewardship. "Human beings are not to worship nature, but neither have they been given a blank check to do whatever they please with it. As the stewards or custodians rather than the owners of creation, they are to care for it and 'guard' it."[28] We may be at the top of the hierarchy of creatures on this earth, but we are not its owners. We are allowed to use it as trustees, not as tyrants, and we certainly do not have any license to trash the place; quite the contrary, since God cares for all of his creation. As Pope John Paul II proclaimed, "due respect for nature" is a moral responsibility owed to God.[29]

Chuck Barlow considers a passage from the Gospel of Luke: "Are not five sparrows sold for two pennies? And not one of them is forgotten before God. Why, even the hairs of your head are all numbered. Fear not; you are of more value than many sparrows."[30] He finds there "the three principles of theocentricity . . . the value of nature, shown through God's constant care—He does not *forget* the sparrows—and knowledge of its disposition; the hierarchy of value, which allows nature to be used by humankind in proper service to God; and the ultimate task of all creation to serve God. . . . The crucial corollary to these principles is the principle that God holds humankind accountable for its use of nature."[31]

The Greek Orthodox Church has even declared environmental degradation to be a sin. In 1995 it convened an international conference on science, religion, and the environment. After the conference, the church "expanded its definition of sin to include sins against nature, such as forcing species into extinction or degrading the atmosphere."[32] (As we will see, one of the founders of contemporary conservatism some fifty years ago called our degradation of nature a sin.)

A key word from the Scriptures that has often been misinterpreted is the term "dominion." "Then God said, 'Let us make man in our image, after our likeness; and let them have dominion over the fish of the sea, and over the birds of the air, and over the cattle, and over all the earth, and over every creeping thing that creeps upon the earth.'"[33] This passage has been used to justify doing whatever we wish with the planet, no matter how much damage is caused to it.

But "dominion or mastery does not imply tyranny and spoliation; it is possible to rule wisely."[34] We are still not the owners of creation. We must exercise our dominion in the way in which a trusted servant would manage the property of his or her master. "'Man has dominion over the "lower" orders of creation, but he is not sovereign over them. Only God is the sovereign Lord, and the lower orders are to be used with this truth in mind. Man is not using his own possessions.' . . . Just as an owner might evict irresponsible tenants, so God would be displeased by a failure to care properly for the earth, for his creation."[35]

Indeed, the word that is translated "dominion" apparently does not have any connotations that would allow tyrannizing over subject people or destroying property. Lloyd Steffen, in an analysis of the biblical term, finds "no justification for equating dominion with domination. In fact . . . dominion originally served justice, reflected relations of intimacy and peaceableness, and conformed to [what we today would call] an ethic of ecological responsibility."[36]

Overall, as James Nash notes, "During the last three decades, a growing number of scholars, from 'progressive evangelicals' . . . to process theologians . . . have been discovering or rediscovering the ecological potential in Christian convictions. In its central affirmations, not simply in its peripheral elements, the Christian faith offers a solid, ultimate grounding for a strong environmental ethic."[37]

Clearly, many Christians and Jews are in fact environmentalists, and the doctrines of Christianity and Judaism provide a solid foundation for environmental protection. But there is a third dimension to the erroneous belief that environmentalism and Christianity are incompatible and that is a historical one. Some have argued that the way in which western Christianity developed has led to many of our ecological problems today. The most famous instance is an essay entitled "The Historical Roots of Our Ecologic Crisis," by the noted medieval historian Lynn White Jr.

White traces the origins of modern science and technology to the central Middle Ages. "The distinctive Western tradition of science, in fact, began in the late 11th century. . . . Before the 11th century, science scarcely existed in the Latin West, even in Roman times. From the 11th century onward, the scientific sector of Occidental culture has increased in a steady crescendo." White finds the (separate) origin of modern technology in the late tenth century when men in western Europe began to apply water power to industrial processes other than milling grain. In the twelfth century, men learned how to harness wind power. From this point, "the West rapidly expanded its skills in the development of power machinery, labor-saving devices, and automation."[38]

White claims that this development, culminating in the modern phenomenon of enormous technological power, was possible because western Christianity taught that nature and all the creatures and things in it existed only "to serve man's purposes." "Man shares, in great measure, God's transcendence of nature. Christianity . . . insisted that it is God's will that man exploit nature for his proper ends. . . . By destroying pagan animism, Christianity made it possible to exploit nature in a mood of indifference to the feelings of natural objects. . . . Man's effective monopoly on spirit in this world was confirmed, and the old inhibitions to the exploitation of nature crumbled." As a result, the Western world believes that people are "superior to nature," and so we have become "contemptuous of it, willing to use it for our slightest whim." Out of that attitude grew modern applied science and technology, which have created the environmental crises we face today and for which "Christianity bears a huge burden of guilt."[39]

White's essay created a considerable controversy, but it persuaded too many environmentalists who, in the words of Carl Pope, executive director of the Sierra Club, simply "considered the case closed."[40]

But many scholars have rejected White's very limited concept of Christianity and its western tradition. The great conservative scholar Leo Strauss traces the source of the modern attitude which White condemns precisely to the western *abandonment* of our biblical heritage.[41]

Some environmentalists have recently taken another look at White's original argument, and discovered that they have overlooked an important point he makes. Carl Pope confesses that "I reread White's essay—and discovered to my chagrin that I and many others had badly misread him."[42]

What they had overlooked is that in searching for a solution to the environmental devastation that he blames on western religious thinking, White does not reject Christianity. He himself was, in fact, a Christian. He argues that if the roots of our ecological troubles are religious, "the remedy must also be essentially religious." Although he believes that the western religious development has led to our problems, he also realizes that Christianity is not monolithic. He then turns to one of the most important Christian saints for his model and proposes that St. Francis of Assisi be adopted "as a patron saint for ecologists."[43] In 1979 Pope John Paul II did just that.[44]

In sum, the Judeo-Christian religions offer a solid foundation for environmentalism. Indeed, their teachings, if taken seriously, virtually require us to be protectors of this earth. Max Oelschlaeger believes that the churches and synagogues may even offer the last remaining hope for avoiding ecological crisis. In a detailed analysis of the complete spectrum of Judaic and Christian theologies from fundamentalist to liberal, he argues that the churches and synagogues are the last remaining places in America where a sizeable number of people may be willing to listen to messages in favor of stewardship of God's earth.[45]

So, we may dismiss this misguided notion that our religious tradition is somehow or other inconsistent with environmentalism.

"ENVIRONMENTALISM IS JUST GLOOM AND DOOM"

Another error conservatives often make is to reject environmentalists as mere prophets of gloom and doom. These conservatives tend to reject all warnings about environmental problems because of the frequent predictions that they will all lead to catastrophe and apocalypse.

It is quite true that environmentalists have used this tactic far too often. They seem to think that the only way to get the attention of an ignorant and apathetic public is to use scare tactics and fear appeals. And the only thing the mass media will report is high drama so, they think, they must make pollution and environmental degradation seem as cataclysmic as possible.

Such hyperbole may in fact be a poor rhetorical strategy. But it is passing strange that conservatives would object to such doomsday rhetoric because they use it constantly themselves. One thinks, for example, of some titles of classics from an earlier day: Friedrich Hayek's *Road to Serfdom* and James Burnham's *Suicide of the West*. Richard Weaver wanted to title his first book *The Fearful Descent*, but the publisher vetoed it. The pages of conservative journals have long been full of predictions that our civilization and our country face ultimate doom from a dozen or so causes: creeping socialism, communist infiltration, meddling bureaucracy, declining "family values," and so on. Indeed, William Rusher, longtime publisher of *National Review*, recently observed that one characteristic of the conservative movement in general is an "appetite for doom and gloom."[46]

Likewise, industrialists frequently claim that proposed environmental safeguards will have catastrophic consequences and doom America's economy forever. To take just one example, during the debate over the Clean Air Act of 1970, Lee Iacocca, vice president of Ford at the time, claimed that the new regulations would force automobile production to cease in the United States after January 1, 1975, and cause permanent damage to our economy.[47] If gloom and doom rhetoric is grounds for automatic rejection of an entire political position, then conservatives ought also to reject out of hand both their own conservatism and the anti-environmental stance of the business world.

Focusing on poor rhetorical strategies of environmentalists is nothing more than a distraction. It is time for both groups to look beyond that to the very real ecological problems our industries and our lifestyles have created for all of us.

Environmentalists may be realizing that much of their rhetoric has been counter-productive. Dave Foreman now wisely advises: "Don't exaggerate, ever. Things are always plenty bad enough to say it straight, and any exaggeration is going to come back to haunt you."[48] That is sound advice for both environmentalists and conservatives.

We have now looked at the major accusations made against environmentalists. But it is a standard tactic of current demagoguery to accuse environmentalists of all sorts of outlandish things. Whenever anyone asserts that "environmentalists want" something ridiculous, we should always demand to know: "*What* environmentalists want that? Name them." And if the accuser cannot find it in the agendas of the major, mainstream environmental groups, we can be very sure that the generalization is dead wrong—and it would be prudent to wonder what the accuser is really up to.

"CONSERVATIVES SHOULD BE 'FOR BUSINESS'"

This brings us to one of the most important misconceptions of all, which has done enormous damage in leading conservatives to violate their principles and oppose environmental protection. This is actually a fundamental misconception about conservatism itself: far too many conservatives seem to believe that above and beyond everything else, they are supposed to be "for business." Almost all policies to reduce pollution or prevent further degradation of our planet will increase someone's cost of doing business, so the business world opposes virtually all such proposals. And far too many conservatives have almost automatically (and blindly) backed the business interests.

This misguided notion that conservatives must be "for business" has long plagued us. Clinton Rossiter rightly calls it "the Great Train Robbery of American intellectual history"[49] and principled conservatives have had to combat it from the first. Russell Kirk notes that very early "an impression began to arise that the new industrial and acquisitive interests are the conservative interest," and he laments that conservatives have "never fully escaped" from this "confusion."[50] Neither the traditionalist nor the libertarian school of thought will support this fallacious belief—quite the contrary. As Peter Viereck says, the goal of conservatives is "to conserve the humane and ethical values of the west rather than the economic privileges of a fraction of the west."[51]

The traditionalists have little sympathy with the mentality of the business world. As Stephen Tonsor insists: "The leaders of the new conservatism are not now, nor will they be, identified with the American business community." Conservatism is properly defined "in religious and moral terms."[52] Traditionalists are concerned primarily with the cultural and spiritual health of our country, and the ethos of the modern business world is in many ways detrimental to both. Richard Weaver warns that in the business mentality "all the virtues are subordinated to successful gain-getting," and this "does constant damage to . . . the spirit which creates culture. . . . Never before in history had this type of person formed a class enjoying social prestige." The fact that it does so now is evidence of our cultural decline.[53]

Russell Kirk worked for a short time in Ford's River Rouge plant, and there he learned to have no love whatsoever for modern industrialism. He rejects the false notion that conservatives "must align themselves with the businessmen" and contends that conservatives "never had much sympathy with industrial aggrandizement." In a devastating critique of the business mentality, Kirk asks, "How many industrialists and financiers take any interest in general ideas? How many know anything about politics? How many, indeed, are really conservative?" He estimates that "in the whole American nation, perhaps, there are not a hundred important businessmen who take an intelligent interest in the problems of modern society. . . . For a long while, the American man of business, generally speaking, has been intent upon getting and spending to the exclusion of almost every cultural and social interest . . . and remains densely ignorant of the nature of true conservatism. . . . The American businessman is inordinately vain. . . he has reason to be ashamed of his record as a cultured man. . . . American businessmen are Philistines," ignorant of general ideas. And Kirk concludes, "A conservative order is not the creation of the free entrepreneur."[54] He specifically warns us about "conservatives" who want to govern by "placating certain powerful interests" and "assure us that great corporations can do no wrong." Such "time-servers," he contends, "cannot govern well."[55]

Thus when businesses and industries object to being required to fix the damage they do to society because it will lower their profits, traditionalists have no sympathy whatsoever.

Nor do libertarians favor giving businesses whatever they want. Quite the contrary. The libertarian philosophy supports free markets. But most environmental problems are instances of "market failure," which are profitable to some people but which libertarians would want to correct.

Libertarians are in favor of the free market, but it has long been recognized that businessmen and businesswomen hate the idea. Of course, they pay lip service to it and praise it in the abstract. But, as Milton Friedman says, there is an "inevitable tendency for everyone to be in favor of a free market for everyone else, while regarding himself as deserving of special treatment."[56] Francis Graham Wilson observes that, regardless of what economists say about the theory of capitalism, "the day-to-day exponents of modern industrialism have been generally in favor of having the government on their side."[57] Adam Smith himself saw that whenever businessmen got together they invariably conspired to fix prices.[58] Today circumstances have changed considerably and what businessmen and businesswomen typically want are subsidies from the government, often direct payments from taxpayers or indirect subsidies in the form of protection from imports.

In the environmental area, what businesses usually want is to avoid paying for as many of the costs of production as possible, to "externalize" their costs, typically in the form of pollution. That is, they want to avoid the costs of acting responsibly and cleaning up the wastes they produce. By polluting, they force innocent victims to bear those costs, and that is a clear instance of "market failure," which libertarians always want to correct: a fundamental principle of the free market is that producers and consumers should pay *all* costs of their production and consumption.

If the market is to work properly, policies must be devised to "internalize externalities" to make the polluter pay. It is the job of conservatives to establish market institutions in which businesses are forced to compete and act responsibly, whether they want to or not. As the *Economist* contends, it is a fundamental mistake of the political right to favor businesses instead of competitive markets.[59] Both environmentalists and conservatives need to realize that it is a violation of conservative principles to let businesses get away with whatever they want to do.

These misconceptions have badly affected conservatives' stand on environmental issues. But there is one more myth that has much wider impact, and it deserves a chapter for itself.

2

The "Environment Versus the Economy" Myth

THERE IS A PERSISTENT MYTH that cleaning up the environment and protecting the earth hurt the economy. All too many people, and not just conservatives, believe that there is a direct trade-off between, for example, pollution control and growth or between "jobs and owls."

This myth is perpetuated by the general tendency of business and industry to give enormously exaggerated estimates of the total costs and consequent damage to the economy that will happen if practically any proposal is adopted that would make them behave responsibly and clean up their collective acts. We have already seen Lee Iacocca's prediction that the proposed Clean Air Act of 1970 would completely shut down automobile production in the United States.

In the late 1980s, when Congress was considering acid rain control legislation, the National Association of Manufacturers predicted that it "would cause serious and lasting damage to the economy" and "achieve only the dubious distinction of moving the United States toward the status of a second-class industrial power by the end of the century."[1] Well, the legislation was passed, sulfur dioxide emissions have been greatly reduced, and the end of the century is behind us. The NAM, presumably, now considers us a second-class economy.

More recently, the auto industry tried to fend off proposals to require them to make more fuel efficient cars by raising the CAFE (corporate average fuel economy) standard to forty miles per gallon. They predicted that it would devastate the industry, putting 300,000 auto workers out of their jobs. But what their "study" did was simply add up all of their employees currently making cars that get less than forty miles per gallon and assume that every single one of them would lose his or her job! It assumed, in other words, that the industry would not even attempt to build a single new car that met the proposed gas mileage standard.[2]

The examples of businesses grossly exaggerating costs of cleanup could easily be multiplied many fold, so it is hardly surprising that the myth of a direct trade-off between the environment and the economy is widespread.[3]

Politicians all too frequently adopt policies based on this myth. The Bush administration, for example, thought that environmental regulation was a major roadblock to economic growth. The White House Council on Competitiveness was established, chaired by Vice President Dan Quayle, to veto environmental regulations, among others. At the state level, politicians often oppose stronger environmental protection for fear of losing economic development to states with lower standards.

With the industries' gloom-and-doom exaggerations and the politicians' rhetoric, it is not surprising that many Americans are personally worried. According to a *Wall Street Journal* poll, fully one-third of all workers are afraid they may lose their jobs because of environmental regulations.[4]

This erroneous belief—that protecting the environment hurts the economy—must be rejected on two levels. As we will see in Chapter 3, principled conservatives are not materialists. Consequently, even if it were true that protecting the environment would hurt the economy, as an objection to cleaning up pollution or conserving resources, it would have little or no force for conservatives. Several basic principles would give precedence to environmental protection over producing more baubles to consume and discard.

But fortunately for the hedonists in our society, the belief is simply false: dozens and dozens of studies have clearly shown that protecting the environment does not hurt the economy. I will first look at some studies that examine the overall, general impact of environmental protection on the economy, and then I will turn to studies addressing more specific questions, such as the impact of pollution reduction on competitiveness, profitability, industrial flight, and employment.

OVERALL, DOES ENVIRONMENTAL PROTECTION HURT THE ECONOMY?

The Organization for Economic Cooperation and Development (OECD) has addressed this question for its member countries (the industrial democracies, including the United States). One study reports that total spending on pollution abatement and control, both public and private, in most member countries has grown since the 1980s. In the 1990s these expenditures generally amounted to 1–2 percent of gross domestic product (GDP), which they predict will grow to about 2.5 percent of GDP in some countries in the next decade. Total environmental expenditures, which also include things like costs of nature conservation, resource protection, and urban amenities, are somewhat larger than this. As for the effects of these expenses on the overall economy, they conclude that "there is no evidence that the growth of environmental expenditure in Member countries will have created . . . serious macroeconomic problems."[5] Another OECD report summarizes the results of research in several member countries, including the United States, "to assess the impacts of environmental programmes on key macro-economic variables, such as GDP, inflation rates, employment levels, and the balance of payments. These studies have generally concluded that the macro-economic effects of environmental policies are quite *small*. Thus environmental programmes are expected to have *little or no effect* on the overall competitiveness of most countries."[6]

The OECD is conducting environmental performance reviews for member countries, and the one for the United States was completed in 1996. They report that total environmental protection expenditures for 1993 probably exceeded 2 percent of GDP. Although this level of spending on the environment is "relatively high," they find "no evidence that the economy has been adversely affected as a whole by strong environmental protection policies."[7]

But what about conditions at a lower level, *within* the United States? Some states have much stricter environmental standards than others. Do they have weaker economies as a result? Two detailed studies provide answers to this question.

In the first, Stephen M. Meyer at the Massachusetts Institute of Technology selected two sets of indicators, one to measure each state's economic strength and one to assess each state's environmental efforts. Both were measured over nearly two decades, from 1973 to 1989. By rating the states on each set and comparing the two lists, Meyer tests what he calls the "environmental impact hypothesis," that is, the assertion that strict environmental protection hurts economic development. The results were released in 1992. The overall conclusions are decisive: "States with stronger environmental policies did not experience inferior rates of economic growth and development compared to states with weaker environmental regulations. In fact the converse was true: states with stronger environmental policies consistently out-performed the weaker environmental states on all the economic measures. . . . The environmental impact hypothesis, while theoretically plausible, has no empirical foundation."[8]

More specifically, Meyer divided the states into three groups on the basis of their environmental strictness: strong, moderate, and weak. He then compared the groups for their economic growth, employment growth, construction employment growth, manufacturing labor productivity, and overall labor productivity. If the claim of a trade-off between environmental protection and economic strength were true, states with stronger environmental policies would have slower economic growth and therefore slower employment growth overall, slower construction employment growth (if businesses flee to environmentally weaker states), lower labor productivity in manufacturing (if industries have to divert too much investment into pollution control), and lower overall labor productivity (if strict environmental laws result in unemployment among highly skilled factory workers and replacement of those jobs with lower skilled service jobs). "The data failed by a wide margin to support the environmental impact hypothesis across all five indicators. In fact environmentalism was found to be positively associated with four of the five economic growth variables. . . . The environmentally strong states outperformed the environmentally weak states by substantial amounts."[9]

A couple of years after Meyer gathered his data, the Institute for Southern Studies conducted a similar investigation. Researchers also ranked all fifty states on two sets of indicators, economic and environmental.[10] Their results completely corroborate Meyer's: "States with the best environmental records also offer the best job opportunities and climate for long-term economic development. . . . The states that do the most to protect their natural resources also wind up with the strongest economies and best jobs for their citizens. . . . The best stewards of the environment also offer workaday citizens the best opportunity for prosperity."[11]

The studies examined so far consider the effect of environmental standards on the overall strength of the economy. Other studies let us get more specific.

DOES ENVIRONMENTAL PROTECTION HURT COMPETITIVENESS?

It is often asserted that environmental protection reduces the competitiveness of businesses and industries. For example, a 1992 survey of industry executives found that 73 percent believe that mandated investments in environmental programs hamper their companies' efforts to improve their competitiveness.[12] If this really is the case, individual companies might become less competitive, and the whole country might become less able to compete in international markets.

The economic evidence, however, indicates that this has not happened, that environmental protection has not adversely affected the competitiveness of American business. Indeed, in some cases it appears that environmental regulations have enhanced the competitiveness of industry. According to Roger Bezdek, among U.S. industries "many of the sectors subject to the most stringent environmental regulations—including chemicals, plastics, synthetics, fabrics, and paints—have become the most efficient and have actually improved their international competitiveness."[13]

Some countries, in fact, are using strict environmental standards as a policy tool to enhance the competitiveness of their industries. The *Economist* reports that "Japan's attack on dirty air at home gave it a competitive edge in the world's markets for cars and air-pollution equipment. Now the government hopes to do much the same with global pollution," for example, by pressing companies to reduce carbon dioxide, the main greenhouse gas.[14]

Michael Porter of the Harvard Business School and Claas van der Linde of the International Management Research Institute in Switzerland have championed the view that strict regulations cause companies to innovate and become more competitive. They believe that frequently it is external pressure that leads a company to examine its product designs and production processes and look for cheaper, more efficient ways to produce. Environmental regulations, they contend, have been an important stimulus to that kind of innovation, which has made many companies more competitive. Their evidence is almost entirely from case studies, but they have found a very large number of such instances, so their claim deserves serious consideration.

The Porter and van der Linde thesis applies particularly to pollution control regulations. "Like defects, pollution often reveals flaws in the product design or production process. Efforts to eliminate pollution can therefore follow the same basic principles widely used in quality programs."[15] "Properly designed environmental standards can trigger innovations that lower the total cost of a product or improve its value. Such innovations allow companies to use a range of inputs more productively—from raw materials to energy to labor—thus offsetting the costs of improving environmental impact. . . . Ultimately, this enhanced *resource productivity* makes companies more competitive, not less."[16]

To support these generalizations, Porter and van der Linde offer a large number of examples. Here I can summarize only a few as illustrations.

- A study of ten manufacturers of printed circuit boards revealed that they adopted thirty-three major process changes, thirteen of which related to pollution reduction. Of the thirteen, twelve resulted in cost reduction, eight in quality improvements, and five in increased productive capacity.[17]
- A Dow Chemical complex in California uses caustic soda to scrub hydrochloric gas, producing a wide range of chemicals. The wastewater used to be stored in evaporation ponds, but new regulations required Dow to close the ponds. So the company redesigned its production process, reducing the use of caustic soda and decreasing caustic waste by 6,000 tons per year and hydrochloric acid waste by eighty tons per year. Dow also captured a portion of the waste for reuse as raw material in other parts of the plant. The cost of the innovation was only $250,000, but it saves the company $2.4 million per year.[18]
- The 3M company has had considerable experience with innovations that reduce pollution and save money. For example, it used to produce adhesives in batches that were then transferred to storage tanks, but one bad batch ruined the entire contents of a tank and resulted in expensive hazardous waste disposal. The company developed a new technique for rapid quality tests on new batches, which reduced hazardous wastes by 110 tons per year at almost no cost, yielding savings of over $200,000 per year.[19]
- Because of new environmental standards, Ciba-Geigy examined wastewater streams at a dye plant in New Jersey. It made two changes to the production process that not only reduced pollution but also increased process yields by 40 percent. Annual cost savings were $740,000.[20]
- When the Montreal Protocol required chlorofluorocarbons to be phased out, companies that used them for cleaning agents looked for alternatives.

Raytheon found an alternate cleaning agent that it could use in a closed-loop system. Average product quality improved while operating costs were lowered. Other companies developed technologies that did not require cleaning at all, which also lowered operating costs.[21]

- Most distillers of coal tar in the United States opposed regulations that required substantial reductions in benzene emissions. The only available solution at the time was to cover tar storage tanks with expensive gas blankets. But Aristech Chemical Corporation developed a way to remove benzene in the first processing step, which saved $3.3 million.[22]
- The Robbins Company, a jewelry manufacturer, was almost shut down because it was violating water pollution standards. Then it adopted a closed-loop, zero-discharge system for handling water used in plating. The new system produced water that was much cleaner than city water and resulted in higher-quality plating and fewer rejects. "The result was enhanced competitiveness."[23]

Porter and van der Linde present many more such examples. Although this evidence is anecdotal, the accumulation of cases is impressive enough that their thesis must be taken seriously. "A solid body of case study evidence . . . demonstrates that innovation offsets to environmental regulation are common." Moreover, "there is no countervailing set of case studies that shows that innovation offsets are unlikely or impossible."[24]

They recognize that these examples "do not prove that companies always can innovate to reduce environmental impact at low cost. However, they show that there are considerable opportunities to reduce pollution through innovations that redesign products, processes, and methods of operation. Such examples are common in spite of companies' resistance to environmental regulation and in spite of regulatory standards that often are hostile to innovative, resource-productive solutions."[25]

The OECD has found the evidence for innovation stimulation to be persuasive. In its review of U.S. environmental performance, the OECD finds that "several industries have gone through considerable changes in response to environmental concerns. . . . Industries that adapted to new environmental requirements were strengthened by this process. The chemical industry, car manufacturers and the pulp and paper industry are examples." The OECD draws a general conclusion from these examples: "Whenever new standards have been introduced at a pace at which industry can adjust, *environmental protection and competitive advantages have gone hand in hand*."[26]

So far environmental regulation has not hurt U.S. economic competitiveness overall and may even have enhanced it in many specific instances. This has happened in spite of the fact that our regulatory regime was not designed to spur innovation. Porter and van der Linde offer some recommendations for change.

Environmental standards could foster innovation offsets that come from new technologies and new production processes by following three principles. First, the standards "must create the maximum opportunity for innovation, leaving the approach to innovation to industry and not the standard-setting agency." In other words, regulatory agencies should set performance standards and let each business or plant determine how to meet them. "Second, regulations should foster continuous improvement, rather than locking in any particular technology. Third, the regulatory process should leave as little room as possible for uncertainty at every stage." A regulatory regime that adhered to these principles would stimulate innovation far more effectively than our present system. In fact, judged by these principles, our current system often "deters innovative solutions, or even renders them impossible."[27] A better regulatory regime, therefore, could produce many more examples of improved competitiveness.

DOES ENVIRONMENTAL PROTECTION CAUSE INDUSTRIAL FLIGHT?

The states are often reluctant to tighten pollution standards for fear that new investment will shun them and locate plants in states with lower standards and existing businesses and industries will move to more lenient states. Likewise at the international level people fear that if U.S. environmental standards are too strict, industries will move to third world countries that are willing to allow high levels of pollution.

This contention has been examined in a number of studies that look at patterns of trade and investment. At the national level, we have seen from the studies by Stephen Meyer and the Institute for Southern Studies that within the United States the common belief is simply not true. States do not win economic advantage by having lower environmental standards. In a comprehensive survey of the economic literature, Jaffe and colleagues summarize other studies reaching the same conclusion.[28] As Meyer says, "Shifts in environmental policy, whether intended to extend environmental control or reduce it, have no discernible effect on state economic performance. . . . States can pursue environmental quality without fear of impeding economic prosperity."[29]

Robert Repetto analyzed investment flows from the United States to third world countries and concluded that they "provide no support for the contention that multinational companies are relocating environmentally sensitive industries in countries with weak regulations."[30] The literature survey by Jaffe and colleagues arrives at a similar conclusion. The third world has slightly increased its overall share of world production of pollution-intensive products, but that is consistent with the general process of development: as countries develop, manufacturing accounts for a larger portion of their economies. "Overall, the evidence of industrial flight to developing countries is weak, at best."[31]

Judith Dean draws a similar conclusion from her review of the economic literature: "More stringent regulations in one country are thought to result in a loss of competitiveness, and perhaps in industrial flight and the development of pollution havens. The many empirical studies that have attempted to test these hypotheses have shown no evidence to support them."[32]

DOES ENVIRONMENTAL PROTECTION CAUSE SLOWER ECONOMIC GROWTH?

We have already seen that Stephen Meyer's study specifically addresses this issue at the state level and finds that this belief is entirely unfounded. Similarly, after reviewing a number of such studies, Roger Bezdek concludes that "recent major empirical studies unanimously reject the hypothesis that there is a negative relationship between environmental protection and economic growth. In fact, when statistically significant relationships are found, they are invariably positive; in other words, the U.S. states and nations of the world with more stringent environmental regulations show the best economic performance."[33]

DOES ENVIRONMENTAL PROTECTION HURT PROFITS?

Environmental regulations require businesses to spend money on pollution control, which might reduce their profits. Robert Repetto puts the question in these terms: "Do establishments with superior environmental performance tend to be more or less profitable than establishments with inferior environmental performance within the same industry?"[34] According to the common belief, environmental performance, measured in emissions per unit of shipments, and profitability should be inversely related.

Repetto used a database generated by the U.S. Census Bureau's Center for Economic Studies that contains detailed information on over 200,000 large manufacturing plants, which he then combined with information from several EPA data systems measuring emissions and the Commerce Department's survey of pollution control expenditures. "The result is a database encompassing thousands of manufacturing establishments . . . and containing detailed information on emissions, production costs, sales, and revenues. Using this database, it was possible to investigate whether firms with superior environmental performance were more or less profitable than their competitors."[35] He selected several sectors with relatively similar product lines that represent a wide range of manufacturing industries and that include sectors with significant environmental impact and relatively large pollution control costs. Within these narrowly defined industrial lines, the environmental performance of plants varies considerably and so, if the standard belief is true, should profitability.

> Looking first at the correlations between manufacturing plants' profitability and their toxic emission (relative to shipments) . . . there is no tendency for higher toxic emissions to be associated with higher profitability. . . . The same general conclusions are borne out by correlations between waterborne emissions and airborne particulate emissions, respectively, and measures of profitability. . . . It is at least equally likely for plants with superior environmental performance to be more profitable. . . . There is no overall tendency for plants with superior environmental performance to be less profitable.[36]

WHAT ABOUT JOBS?

Many believe that stricter environmental protection causes companies to lay off workers. We have seen from the *Wall Street Journal* poll that a third of all American workers apparently fear losing their jobs because of environmental regulations.

Since environmental regulations have virtually no impact on the overall economy, they would be expected to have little or no overall impact on employment. If regulations actually stimulate competitiveness, they might have a positive effect. Moreover, the technology used to meet environmental standards must be produced, and jobs would be created in this new sector. And that is, in fact, what economists have discovered.

According to Robert Repetto, the number of jobs created per dollar of expenditure on pollution control is very similar to the employment created per dollar of sales in American industry in general.[37] Eban Goodstein finds that the overall employment effect of environmental protection is positive: "When the job creation aspects of pollution control policies are factored in, environmental protection has probably increased net employment in the U.S. economy by a small amount."[38] The OECD review of our environmental performance corroborates Goodstein's conclusion: "Over the period 1970–87, the *net employment effect* of federal environmental programmes was calculated to be positive, leading to a reduction in unemployment equivalent to 0.4 per cent of the labour force."[39] "At the national level," as Goodstein concludes, "any claim of a trade-off between jobs and the environment is completely without substance."[40] In an update of this research, Goodstein reaches the same conclusion: "In reality, at the economy-wide level, there has simply been no trade-off between jobs and the environment."[41]

But those findings relate to total employment. Do environmental regulations cause significant dislocations at the local or plant level? How many jobs are lost because of environmental protection policies? An answer to these questions comes from data the Department of Labor has been gathering. Since 1987, the department has collected information on layoffs that idle more than fifty workers. Their database is built from an annual survey of employers who are asked to list the primary cause of layoffs. Three-fourths of all U.S. firms that employ more than fifty people are included in the database. According to the employers' own responses, environmental protection spending "accounted for less than one-tenth of 1% of all mass layoffs nationwide. . . . 1,300 lost positions per year on average could be partially attributed to environmental regulation."[42] An update of this survey covering 1995 to 1997 concludes that about 1,500 workers per year were laid off because of environmental and safety regulations.[43] The recent merger of Exxon and Mobil will by itself cost about seven times that many jobs.

According to Goodstein, the assertion that environmental protection measures result in major job losses "has no basis in reality."[44] Environmental protection has made a small contribution to total employment in our country and it has not caused significant numbers of workers to lose their jobs. The evidence is decisive. As Goodstein says, "One should be wary of any statement preceded by 'All economists agree.' . . . But, in this case, there seems to be universal accord that, on an economy-wide basis, the 'jobs versus the environment' debate is based purely on myth."[45]

At this point we should take a brief look at the environmental protection industry itself because it has created new jobs and should create many more in the future. Environmental protection has been one of the most rapidly and consistently growing industries in the United States over the past two decades. Moreover, it has proven to be virtually recession proof. Between 1970 and 1992 the industry expanded sixfold, a sustained real average growth of almost 9 percent a year. The GDP growth, by contrast, was between 2 and 3 percent. And when the economy was suffering from the oil price shocks of the late 1970s, the environmental protection industry grew by over 60 percent. In the severe recession of the early 1980s, the industry expanded by some 22 percent.[46] This industry is, of course, stimulated by strict environmental standards. As the *Economist* reports, America's fast-growing environmental service industry is centered in California, which has "the toughest environmental rules in the world."[47]

According to the OECD, in 1992 environmental expenditures had created some 4 million jobs in the United States, which will grow to 5 million by the year 2000. This sector of our economy totaled $170 billion in 1992 and employed 3 percent of the workforce.[48]

The export potential of the environmental protection industry is great. In the early 1990s the worldwide market for environmental protection services and technologies was $370 billion, of which $50 billion took the form of international trade. The United States was the single largest exporter, at $6 billion.[49]

As environmental standards have become stricter in our competitor nations—especially Germany and Japan—the U.S. international position in the industry is threatened. Germany has already surpassed the United States in this export market. Its government is convinced that its environmental regulations, already the strictest of all nations, will stimulate the development of a wide range of new "green" technologies that can be marketed worldwide. Moreover, the Germans "believe that new efforts to curb pollution by boosting efficiency will further reduce operating expenses in their already efficient economy, providing them with a competitive edge over Japan and (especially) the United States." A former aide to the Environment Ministry commented that "what we are doing here is economic policy, not environmental policy." Likewise, according to a representative of MITI, Japanese industries see "an inescapable economic necessity to improve energy efficiency and environmental technologies, which they believe would reduce costs and create a profitable world market."[50]

Why are these two countries threatening the U.S. lead in the international trade of environmental protection technologies? Because "thousands of . . . born-in-the-U.S.A. environmental products were abandoned during the 1980s as the Reagan and Bush administrations, the Congress, and many state officials turned their backs on environmental protection, orphaning technologies that now stand to generate billions . . . of dollars for their new proponents."[51]

It must be emphasized here that throughout this summary of studies on the environment versus economy myth, *none of them has even mentioned the environmental and health benefits of cleaning up our country. These studies have only considered economic effects.* Even in terms of that extremely narrow focus, protecting the environment has not hurt the economy—even though our current regulatory regime is far from optimal. As Roger Bezdek concludes,

> the often-repeated dichotomy of economic prosperity versus environmental protection is basically not an issue. . . . Environmental protection, rather than being a drag on the economy, can represent a major profit-making, job-creating, tax revenue-generating opportunity for this nation. Further, until the debate shifts from artificial tradeoffs, such as "jobs versus spotted owls" and "profits versus energy efficiency," the United States will be at a competitive disadvantage because other nations have already identified environmental technology as a 21st-century winner and are investing heavily in it.[52]

JOBS VERSUS OWLS? THE ENDANGERED SPECIES ACT

Although environmental protection overall has not harmed our economy, the "jobs versus owls" slogan that Bezdek quotes points to one law in particular that requires a careful look. The most contentious of all our environmental laws is, of course, the Endangered Species Act (ESA). I will consider the specific provisions of this act and suggest ways of improving it in the discussion of biodiversity (Chapter 8). Here we are only concerned with the economic effects of the act.

Around the time when the Republicans took over in the 104th Congress (1995), opponents of the Endangered Species Act were using as "evidence" a number of "horror stories," tales that featured poor suffering citizens who were unable to protect their homes or were unable to do something on their land or incurred enormous expenses because of some rat or fly or creepy-crawly thing. Virtually none of these stories could be corroborated. Most of them were obviously so grossly exaggerated that any kernels of truth that might lie behind them had long since been lost in the telling. Some were clearly fabricated out of whole cloth. Nevertheless, politicians and even the media simply told and retold one horror story after another, making no attempt to verify them. In the acrimonious wrangling over this law, evidentiary standards were about as low as they can get.

There were, however, some relatively obscure attempts to determine the truth about several of these horror stories. Since the stories played such an important role in the attack on the Endangered Species Act, we should examine at least a few examples.

One of the most notorious stories was the subject of an investigation by the General Accounting Office (GAO), an investigative agency of Congress. In this case, the Endangered Species Act was blamed for the destruction of some homes in 1993 by a wildfire near Riverside, California. Homeowners were prohibited from disking to clear fire protection areas around their houses because the area was habitat for the endangered Stephens' kangaroo rat. Home owners were allowed to create firebreaks around their houses by mowing or other forms of weed control that did not disturb the ground. According to the county counsel, the local fire department, and the Fish and Wildlife Service, these measures would provide adequate firebreaks.

The fire in October 1993 was one of the biggest of the year, with winds up to eighty miles per hour. It burned some 25,000 acres and destroyed twenty-nine houses. The GAO was asked to investigate charges that some of those houses were lost because the ESA prevented disking of firebreaks. The GAO "found no evidence to support these views. Homes where weed abatement, including disking, had been performed were destroyed, while other homes in the same general area survived even though no evidence of weed abatement was present. . . . Weed abatement by any means would have made little difference in whether or not a home was destroyed."[53] The officials interviewed by the GAO "generally concurred that no type of firebreak could have ensured a margin of safety for homes given the force of the fire."[54] Despite the GAO's published study, this story resurfaces occasionally.

In 1995 the Fish and Wildlife Service (FWS), which administers the act, published *Facts About the Endangered Species Act* as part of a packet of materials for the press. It lists thirty-four of these horror stories and gives the FWS response to each. I will summarize just a few examples.

The Stephens' kangaroo rat was also blamed for other things, for example, costing home owners nearly $2,000 each in fees to pay for habitat to be preserved for the rat. It was also alleged that these fees caused some farmers to lose half of their acreage. The FWS replied that this fee in actuality is only a fee *per acre* on *new development*, or about $200 on a $95,000 home. It is not a fee on current uses of land.

Another tale claimed that the Fish and Wildlife Service was going to declare some 20 million acres in thirty-three Texas counties as critical habitat for the golden-cheeked warbler. FWS responded that there are fewer than 800,000 acres of potential warbler habitat in the entire state, and that Secretary of the Interior Babbitt announced that designation of critical habitat would not be necessary if habitat conservation plans were put into place. Work is now proceeding on the plans.

Another allegation is that FWS sues private landowners in Texas if they try to control cedar on their property. But the FWS has repeatedly said that control of cedar and ongoing ranching practices do not harm the habitat of the golden-cheeked warbler.

In testimony before a Senate committee, a representative of a cattlemen's association claimed that a widow near Austin, Texas, was threatened with prosecution by FWS because she wanted to clear brush from a fencerow. FWS responds that the woman had been advised that clearing a thirty-foot-wide, one-mile-long fencerow might harm endangered songbird nesting habitat. But after meeting with her and assessing the ground, FWS allowed her to clear the fencerow.

Another choice story alleges that FWS forced domestic geese in Utah to vomit to see if they had been eating endangered Kanab ambersnails. Their owner was supposedly threatened with a fine of $50,000 per snail eaten by a goose. In actuality, some geese were removed from a pond inhabited by Kanab ambersnails, but none was forced to vomit and no one was ever threatened with a fine for snails consumed by the geese.

These stories demonstrate gross exaggeration of a kernel of fact. Other stories must have been fabricated. For example, one claims that a farmer in Michigan was told that he would be allowed to return to farming if he gave the government a square mile of his property and a mitigation fee of $300,000. The farmer refused and was fined $300,000. FWS checked with all the relevant offices in Michigan and found no record of any such case.[55]

Another tale claims that thousands of farm acres were lost to flooding in Monterey County, California, for conservation measures to protect a species of salamander. But there were no conservation measures of the sort there because the salamander does not even live in that area.[56]

These are just a few examples of the kinds of tales that were being used as "evidence" by opponents of the Endangered Species Act in the mid-1990s. Such horror stories merely deflect attention from serious analysis of the act and its economic impact.

In fact, very few projects are ever stopped by the ESA. A large number of federal agency projects and private developments require federal permits. In these cases, there is a required consultation procedure. The FWS informally reviews the proposals to see if they might affect endangered species. (The National Marine Fisheries Service has jurisdiction over fish that spend all or part of their lives in the sea, but the procedure is similar.) Most proposals pose no problems and are quickly approved. In the relatively few cases that might involve some adverse effect on an endangered species, there is a formal review in which the FWS investigates the proposal's potential impact in greater detail and issues a biological opinion. If it finds that there will be an adverse effect on an endangered species, FWS suggests modifications or mitigations that would make the project acceptable. "Instead of stopping projects, consultations usually bring about agreements to change them so that their effects on endangered species are reduced or eliminated."[57] Only a miniscule fraction of reviewed proposals actually get cancelled. As Hank Fischer of Defenders of Wildlife says, "The strength of the Endangered Species Act has always been in modifying developmental activities, not in stopping them."[58]

In fact, it seldom makes modifications. Two important studies provide some illuminating statistics. In 1992 the GAO examined consultations from 1987 through 1991. There were 16,161 informal reviews and 2,050 formal ones. Some 89 percent of all consultations were resolved informally and the projects proceeded as planned. Of the formal reviews in which the agencies issued biological opinions, over 90 percent concluded that the proposed action would not hurt an endangered species. Out of all those proposals, over a period of five years, in only 181 cases did the agencies conclude that the project would harm an endangered species. And in 158 of those cases they were able to suggest "reasonable and prudent alternatives" that would allow the projects to proceed. In only twenty-three cases was it not possible to find "reasonable and prudent alternatives."[59]

The GAO only examined consultations for which the FWS and NMFS had produced written documents. But most consultations under the ESA do not result in any written document. So the World Wildlife Fund obtained records of *all* consultations, written and oral, and updated the statistics through 1992. In that six-year period there were 95,185 informal consultations and 3,052 formal ones. Of the 3,052 formal reviews, in only 360 cases did the agencies find that the proposal would jeopardize an endangered species.

Even this small number is inflated because two federal agencies asked for separate opinions on items that would normally be reviewed together on a programmatic rather than an individual basis. Of the 360 jeopardy opinions, 152 were part of the EPA's pesticide registration program and all were allowed to proceed after the EPA agreed to suggested labeling requirements. (The GAO did not include these in its figures because of their unusual nature.) Fifty-two of the jeopardy opinions concerned Bureau of Land Management (BLM) timber sales in the spotted owl region. Eight of the sales proceeded with modifications. In a political action that received a lot of publicity, the BLM appealed the other forty-four to the "God Squad," the cabinet-level committee that can exempt projects from the ESA. The God Squad exempted thirteen of the sales and the BLM ultimately withdrew all of them. If these two agencies had asked for the normal programmatic reviews, there would have been only 158 jeopardy opinions in the six-year period—out of over 95,000 proposals.

Almost all of the 360 proposals that resulted in jeopardy opinions proceeded with modifications. FWS identified only fifty-four projects that were finally withdrawn or blocked (forty-four of them were the BLM timber sales) and NMFS blocked one project. If the BLM had made the normal consultation request, the total number of projects blocked by the ESA would have been twelve—two per year. That is, depending on how you count the timber sales, about six-hundredths of a percent (.06 percent) or one-hundredth of a percent (.01 percent).[60]

Thus virtually all of these development proposals proceeded. John Sawhill, head of The Nature Conservancy, provides this perspective: "A developer faced a greater chance . . . of having an airplane crash into something he built than having a project stopped by the Endangered Species Act."[61]

Some industry critics claim that even if the consultation process almost never stops a project, it results in long delays that increase the cost. In 1987 the GAO investigated this charge in relation to western water projects and determined that any such effects were minor. Other events occurring at the same time, such as difficulties in obtaining financing, had more significant effects than the ESA consultation process.[62]

So few projects ever run into problems with the Endangered Species Act that the overall economic impact of the act is miniscule. Some interesting data were compiled by Stephen Meyer of MIT in a study entitled *Endangered Species Listings and State Economic Performance.* He estimated the impact of endangered species listings on state economic development between 1975 and 1990, the entire lifetime of the ESA for which complete data were available. He used two standard measures of state economic performance that indicate development activity: growth in construction employment and growth in gross state product. The impact of the ESA was measured by the number of species listed per state. Meyer realizes that the latter is an imperfect indicator.

> Some listed species range over very small habitats. Habitat for the Tecopa Pupfish was under an acre. Others, such as grizzly bears require hundreds of thousands of acres. Thus, two states with a single listing each may experience very different impacts. Then too many listed species have overlapping territories. For example, the area designated as habitat for the California Gnatcatcher contains some 37 other endangered species. Thus, it is likely that the cumulative impact of these 38 listings may be substantially less than the sum of their parts. . . . Unfortunately, data are not yet available to allow us to measure endangered species "burden" directly by land restrictions.[63]

Meyer used the data for the forty-eight contiguous states, broken down into three periods of five years each: 1975–1980, 1980–1985, and 1985–1990. For construction employment growth, "none of the patterns in any of the time periods support the assertion that endangered species protection results in measurable reductions in state economic performance. In fact there seems to be a modest *increasing* (positive) trend during the first two periods in construction employment growth as the number of listed species rises, and no trend in the last period. If endangered species listings are 'trashing' state economies there is no sign of that impact on construction employment." Meyer then examined gross state product growth, along with endangered species listings. "Here again the pattern predicted by critics of the Endangered Species Act fails to appear. There is no trend of declining economic performance as species listings increase. Instead all three periods show a modest increasing rate of gross state product growth associated with increasing numbers of species listings." The results of these two comparisons "strongly contradict the argument that endangered species listings impede state economic growth and development."[64]

Meyer concludes, "The one and a half decades of state data examined in this paper strongly contradict the assertion that the Endangered Species Act has had harmful effects on state economies. Protections offered to threatened animals and plants do not impose a measurable economic burden on development activity at the state level. In fact the evidence points to the converse. The combination of robust development and population migration accelerates the rate of endangered species listings."[65]

Listing of endangered species may, of course, have impacts on economic activities at the local level or on a particular developer. But "the economic effects of endangered species listings are so highly localized, of such small scale, and short duration that they do not substantially affect state economic performance in the aggregate."[66]

At the end of the study, Meyer relates his findings to the policy level. "The evidence is clear: Based on the actual economic experience under the Endangered Species Act weakening the Act will not spur job creation and economic growth. It will not launch poor rural or western communities on the road to prosperity. It will not save overextended developers from bankruptcy. If 'growing the economy' is the top priority of government then we should focus on policy options that can make a difference."[67]

Although the overall economic impact of protecting endangered species is for all practical purposes nonexistent, at the local level it may on occasion have a significant impact. But even then, the effect of the ESA can be much greater on the decibel level and blood pressure of its opponents than on actual economic performance. The ESA often may be used as a scapegoat to justify avoiding the real sources of economic problems. And the best illustration of this is the most notorious case of all: the northern spotted owl. The "jobs versus owls" slogan that got so much publicity, and was even picked up by George Bush in the 1992 campaign, was basically a smokescreen to hide the real issues.

The real issue never was "jobs versus owls." It was, and still is, the protection of the last remnants of old-growth forest in the Pacific Northwest, which once covered some 25 million acres of Washington, Oregon, and northern California west of the Cascades. Since they contain the most valuable timber, they have been heavily logged for many years and now only a small fraction remains. The best and most recent study, done in 1991, found that only 2 million acres—8 percent—were left.[68] There is less than that today. And much of it has been fragmented by clear-cuts, so its value as habitat is considerably reduced for old-growth dependent species.

With so little of the old-growth forest left, environmentalists tried to protect at least the portion remaining in significantly large, unfragmented blocks. They met fierce resistance from loggers and the timber industry. President Bush wanted a "balanced" solution, but obviously any possibility of balance had long since vanished in favor of logging. And, frankly, if the timber industry is so incompetent that it cannot survive on the enormous forestland base it has already been given, it is difficult to believe that sacrificing the last of the ancient forest would save the companies and the logging jobs. In fact, the industry has long been dominated by cycles of boom and bust, and by a "cut and get out" mentality that has left impoverished and unstable communities in its wake.[69]

There is no doubt that timber workers often have hard lives. James Elias interviewed laid-off timber workers and their families, and found fear, even despair, about their future. For example, one worker told him,

> When I worked at the mill, we always had ups and downs. The way I figured it, we all worked there so I didn't panic through the hard times. We were all in the same boat. . . . Now that it really looks like the mill's not going to be there anymore, a lot of people have panicked and left. A lot of people are real distraught. . . . I know a family with five kids, they had a beautiful home, and they lost everything. . . . They've worked hard all their lives. Now there's nothing left.[70]

Blame the spotted owl? Blame the environmentalists? No. This interview was conducted before the spotted owl even went to court.

The history of the timber industry and timber jobs in the Northwest since World War II reveals some important trends that provides the proper context for considering the "jobs versus owls" issue. The supposed golden era of logging extended from the end of World War II to the mid-1960s. Timber cut from private lands and the national forests increased dramatically. Yet the number of jobs nationwide in logging and milling dropped from 572,000 in 1947 to just 342,000 by 1964.[71]

In Oregon, more than a third of the state's large sawmills closed between 1948 and 1962. Even more dramatic was the failure of smaller sawmills—over 85 percent were shut down during the same period.[72] After the recession of the early

1980s, Oregon timber production expanded once more and set new records in 1986 and again in 1987. But these production levels were achieved "with fewer mills and fewer workers per mill. The industry [in 1989] is operating only seventy-five percent as many mills as it did in 1979, and it employs only two-thirds as many workers as it did then. In addition, this reduced work force is being paid less. . . . The health of the wood products industry . . . seems to have been maintained. But doing so has had serious economic and social consequences at the local level. Fewer communities . . . have mills; the mills that continue to operate have fewer workers; and those still working have less income."[73]

In the Pacific Northwest as a whole, in the late 1970s there were 534 timber mills employing almost 200,000 workers. They produced approximately 11 billion board feet of lumber per year. By 1988, production had increased to 16.5 billion board feet. But because of automation, fewer than 160,000 workers were employed, and the number of mills had declined to 453.[74]

Even before the "golden" days after the war, experts were warning that the rate of timber cutting was not sustainable.[75] By the late 1980s the old-growth forest was almost gone. Jobs in the timber industry had been vanishing for decades. Timber workers had been laid off at a high rate although the already unsustainable cut was going up and up. Another person Elias interviewed told him, "You talk to the old timers here and they'll tell you that [the timber company] said they had enough timber to last them for the next hundred years. Then they got greedy in the last four or five years and literally raped the ground, took all the trees and did very little planting. . . . Now this happens [the mill closes]. Sure, I'm disgusted."[76]

And *all* of that happened *before* the spotted owl had its day in court.

Against that background in the Pacific Northwest—decades of unsustainable logging, thousands of workers being laid off and many mills closing even while the total cut was increasing, social instability and financial depression in many rural communities—enter the elusive and reclusive northern spotted owl.

The spotted owl case had its origins in the 1970s, when scientists began to pay serious attention to the study of old-growth forests. In earlier years it was generally believed that the ancient forests of the Pacific Northwest—with trees 250 years old and more—were decadent because trees that old grow very slowly. Those forests were also believed to be "biological deserts." But when subjected to serious study, the ancient forests proved to be anything but biological deserts. They are exceedingly rich and complex ecosystems, with many species dependent on them for all or part of their life cycle.

It just so happened that one of the first studies of old-growth forests focused on the northern spotted owl. Scientists quickly perceived that logging practices were threatening the owl's existence, and the bird became an issue in the forest-planning process. Throughout the 1970s and early 1980s additional studies kept raising estimates of the amount of old growth needed by the spotted owl. The timber industry became alarmed and did its best to reduce the amount of timber that would be reserved to protect the bird. The Forest Service tried to accommodate all sides through its elaborate planning process and did its best to devise owl protection plans that would not significantly reduce timber sales.

For something over fifteen years, a long series of committees and study teams reviewed the scientific data and made recommendations for forest plans. One proposal after another was rejected—by environmentalists if they thought it gave insufficient protection to the bird and by industry if they thought it would reduce timber sales.

Throughout this long process, the Endangered Species Act played no role. The question was over the Forest Service's legal requirement under the National Forest Management Act (NFMA) to manage the national forests for wildlife and other multiple uses, not just timber. The Fish and Wildlife Service stayed firmly planted on the sidelines as long as possible. Finally, environmentalists formally asked the FWS to declare the owl an endangered species. The FWS refused and the environmentalists sued. The court ruled that the FWS decision had been arbitrary.[77] The agency reconsidered as ordered by the court and declared the owl "threatened." Environmentalists sued again because even though the FWS had declared the owl's existence threatened, it had not designated the required habitat for it to survive. Again the court agreed with the environmentalists and ordered FWS to designate critical habitat for the owl.[78]

Separately, suit was brought against the Forest Service under the NFMA to force the agency to establish standards to ensure the viability of the spotted owl. The district court stopped all old-growth timber sales on national forests in the spotted owl region until the Forest Service could revise its plans to protect the owl.[79]

This was basically the situation when President Clinton took office. He convened his much publicized Forest Summit in Portland, and subsequently his administration released its Option 9 management plan for the region. Old-growth reserves would be set aside within a matrix of forestland with different levels of timber management. Timber sales would be much reduced from the levels of the 1980s, but some old growth would be cut. Eventually the court accepted the new plan, and that is basically where we are today.[80]

The spotted owl case probably contributed to the pain in some rural communities in the Northwest. There is no doubt that the poor owl suddenly became the scapegoat for many economic woes that were, as we have seen, already far advanced. We are now nearly a decade removed from the injunction against timber sales, so perhaps we have sufficient perspective to see what the effects of the case really were.

At the outset, the wood products industry predicted that saving the owl would result in enormous losses. In Oregon, it predicted, some 100,000 jobs would vanish.[81] One industry source estimated that 147,000 jobs would be lost in the Pacific Northwest.[82]

Environmentalists' economic analyses, by contrast, predicted far fewer losses. For example, the Wilderness Society estimated that no more than 9,000 jobs would be lost by the year 2010, most of which would be regained by 2040 as the industry adapted to a second-growth timber base.[83]

After several years, it appears that the environmentalists were far closer to the mark. Overall, the regional economy suffered no disaster. Indeed, the Northwest was the fastest-growing region in the entire country, with unemployment rates at their lowest in twenty-five years.[84]

In Oregon, the wood products industry lost only 4,500 jobs, spread out over five years—not the 100,000 the companies had warned would go—at a time when the state was creating new (nontimber) jobs at a rate of some 50,000 every single year. In large part, the industry was able to adapt to processing smaller-diameter second-growth trees from private farms. It became more efficient, finding uses for previously discarded material. And it also succeeded in increasing local value added, such as making cabinets and door and window frames instead of just exporting lumber.[85]

For the Pacific Northwest as a whole, we should look at employment *trends*, not just employment numbers. The timber industry, we have seen, had been shedding workers for decades, even when it was cutting more trees and producing more board feet of lumber, and more jobs would be eliminated in the future. As Judge William Dwyer concluded, based on the evidence presented to the court in the timber sale injunction case, "Job losses in the wood products industry will continue regardless of whether the northern spotted owl is protected."[86] So Freudenberg and colleagues contend that the proper approach is to ask, Did the spotted owl case accelerate the *rate* at which employment in the industry was decreasing? If protecting the spotted owl did indeed harm employment in logging and milling, then the rate at which the industry was losing jobs would have increased.

Freudenberg and his colleagues found nothing of the sort. Their statistical analysis of employment trends in logging and milling led them to the following conclusion: "We find that the 1989 listing of the spotted owl has no significant effect on employment—not even in the two states where the debate has been most intense [Oregon and Washington]."[87]

> Despite the widespread and apparently heartfelt conviction that the jobs of rural loggers and primary wood processors in the Pacific Northwest are being endangered by federal protection of the spotted owl, and by environmental protection more broadly, quantitative analysis provides a very different picture. Based on . . . analyses of the best available data on employment in logging and milling, whether in the Pacific Northwest or in the nation as a whole, there is simply no credible evidence of a statistically believable job-loss effect. This conclusion holds . . . whether the focus is on the period associated with spotted-owl protection in 1989–1990 or on the period since "Earth Day" and the passage of the National Environmental Policy Act in 1969–70.[88]

And the authors conclude, "There is simply no quantitative evidence of any statistically credible increase in job losses associated with the federal listing of the northern spotted owl as a 'threatened' species."[89]

Efforts to protect the spotted owl have not produced the economic devastation the industry and some conservative politicians and pundits predicted. Indeed, in the case of the politicians and pundits, much of their "concern" for timber jobs amounted to "crocodile tears" at best—the sort of "selective indignation" that the *National Review* in its early years attributed to the liberals. These conservatives unleashed their wrath against environmentalists over the owl. But (except for a small group of populists) one never sees from them that same sort of indignation about the thousands and thousands of people laid off by corporate mergers or downsizing, even when the companies are earning high profits. One never sees them attacking, say, General Motors, AT&T, or IBM for shedding so many workers—many more than were ever in jeopardy in logging. *The Economist* has reported over the past few years on studies of downsized firms, which found that in many cases the companies were no more profitable after laying off all those employees than they had been before. Many of those *real* job losses and the personal sufferings of laid-off workers were apparently all for naught. At least in the case of the spotted owl we have saved some of the little old-growth forest that still remains—we came out of that case with something positive to show for it.

In actuality, we achieved a great deal more, even in narrowly economic terms. Protecting the spotted owl and its forests needs to be considered in the context of the entire Pacific Northwest economy, not just the timber industry. When we look beyond that one extractive industry and its very loud lobbyists, we find that protecting the ancient forests has considerable positive value. Cost-benefit studies have concluded that the benefits of preserving the owl and its old-growth forests far outweigh any costs. At the height of the controversy, Jonathan Rubin, Gloria Helfand, and John Loomis, using standard contingent valuation (CV) methodology, concluded that in the long run the economic benefits of preserving the forests exceeded the costs by nearly three to one.[90]

Another cost-benefit study was conducted by Daniel Hagen, James Vincent, and Patrick Welle. Also using CV methodology, they calculated benefit/cost ratios "using 'best' and 'lower-bound' estimates of the benefits of preservation. Under all combinations of assumptions, the estimated benefits exceed the costs of the conservation policy."[91] The benefit-to-cost ratios of preserving the old-growth forests and the spotted owl were at least 3.5 to 1, ranging up to 14 to 1 under a very conservative set of assumptions. Under one assumption about future income, the benefit-to-cost ratio for preserving the owl would be 42.5 to 1![92]

The authors recognize that contingent valuation only measures some of the economic benefits of preserving the ancient forests. Their CV survey does not measure "use values associated with commercial fisheries, stream-flow maintenance, and water quality, which would have to be estimated separately."[93] At least some of those uses would then have to be added to the benefits of preservation because logging reduces these values by increasing erosion that damages streams and rivers as habitat for fish and that increases the costs to cities and towns that get their water from that watershed. Logging also reduces the value of the area for recreation. But even *without* calculating these additional benefits of preservation, considering only the benefits found using the CV methodology, logging the last old growth is clearly, as the authors put it, "a value-reducing activity."[94]

When the economy of the Pacific Northwest is considered in its totality, the importance of that conclusion becomes obvious. Preserving the environmental amenities of the region is far more important for its economic health than cutting down more trees. In December 1995 a group of thirty-four economists from the region issued a "consensus report" entitled *Economic Well Being and Environmental Protection in the Pacific Northwest*, under the editorship of Thomas Michael Power of the University of Montana. Their analysis of the region's economy and the factors critical to its strength leads them to conclude that environmental protection is vital. "The region is successfully navigating from being dependent on a few extractive industries to having a modern, widely diversified economy. . . . As quality of life becomes more important to the region's economy and natural-resource extraction becomes less important, a shift is taking place in the economic role that natural resources play." Especially in the Pacific Northwest, "a healthy environment is a major stimulus for a healthy economy. . . . *In short, the Pacific Northwest does not have to choose between jobs and the environment. Quite the opposite: A healthy environment is a major stimulus for a healthy economy.*"[95]

After considering the major changes taking place in the structure of the regional economy, the economists conclude, "*Policies and actions that significantly diminish the natural environment may threaten this region's economic future and should be undertaken only after careful deliberation shows that they are worthwhile.*" In particular, "*in many instances, the highest-value use of a forest, river, or other resource will be to protect and enhance it, because this will strengthen one set of forces that is creating new jobs and higher incomes.*" And they warn that "*reversing . . . efforts to protect our environment could impose serious economic harm on the region.*"[96] They explicitly reject the notion that there is a trade-off between environmental protection and jobs. In reality, the issue is "jobs-versus-jobs." "*The new jobs and income that are vital to the region's economic future will depend more on the protection of [the unique natural resources] than on their degradation.*"[97]

According to economics professor W. Ed Whitelaw of the University of Oregon, "we have passed the point where the forest standing yields more jobs and higher income than cutting it down."[98] Indeed, even during the height of the spotted owl controversy, the economy of the Pacific Northwest hardly skipped a beat. It was the fastest-growing region in the entire country. "Far from being a period in which the Pacific Northwest region has been 'outta work for every American,' the period since the listing of the spotted owl has actually been one of soaring job growth in the Northwest."[99] As Power comments, "The jobs-versus-environment folks were predicting a new Appalachia. The opposite has happened."[100] Oregon's unemployment rate, for example, in 1996 was a full percentage point below the national average. Even in rural areas, "a common refrain heard on the Main Streets of the state's timber towns is that anyone who wants work has it."[101]

How could this be—this discrepancy between economic reality and the gloom and doom predicted by the anti-environmentalists? The timber industry, of course, had a vested interest in exaggerating the effects of preserving the remaining ancient forests. And public perceptions often seemed to go along with the picture painted by the industry's propaganda, even though it bore virtually no relation to reality.

Thomas Power explains that the common view of a region's economy is often based on folklore that originated long, long ago, and this is clearly the case with the Pacific Northwest. When we think of the upper left corner of the United States, we almost automatically think of forests and logging and timber towns. But this picture is now grossly outdated. Today the entire extractive industry constitutes only a tiny fraction of the regional economy. The entire wood products industry, which is specifically at issue here, provides less than 3 percent of the jobs in the region, and its portion is falling. Even in the timber districts within the Northwest, the wood products industry provides only 5 percent or less of total personal income.[102]

CONCLUSION

In sum, neither environmental protection in general nor the Endangered Species Act in particular has had the devastating impact on our economy that their opponents have claimed. On the positive side, we have cleaner air and water and land to show for it (although the cleanup is far from done) as well as a significant amount of habitat preserved for the future and for other creatures.

Since we are concerned with both misconceptions and principles, it is important to stress that even if some measures to protect the environment would reduce economic growth, for a principled conservative that would be largely irrelevant. Conservatives, we will see shortly, are not materialists, and the principles involved in protecting the environment are far more important than a few more baubles to consume and discard.

The misconceptions and fallacies we just examined have kept conservatives from giving serious consideration to environmental problems and policies. They should reverse this tendency. Environmentalists are not inherently left-wingers and environmentalism is not an anticapitalist conspiracy. Efforts to protect our earth do not go against Western culture or Christianity—quite the contrary. Gloom and doom rhetoric, used so conspicuously by conservatives themselves, should no longer distract us from consideration of our very real environmental problems. Conservatism as a political philosophy is in no way a capitulation to whatever the business world wants—quite the contrary. And the supposed trade-off between the environment and the economy is a myth.

None of this is to say that current laws and regulations cannot be improved, that we have an ideal regime, economic or environmental. There are many improvements that could be made, new policies based on conservative principles that would be more effective in cleaning up the environment, preventing pollution, conserving resources, and preserving our natural heritage. And we have barely begun to confront some daunting environmental problems that will require conservative policies to solve them.

But before we can consider those problems and policies, we must examine some fundamentals of the conservative political philosophy, to see what principles conservatives should apply in selecting environmental policies.

3

Nine Conservative Principles

HAVING DISPOSED OF SOME MYTHS and fallacies about environmentalism that have badly confused the public debate, I now turn to the positive side: the principles of the conservative political philosophy that are relevant to environmental issues. These principles clearly support conservation, protection of our environment, and preservation of our natural heritage.

As noted in the introduction, the conservative political philosophy is not a single, unified doctrinal system. Rather, American conservatism embraces a spectrum of ideas on politics and economics that ranges from free market libertarianism on the one side to Burkean traditionalism on the other. The typical conservative intellectual falls near the middle of that spectrum, borrowing from both sides. Although almost no one likes the word, the "fusionism" advocated by Frank Meyer represents the typical conservative thinker's position. The differences among conservatives thus are largely differences in emphasis on principles that they all share. This allows us to examine some of those principles and to synthesize from them a conservative approach to environmentalism.

The principles I will consider have been selected because of their environmental implications. I do not mean to imply that these are the only principles of conservatism; I am not attempting to describe the entire conservative philosophy. The principles I have chosen are acceptable across the conservative spectrum, with more or less emphasis from various points on that spectrum:

1) Conservatism is not materialistic.
2) Freedom of the individual is extremely important.
3) An ever present corollary of freedom is responsibility.
4) Private property is a fundamental social institution.
5) The free market is a fundamental social institution.

In the responsible exercise of our freedom, three principles should govern our thinking and our actions and our policies:

6) Piety, especially piety toward nature, should be our governing attitude.
7) Society is intergenerational.
8) Prudence is the most important political virtue.

Finally, in applying their principles,

9) Conservatives are not ideologues; conservatism is not an ideology.

The first and last principles do not have immediate or direct policy implications, but they pervade all conservative thought, and their indirect influence on thinking about environmental problems will become clear. These principles are the ones conservatives violate when they take anti-environmental stands.

In explaining these principles, I will rely on rather extensive quotations from leading thinkers, especially from the "founding fathers" of contemporary conservatism. It is important to see how they developed these principles in their own words.

CONSERVATISM IS NOT MATERIALISTIC

Conservatives are not materialists. They believe that there are many more important things than production and consumption. The emphases of the two schools at the ends of the conservative spectrum are quite different, but they share this common premise.

The traditionalists are most emphatic in rejecting materialism as an end for either the society or the individual. As William Harbour observes, they "are increasingly critical of a society they see clearly marked by . . . a materialistic culture which reduces human choices to a hedonistic calculus."[1] Since religion plays a central role in the lives of most traditionalists, they remind us that we cannot serve both God and mammon, and that greed and avarice are still sins.

Traditionalists are primarily concerned with the health of our spirits and our culture, where they find great cause for concern. As the great historian Christopher Dawson wrote, "we now realize that a civilization may prosper externally and grow daily larger and louder and richer and more self-confident, while at the same time it is decreasing in social vitality and losing its hold on its higher cultural traditions."[2] Traditionalists generally agree with T. S. Eliot's famous judgment: "We can assert with some confidence that our own period is one of decline; that the standards of culture are lower than they were fifty years ago; and that the evidences of this decline are visible in every department of human activity."[3] Eliot wrote that in the late 1940s, and the concern is still with us today, perhaps more so than ever. Witness, for example, William Bennett's *The Index of Leading Cultural Indicators,* which he contends demonstrates that "over the past three decades we have experienced substantial social regression."[4] And the major cause of decline, traditionalists contend, is materialism.

If any single individual can be credited with starting the postwar conservative intellectual movement in America, it is Richard Weaver. His first book, *Ideas Have Consequences*, published in 1948, was the first shot fired in the conservative attack on the liberal establishment. This book is, in Frank Meyer's judgment, "the *fons et origo* of the contemporary American conservative movement."[5]

Weaver contends that the basic cause of our cultural and spiritual decay may be found in a change in our conception of humanity: "Man created in the divine image, the protagonist of a great drama in which his soul was at stake, was replaced by man the wealth-seeking and -consuming animal."[6]

America, Weaver believes, has become a "mass plutocracy," which is an unhealthy state of mind for the whole society, not just the wealthy. People believe that "money is the axis on which the world turns." Americans trust that money will "bring security and happiness almost automatically."[7] We end up with a "spoiled child psychology" that infects the whole society.[8]

This modern belief about the nature of humanity is relatively recent, and historically it is very unusual. "Non-materialist views of the world have flourished for most of our history, have inspired our best art and held together our healthiest communities. This is, indeed, the 'natural' view."[9]

The materialism that characterizes modern America is, Weaver contends, an unmistakable indication of cultural and spiritual decline. The modern conception of human beings as essentially producers and consumers is "symbolic of spiritual decadence."[10] All the technological marvels of which modern people are so proud may be merely "a splendid efflorescence of decay."[11] The temptations and distractions they place constantly before us "may make civilization more rather than less difficult of attainment."[12] We need to remember that "there is no correlation between the degree of comfort enjoyed and the achievement of a civilization. On the contrary, absorption in ease is one of the most reliable signs of present or impending decay."[13]

Weaver was a Southerner and spent his entire professional life as a student of the culture of the Old South, looking always at its strengths and weaknesses for the lessons we might learn from them. Weaver realized full well, of course, that "there cannot be a return to . . . the Old South under slogans identified with them."[14] But he believed there are valuable lessons to be learned. "The Old South may indeed be a hall hung with splendid tapestries in which no one would care to live; but from them we can learn something of how to live."[15] What Weaver believed he had found in the Old South, despite all its faults, was "*the last non-materialist civilization in the Western World*." And the Old South therefore confronts us with a challenge: "to save the human spirit by re-creating a non-materialist society."[16] Whether he is right or wrong in his historical judgment about the antebellum South, the point for conservative philosophy is clear.

Russell Kirk did more than anyone else to establish Burkean traditionalist conservatism as a respectable and important political philosophy for our time. He decisively rejects the modern "reduction of human striving to material production and consumption."[17] If Burke could see us today, he "never would concede that a consumption-society . . . is the end for which Providence has prepared man."[18] Kirk wrote in 1957 that a very real danger for our society is that "we may become pigs in the sty of Epicurus."[19] Perhaps by now we have long since been reduced to that level. And a very low level it is, for "the man bent upon gratifying his appetites is servile, however rich he may be."[20]

Kirk condemns our materialistic, consumption-maximizing society throughout his works, but perhaps nowhere better than in his book *Prospects for Conservatives.* In a chapter entitled "The Question of Wants," he writes:

> Every year, we become still more the slaves of creature-comforts and effortless amusements. Our whole economy . . . is calculated to endorse and increase this appetite for material goods, and we are urged to consume everything, at the greatest possible rate . . . Tocqueville believed that a gross materialism was the greatest menace to the American democracy; and Orestes Brownson denounced the American knack for creating new wants as an impious and ruinous passion, encouraging novel appetites that never can be truly satisfied. Our real wants, he said, are those of the spirit; but those we neglect, spending our days in a meaningless pursuit of mundane pleasure which, in the end, must betray us. . . . Now the thinking conservative, I believe, ought to contend with all his strength against this particular degradation of the American mind and heart. The conservative knows that material production and consumption are not the purpose of human existence.[21]

According to Kirk, "the conservative thinker does not believe that men are made happy by creating and stimulating new wants. He does not believe that men are made happy by attempts to satisfy to repletion every physical craving. . . . [Therefore], the American conservative will endeavor to exert some intelligent check upon material will and appetite. If, indeed, as the economic optimists declare, we are in an age of plenty, then we ought to begin to employ these material endowments after a fashion which will bring our wants into conformity with truly human needs."[22]

The traditionalists' favorite economist is Wilhelm Röpke, who served as an adviser to Ludwig Erhard, guiding the postwar economic policy of West Germany that created the "German miracle." Röpke advocates the free market, but he always insists that economics and production and consumption have only limited and subordinate roles to play in life. And he severely criticizes the modern West for "squandering their spiritual patrimony, in pursuit of a higher standard of living." He claims that the result is "a cultural catastrophe." "Man simply does not live by radio, automobiles, and refrigerators alone, but by the whole unpurchasable world beyond the market, the world of dignity, beauty, power, grace, chivalry, love, and friendship."[23]

Röpke insists that we need "to confine the acquisitive instinct to socially tolerable forms."[24] He believes that utilitarianism, economism, and materialism have a "stultifying effect on life. . . . A society which concentrates on material gains will be at once immensely productive and immensely sterile, satiated and hungry, busy and enormously bored."[25] Röpke condemns the "cult of the standard of living," which is "a disorder of spiritual perception of almost pathological nature, a misjudgment of the true scale of vital values, a degradation of man not tolerable for long."[26] Believe it or not, that comes from the pen of an economist! Were there more like him, Edmund Burke might not have placed economists in the same category with "sophisters"![27]

Traditionalists thus are outspokenly antimaterialistic, as might be expected. But libertarians also, from a very different perspective, are likewise not materialists. They place enormous emphasis—sometimes nearly exclusive emphasis—on the free market economy. But for most libertarians, this is a means to an end, and the end is personal freedom, not production or consumption. As William Harbour explains, "the 'libertarian' Conservative argues that the crucial question to ask is whether or not capitalism helps to create some of the conditions for human freedom by creating a great area in life that is outside the coercive powers of the state."[28]

Murray Rothbard identifies the notion that libertarians are hedonists as a myth that he believes must be refuted. He responds that libertarianism is not a way of life; rather, "it offers liberty, so that each person is free to adopt and act upon his own values and moral principles." Libertarianism is a political theory, not a moral theory or a lifestyle. "What a person *does* with his or her life is vital and important, but it is simply irrelevant to libertarianism."[29] He is glad that the free market offers a wide range of choices for consumers and raises their standard of living. But when he presents a list of the desirable consequences of the free market, raising standards of living and satisfying consumer wants are placed last. At the top of the list are individual freedom, general mutual benefit, mutual harmony, and peace.[30]

Likewise, Tibor Machan denies that libertarianism is materialistic. "A preoccupation with material wealth is supposed to be implicit in libertarianism. But there is no requirement in that political doctrine to the effect that human beings ought to strive for material wealth, even if ordinary prudence would be expected from anyone as regards his material needs and wants."[31]

Machan offers a defense of free market capitalism that is based on its moral superiority to other political systems.[32] He develops this argument in explicit contrast to the more common defense based on its productivity of material wealth. He believes that "a moral argument for the free market system is available and is needed to supplement, and sometimes even to counteract, what economists have to say in support of capitalism."[33] For Machan, the economic justification for capitalism as superior at creating wealth is inadequate: "It presupposes the superior value of material prosperity, something that is by no means necessarily true."[34] Like Rothbard, Machan places capitalism's productivity low on his list of the advantages of capitalism. "What counts the most, what is centrally significant about this political economic system, is that it enables individuals to live a morally dignified life, to be in maximum command of their own existence in whatever conditions of existence they happen to be born into."[35]

Machan believes that the label "capitalism" for the free society is misleading because it suggests that the primary purpose of that system is economic. "No doubt, prosperity is vital to a morally good human life, but it isn't its major objective. Human happiness is. And such happiness is by no means attained by being merely prosperous."[36]

That conservatives on principle are not materialists pervades all their thinking. It has important if indirect implications for solving environmental problems because it means that the standard objections to controlling pollution and conserving resources—that they will reduce profits or slow economic growth—have little or no force (even if they were true). Those standard objections only carry weight with people whose motives are materialistic. Machan, for example, uses what should be, even for pseudoconservatives, a currently compelling analogy: "To permit [polluting] production to continue on grounds that this will sustain employment would be exactly like permitting the continuation of other crimes on grounds that allowing them creates jobs for others."[37]

FREEDOM OF THE INDIVIDUAL IS IMPORTANT

Whether approached from the traditionalist or the libertarian perspective, personal freedom is a primary value for all conservatives. For libertarians, securing the freedom of the individual is the sole legitimate purpose of politics and government. According to Tibor Machan, "liberty is the paramount value to be sought by way of politics."[38] Murray Rothbard contends that "government is distinguished from every other group in society as being the institution of organized violence. Libertarianism holds that the *only* proper role of violence is to defend person and property *against* violence, that any use of violence that goes beyond such just defense is itself aggressive, unjust, and criminal," even when used by government.[39]

Traditionalists would not go to that extreme, for they believe that there are many other proper functions for government. Moreover, the traditionalists are very concerned about the relationships between freedom and order. Libertarians stress only one of these terms, with results which traditionalists often contend are simply utopian. Traditionalists contend that both are necessary: proper order is a prerequisite for freedom, and freedom is not to be confused with license. It is within the dialectic between these polar terms that political life must be lived and institutions organized. Russell Kirk argues that "there cannot be genuine freedom unless there is also genuine order in the moral realm and the social realm. . . . Although there cannot be freedom without order, in some sense there is always a conflict between the claims of order and the claims of freedom."[40]

Order must be inner as well as outer. As character has declined, that inner order necessary for a proper exercise of freedom has broken down. And with it, freedom itself has dwindled. "The less control survives within private life and local community, the more control will be exerted by the centralized state, out of necessity."[41] If our cultural and spiritual health were restored, we could regain the foundations for freedom.

Traditionalist conservatives, unlike libertarians, believe that the state has many proper functions, even when cultural standards are sufficiently internalized to provide proper order. The optimal combination of freedom and order would allow for what British political philosopher John Gray calls the "autonomy" of the person. Gray was a libertarian who became a traditionalist, so his concept of autonomy includes a substantial element of freedom for the individual, but freedom that is exercised within a rich cultural context full of public goods, many of which will be provided by government. His concept of autonomy is worth a closer look.

Gray's book *Beyond the New Right* is a critique of libertarian, pure free market theory from the perspective of a Burkean conservative. Gray contends that "the principal defence of liberty is not in its promotion of efficiency or productivity but in its contribution to the autonomy of the individual."[42] But an autonomous person is not an atomistic, isolated individual, not a "free-floating sovereign self." An autonomous person "will typically emerge from strong and stable communities and will remain embedded in them."[43]

Consequently, one of the "preconditions of autonomy" is "a broad diversity of institutions aiming to provide the conditions of autonomous action, as well as a rich and deep common culture containing choiceworthy options and forms of life."[44] If autonomy is to be meaningful and valuable, it "requires not only capacities for choice on the part of the individual but also a span of worthwhile options in his or her cultural environment. In the absence of this, autonomy wanes, and the lives of individuals become the poorer, however many choices they make."[45] "Conservative individualists, unlike their liberal and libertarian counterparts, recognise that the capacity for unfettered choice has little value when it must be exercised in a public space that . . . is filthy, desolate and dangerous. The exercise of free choice has most value when it occurs in a public space that is rich in options and amenities, and its value dwindles as that public space wanes."[46]

This rich public space does not emerge automatically, nor is it the independent creation of each individual. "An individualist order is not free-standing, but depends on forms of common life for its worth and its very existence. Equally, autonomy is valueless if it is exercised in a community that is denuded of the inherently public goods which create worthwhile options and which thereby make good choices possible."[47] The public goods and amenities that are part of this environment will often have to be provided by the government.

The free market has a critical role for autonomy, but not the exclusive role attributed to it by the libertarians. Gray does not believe "that the institutions of the market have any unique role in promoting and enhancing individual autonomy. That value is also promoted in voluntary associations—families, churches and many other forms of life—in which market exchange is peripheral. The claim is not that market institutions alone promote autonomy in society, but rather that in their absence people will be denied autonomy in a vital part of their lives—the economic dimension in which they act as consumers and producers."[48]

Thus, although both schools of conservative thought value freedom, there are some considerable differences in the philosophical foundations they offer for liberty and in the emphasis they put on this value in relation to other principles and values. Nevertheless, both schools of thought believe that the modern collectivist welfare state encroaches far too much on our rights and liberties. And both schools of thought contend that the free market should be expanded as one way of enhancing our freedom.

But freedom does not just mean freedom from governmental encroachment; it also means freedom from imposition by other people. In the libertarian context, Tibor Machan explains, "each person would be completely free of others' intrusions or could count on legal sanctions when such intrusions occur."[49] Murray Rothbard similarly claims that "everyone should be free of violent invasion."[50] While the traditionalists might not argue in such absolute terms, nevertheless they would accept the general premise.

And here we find some major implications for environmental policies. For example, pollution infringes on the freedom and rights of all of its victims. It imposes costs on them that should not be theirs to bear. Those costs can be monetary, as when people have to pay more to clean up their water because of pollution from factories upstream. All too often the costs also take the form of pain and suffering. To take a recent example, scientific studies have concluded that emissions of small particulates and ozone still cause tens of thousands of premature deaths each year and hundreds of thousands of cases of respiratory distress, especially among children.[51] Or consider a legacy of the chemical industry: an environment so pervaded by persistent toxic chemicals that they can now be found in the bodies of everyone on the entire planet. As Gordon Durnil asks, "Is not the insidious invasion of our bodies by harmful unsolicited chemicals the most flagrant violation of our individual rights?"[52]

All forms of pollution are violations of the conservative principle of freedom. As Tibor Machan emphatically contends, "*capitalism requires that pollution be punishable as a legal offense that violates individual rights.*"[53] So, for principled conservatives, appropriate policies to reduce pollution are a means of protecting not only our health but also our freedom.

RESPONSIBILITY IS A COROLLARY OF FREEDOM

Closely and inherently connected with freedom in the conservative philosophy is its corollary, responsibility. As Friedrich Hayek argues, "Liberty not only means that the individual has both the opportunity and the burden of choice; it also means that he must bear the consequences of his actions and will receive praise or blame for them. Liberty and responsibility are inseparable."[54]

Traditionalists would go much farther than libertarians in claiming that we have certain duties to society. Thus Russell Kirk contends that there is virtually a one-to-one relationship: "Every right is married to a duty; every freedom owes a corresponding responsibility."[55] As William Harbour elaborates, traditionalist conservatism

> advances a view of human conduct which emphasizes the idea of accepting and acting upon extensive responsibilities. These responsibilities are not regarded as matters of pure choice, matters that can be taken up and abandoned at whim depending upon the mood of the individual. Rather, these responsibilities are set by man's relationship to God and the relationships (many of which are not voluntary) that men develop with each other in different societies. . . . Conservatism distrusts talk about freedom that gives exclusive stress to notions of rights and the claims that individuals make against society while it ignores the notion of responsibility.[56]

Libertarians place much more stress on freedom, and much less on responsibility to society. But both would agree that we must act responsibly toward one another and both would insist that we must take responsibility for our own actions and their consequences. Both schools of thought would accept Midge Decter's claim that an essential conservative belief is "the taking of responsibility for what one does and what one is."[57] And there, of course, is where this principle impacts environmental policies.

Just as pollution infringes on the freedom of its victims, so it is, by its very definition, an attempt by the producer to escape from being responsible for his or her actions. If a manufacturing process produces wastes, it is the responsibility of the producer to dispose of them properly. By emitting them into the air or dumping them into a river, the producer is simply attempting to avoid being responsible for his or her actions. And for a conservative this has much greater implications than the economist's concern for "inefficiency" resulting from "negative externalities," a point that will be considered later. Responsibility is a moral issue and irresponsibility is a moral fault. As Gordon Durnil says, "We conservatives bemoan the decline in values that has besieged our society. . . . Why then should we not abhor the lack of morality involved in discharging untested chemicals into the air, ground, and water to alter and harm, to whatever degree, human life and wildlife? As a conservative, I do abhor it."[58]

This principle lies behind such recent policies in the United States as welfare reform (to make people take responsibility for their own lives) or "getting tough on crime" (to make criminals take responsibility for the damage they do to society). Unfortunately, when faced with environmental issues, somehow or other far too many of our conservative politicians and pundits simply abandon this principle. Gordon Durnil observes that the coddling of criminals "really upsets most conservatives. . . . But when the executive of some large conglomerate violates the laws by discharging some onerous substance into the water or air or onto the ground, we [conservatives] pay little attention. . . . A conservative should believe that industry executives, as well as individuals, are responsible for their actions."[59]

Almost without exception, when confronted with evidence that it is harming the public with its pollution, business and industry have attempted to evade their responsibilities. Durnil gives a vivid and depressing description of the various tactics they use, from denial of the scientific evidence, to systematically disrupting dialogue at meetings and hearings to make sure no decisions can be reached, to marshalling their hordes of lawyers and lobbyists to block any measures the government might take.[60] For a recent example, when the EPA, after analyzing hundreds of scientific studies, concluded that current levels of particulate and ozone emissions were still dangerous to the public, the immediate reaction of the industries involved was to pool their money to wage a counterpublicity campaign and to file suit to prevent the EPA from protecting the public by stricter emissions requirements.[61]

But what is even worse is that conservative politicians almost always come to the industries' defense—in complete and total violation of their principles. Consider the recent case of timber companies in the Pacific Northwest. They clearcut forests, even on steep slopes where the completely predictable result will be massive mudslides. In the extremely wet winter of 1996–1997, mudslides wiped out many private homes below clearcuts and killed a dozen people. Afraid of lawsuits from the homeowners, the timber industry in Oregon moved rapidly to evade responsibility for the results of its action and "quietly won indemnification through special state legislation."[62]

Conservatives have long objected to liberals making excuses for criminals and erecting technical and procedural barriers to making them pay for the harm they do to society. Yet these same conservatives turn around and do exactly the same thing for polluters. This hypocrisy clearly offers a powerful rhetorical opportunity that environmentalists should take advantage of more often. For example, in

Arizona, Republican politicians were trying to weaken enforcement of pollution control laws. One Sierra Club activist, herself a Republican, found that the most effective attack was to expose the inconsistency of legislators wanting to be "tough on crime" and then trying to give immunity for environmental crimes. When she pointed out their double standard, "things got real quiet. . . . They say their agenda is conservative, but what it really comes down to is 'line my pockets, please.'"[63]

In similar fashion, most proposals by conservative politicians for "deregulation" are actually a violation of conservative principles. Air and water quality regulation is an attempt, at least, to force polluters to act responsibly and to make the polluter pay. But no principled conservative could simply propose "deregulation" and leave it at that. All attempts to allow polluters to evade their responsibility are violations of basic conservatism. What principled conservatives would do instead is propose *replacing* inefficient and ineffective command and control regulations with *better and more effective* market-based ones, as we will see in later chapters.

PRIVATE PROPERTY IS A FUNDAMENTAL SOCIAL INSTITUTION

Private ownership of property is important for conservatives as a matter of principle. The libertarian position is expressed by Tibor Machan, who contends that people have basic rights to life, liberty, and property. He contends that private property rights are derived "from the principle of the right to life" and they secure for us "our sphere of personal moral authority."[64]

Traditionalists also stress the right to own private property. Russell Kirk lists six fundamental canons of conservative thought, among which is "persuasion that freedom and property are closely linked."[65] According to Francis Graham Wilson, one of the distinguishing characteristics of the conservative mind is its defense of the institution of private property.[66] Richard Weaver goes so far as to claim that the right to own private property is "the last metaphysical right."[67]

Some important implications for environmental policy flow from the notion of property rights. But unfortunately our politicians and commentators have again failed to draw from this principle even the most obvious conclusion. According to Adam Smith, "the first and chief design of every system of government is to maintain justice, to prevent the members of a society from encroaching on another's property."[68] Pollution in any form is a violation of the property rights of all of its victims. If property is to be respected in principle, if property rights are to be enforced, then pollution clearly needs to be prevented as a matter of principle. As Mark Sagoff says, "A society that takes property rights and consent seriously, such as ours, will then at least enact environmental laws that seek to minimize and eventually eliminate pollution. A planned or centralized economy, in contrast, may permit and may even require pollution and any other transfer of property rights, without the consent of the initial owners, as long as the transfer is efficient or the benefits exceed the costs."[69]

Traditionalists also call attention to an important distinction in the ownership of property that has significant environmental implications. There is a vast difference, they claim, between individual, personal ownership of homes and farms and businesses, and the modern phenomenon of corporate ownership.

The environmental implications of this distinction can be seen if we compare, say, a family farm with an industrial agribusiness. The family will probably think of its farm as a patrimony. It may well have been in the family for generations, and they will want it to provide livelihoods for their children and grandchildren. If the economic system will allow it, they will try to protect the land and its productivity for the long-term future. They will probably keep some of their land as habitat for wildlife, if for no other reason than to make the farm a more pleasant place to live.

An agribusiness, on the other hand, will usually treat the land as a mere investment, from which it will insist on the maximum return in the short run. This will probably mean eliminating all amenities on the land, planting from border to border, "mining" the soil, with a maximum application of chemicals regardless of where they end up. And when the land is exhausted and ceases to provide returns at least as great as alternative investments, the company will simply abandon the land, selling out and putting its money elsewhere.[70]

We have a very vivid illustration of this difference in the way the environment is treated in the case of Pacific Lumber. This family-run company owned vast tracts of forests in California, including most of the old-growth redwoods in private hands. Its policy for decades had been to manage its forests for the long-term future. It never used clearcutting and purposely kept its forest ecosystems intact. It was profitable and free of debt. But in 1985 it was taken over by a corporate raider who proceeded to liquidate the forests to pay off the junk bonds that financed the purchase.[71]

Conservative principles are opposed to such abuse of property. According to Clinton Rossiter, we have here a good illustration of the idea that rights entail duties: "No right carries with it greater obligations than the possession of property, which is a legacy from the past, a power in the present, and a trust for the future."[72] The right to property includes the duty to use it responsibly; there is no right to abuse one's property.

Since our Western tradition is based on Judeo-Christian doctrine, biblical teaching is relevant here. Chuck Barlow finds that the Old Testament denies any right to destroy property, one's own or another's, because of our duty to God for stewardship of His creation. There is "a general principle against needless destruction of nature. . . . No exception exists for 'wise use' of private, as opposed to communal or public, property that would amount to reckless destruction. 'One is not permitted to destroy one's own property any more than he is permitted to destroy another's.'" Destruction of property "is an evil because it harms the realm of God and his creation."[73]

Conservatives should design policies to protect the property rights of victims of pollution in all of its forms. They should also design policies of incentives and disincentives that would lead all owners to exercise the kind of careful stewardship normally practiced by individuals and families.

THE FREE MARKET IS A FUNDAMENTAL SOCIAL INSTITUTION

The free market is an institution that allows us to exercise freedom in an important realm of our lives. As such, it is supported by both traditionalists and libertarians. Although they often differ on the proper extent of the market—for example, in the debate over legalizing drugs—both schools of conservative thought agree in general that the market is too restricted today and should be extended. For example, traditionalist political philosopher John Gray claims that the agenda of market expansion is far from exhausted. "There is much farther to go in extending market institutions into hitherto sacrosanct areas, in reducing taxation, inflation and government expenditure, and in privatising industries and services."[74] He is specifically referring to Britain, but his point applies to the United States as well.

The free market is usually friendly to the environment, if it is allowed to work properly. There are two fundamental market principles in particular that would help protect the environment if they were properly institutionalized. Unfortunately, in both cases our conservative politicians have failed to support them.

One principle of the free market, in economic terms, is that negative externalities should be internalized in prices. That is, *all* of the costs of producing a product should be borne by the producer and included in the price the consumer pays. Negative externalities are costs that the producer avoids and innocent victims are forced to pay. All forms of pollution are negative externalities. For a market to work properly, its institutions need to internalize those costs or, in plain English, make the polluter pay.[75] That is, even in terms of free market economics, businesses and industries have no grounds to complain about having to pay the costs of pollution control, no more than about having to pay wages to their employees or to buy their raw materials and supplies. Those costs are properly their costs and no one else's.

We have, of course, met this idea before, under the headings of "freedom," "responsibility," and "property rights." Economists simply add another reason for forcing polluters to clean up: it increases "economic efficiency" in the allocation of scarce resources. Externalities misallocate resources because, in this case, if pollution control costs are not included in the price of a product, its price will be too low and too much of it will be produced. And the corollary of this is that society will have too little clean air and water. In a properly functioning market economy, virtually all pollution of our air and water and land would have been prevented.

Conservative economists usually believe that where it is easy to identify who is hurt and who is benefited by pollution, it would be best to let the parties negotiate a solution, perhaps by agreeing on a compensating payment to the victim of the pollution. If they cannot reach an agreement, the victim could sue for damages. (The ways in which our legal system would render private enforcement of environmental standards unworkable in most cases is considered in Chapter 10.)

But even these economists acknowledge that in many instances there is no way to determine (1) who is causing environmental degradation and by how much or (2) exactly how much damages a particular victim suffers from specific polluters. Air pollution in a large city, for example, comes from thousands or even millions of sources—cars and trucks, factories, utilities, lawnmowers, dry cleaners, and so on. Government, therefore, has a proper role to play in solving these "market failures" and making sure that negative externalities are internalized in prices. Even Milton Friedman concedes that "third-party effects of private actions do occur that are sufficiently important to justify government action."[76]

However, governmental actions also have externalities, or third-party effects, so the burden of proof, Friedman contends, should be on proponents of particular governmental interventions to solve market failures.[77] In making sure negative externalities are internalized, Friedman says, there is "no hard and fast line how far it is appropriate to use government to accomplish jointly what it is difficult or impossible for us to accomplish separately through strictly voluntary exchange."[78] Each instance has to be decided on a case-by-case basis.

Consequently, what conservatives obviously need to do is to design appropriate policies that make the polluter pay in order to correct this "market failure." Bureaucratic command-and-control regulation is an attempt to do that, but an enormously wasteful one that is often far too ineffective. Market-based policies, such as tradable emission permits or deposit and refund schemes for proper disposal of hazardous materials, could do the job more effectively, at far less cost, and preserve flexibility and freedom for producers at the same time. Conservatives should be at the forefront, creating policies of these types to prevent pollution and make the market work more efficiently. Some of these policy options will be discussed in Chapter 4.

A second economic principle is that governmental interference in the market should be reduced to a minimum. Of primary concern here are the numerous governmental subsidies that are responsible for a tremendous amount of environmental damage. The U.S. Forest Service sells trees at prices that do not even cover the costs of the sales. The timber is so cheap that logging is "profitable" even on steep slopes with low-quality trees, places that would never be cut at all if the logging companies had to pay the costs. Our forests are clearcut because that is the easiest way to harvest the trees, even though the damage to the environment is often great. Clearcutting exposes the soil to massive erosion, which then destroys the streams and rivers as habitat and recreation areas. And it often makes replanting problematical or even impossible. American taxpayers are losing hundreds of millions of dollars every year for the privilege of having our forests and streams destroyed, forests that are far more valuable for watershed protection and recreation than they are for lumber and wood pulp. Without the subsidies to logging companies, our national forests and their watersheds would be in much better condition. (Management of our national forests is considered in greater detail in Chapter 5.)

We taxpayers also lose hundreds of millions of dollars every year subsidizing western ranchers, who raise a miniscule portion of America's beef. They overgraze the public range for prices far below the going market rate, leaving much of the land "cow bombed" and the streams beaten to muddy messes. The ranchers get cheap, subsidized water for irrigation from enormously expensive federal water projects whose dams and reservoirs have done so much damage to the rivers and streams of the West. They even get federal agents to kill coyotes for them. Without the subsidies, much damage to the ecosystems in the West would have been avoided. Remove the subsidies and many of these lands and streams will begin to recover.

But when the Clinton administration—Democrats, of all people!—tried to take a few modest steps toward the market (e.g., by raising grazing fees ever so slightly toward market rates) the Republicans denounced them for waging a war on the West![79] Hardly the reaction of principled conservatives. (See Chapter 5 for an analysis of public land grazing problems and conservative solutions.)

These are just two examples in which violating this market principle has done enormous damage to the environment. There are numerous other areas in which similar degradation is due to governmental subsidies. For several years now a coalition of fiscal conservatives and environmentalists has published an annual *Green Scissors* report in which they advocate eliminating several billion dollars from the federal budget that go to subsidizing environmental destruction. In the latest edition, for example, they identified subsidies on such things as coal, nuclear power, dams, recreation, sugar, some specific road and highway projects, flood insurance, and so on—seventy-seven wasteful programs costing $50 billion that are also harmful to the environment.[80]

PIETY IS A VIRTUE, ESPECIALLY PIETY TOWARD NATURE

The writings of conservatives include many references to the virtue of piety or reverence and its close relative, humility. As Edwin Feulner observes, "Piety is common in the conservative movement, among men and women of every religious tradition and, sometimes, none at all."[81] Conservatives believe that piety and its related attitudes define our proper approach toward the mystery of creation. We inhabit, for but a brief moment, a world that we did not create, that we cannot fully understand, and that will be here long after we are gone. Therefore, as Frank Meyer says, "in a deep sense [conservatives] must have piety toward the constitution of being."[82] Likewise, Eric Voeglin, in returning to our Hellenic and biblical heritage, finds that those views "had stressed man's dependence on transcendent being or authority and hence the need for reverence and awe—in a word, piety."[83] And Leo Strauss contends that piety is the first step in seeking comprehension of the world.[84]

Today, this attitude of piety is especially necessary toward nature, and no one has expressed this better or more forcefully than Richard Weaver. In an article titled "The Southern Tradition," published after his death, Weaver contended that

> the attitude toward nature . . . is a matter so basic to one's outlook or philosophy of life that we often tend to overlook it. Yet if we do overlook it, we find there are many things coming later which we cannot straighten out. . . . Nature [is] something which is given and something which is finally inscrutable. This is equivalent to saying that . . . it [is] the creation of a Creator. There follows from this attitude an important deduction, which is that man has a duty of veneration toward nature and the natural. Nature is not something to be fought, conquered and changed according to any human whims. To some extent, of course, it has to be used. But what man should seek in regard to nature is not a complete dominion but a *modus vivendi*—that is, a manner of living together, a coming to terms with something that was here before our time and will be here after it. The important corollary of this doctrine, it seems to me, is that man is not the lord of creation, with an omnipotent will, but a part of creation, with limitations, who ought to observe a decent humility in the face of the inscrutable.[85]

All of our conservative politicians and pundits should study this passage carefully. This is not a description of deep ecology or some New Age nature worship. This is a clear statement applying a fundamental principle of conservatism—piety—to the natural world, and it expresses an attitude that is desperately needed today.

Nature, Weaver contends, is to be revered and respected because it is "original creation." It is a divinely provided order and therefore it is "providential."[86] Nature is "the creation of a benevolent creator" and is therefore good.[87] Since nature is God given, the mystery and transcendence of the Creator can be seen in it; "nature and supernature" appear "in their inextricable involvement."[88] Weaver affirms John Crowe Ransom's contention that "out of so simple a thing as respect for the physical earth and its teeming life comes a primary joy, which is an inexhaustible source of arts and religions and philosophies."[89]

In nature, we find that we are part of a structure of reality that is independent of our own wills and desires.[90] We are a part of nature; nature is "the matrix of our being."[91] But Weaver's piety is not pantheism. It is based entirely on the Judeo-Christian realization that "the earth is the Lord's and the fulness thereof."[92] And that realization has the profound effect of undermining a fundamental tenet of the modern world, that human beings are the center of everything. Piety is "an attitude toward things which are immeasurably larger and greater than oneself without which man is an insufferably brash, conceited, and frivolous animal," which is exactly what modern people have become.[93] Weaver describes his own growth away from that falsehood, as he found himself "in decreasing sympathy with those social and political doctrines erected upon the concept of a man-dominated universe."[94]

Nature is the creation of a power far transcending our own, and we know very little about it. Even with the phenomenal advances of modern science, nature remains largely unknown and mysterious. "The wise student of her still says modestly with the soothsayer in *Antony and Cleopatra* 'in nature's infinite book of secrecy a little I can read.'"[95]

Once we realize the enormous extent of our ignorance about creation, piety comes to us as "a warning voice" that we are mere mortals and must think as mortals. We cannot know everything or control everything. We must recognize our own limitations and accept the contingency of nature, which will give us the protective virtue of humility.[96]

One aspect of the modern world and the modern mentality that Weaver vigorously condemned throughout his life is our unmitigated aggression against nature, "disfiguring her and violating her."[97] "For centuries now we have been told that our happiness requires an unrelenting assault upon [nature]; dominion, conquest, triumph—all these names have been used as if it were a military campaign. Somehow the notion has been loosed that nature is hostile to man or that her ways are offensive and slovenly, so that every step of progress is measured by how far we have altered these."[98] But "if nature is something ordained by a creator, one does not speak of 'conquering' it. The creation of a benevolent creator is something good, and conquest implies enmity and aggression."[99] This aggressive attack on nature is irreverent and consequently nothing less than a sin, from which we can only be absolved by the recovery of the ancient virtue of *pietas*.[100]

Of course, Weaver recognizes that we have to use nature to live and that requires, to some extent, changing it, but we should not force changes in nature simply from hubris or senseless presumption.[101] He accepts John Crowe Ransom's conclusion that in the past societies "have not conceived it in their interest to war continually against nature. Once a reasonable material establishment had been achieved, man turned his attention to humane ends. . . . He concludes a truce with nature, and he and nature seem to live on terms of mutual respect and amity."[102]

Changing nature beyond limited degrees usually causes at least as many problems as it solves. We know so little about nature that our meddling with parts of it almost always produces unforeseen undesirable consequences.[103] That is why old, traditional societies come to terms with the world. They accept the fact that efforts to change nature beyond a certain point will cost more than they will produce.[104]

Weaver's argument that we should approach nature in the spirit of piety is not unique to him. It was shared by the Fugitive-Agrarians who were his mentors at Vanderbilt, and Weaver expanded on John Crowe Ransom's essay in *I'll Take My Stand*, as already noted. Other noted leaders of traditionalist conservatism have also affirmed the importance of our attitude toward nature. T. S. Eliot claims that "religion, as distinguished from modern paganism, implies a life in conformity with nature. It may be observed that the natural life and the supernatural life have a conformity to each other which neither has with the mechanistic life. . . . A wrong attitude towards nature implies, somewhere, a wrong attitude towards God. . . . [We should] struggle to recover the sense of relation to nature and to God."[105]

Likewise, Russell Kirk observes that people in our century are proud of conquering nature. But if we finally succeed, the victory may end in our own destruction. Our pride must be replaced by piety, which includes respect for the natural balance in the world.[106] In his best-known book, *The Conservative Mind*, Kirk condemns our impious aggression against nature: "the modern spectacle of vanished forests and eroded lands, wasted petroleum and ruthless mining . . . is evidence of what an age without veneration does to itself and its successors."[107]

The environmental implications of piety toward nature are considerable and far-reaching. Now that we have "developed" or destroyed almost all natural areas, it is simply a pious duty to preserve the few remaining fragments of wilderness.

One of the most notable aspects of the world we inhabit is the incredible diversity of forms of life that have adapted to fill every niche on the globe. But by our actions we are wiping out whole ecosystems and causing species to become extinct at a rate hundreds or thousands of times greater than natural.[108] Piety demands that we respect all of creation and do everything we can to preserve these unique plants and animals.

Our Judeo-Christian heritage provides explicit support here. God told Noah to save *all* of the different kinds of creatures, not just the ones he thought were cute or the ones he could make a profit from later. Whether that story is interpreted literally or allegorically, the moral is the same. And there is absolutely no reason to assume that God expects any less of us today.[109]

Piety also requires us to be far less destructive of nature in producing our goods and services. As Richard Weaver concludes, the destruction we have wrought on nature is nothing less than "sin."[110] For example, clearcutting forests is the easiest way to harvest them, but it is impious in the extreme. It is enormously destructive, leaving a wake of devastation like a battlefield from World War I. It is perfectly possible to get the timber we need profitably by acting in harmony with nature, preserving forests as complete ecosystems, which protects the soil and streams and animals as well. Likewise, Wendell Berry's indictment of strip mining as it has been practiced in Appalachia points to another area of our economy which piety demands that we change.[111] And several other areas of our economy would have to be changed as well.

SOCIETY IS INTERGENERATIONAL

Another fundamental principle of conservatism is that society is intergenerational. In the words of Edmund Burke, "it is a partnership in all science; a partnership in all art; a partnership in every virtue, and in all perfection. As the ends of such a partnership cannot be obtained in many generations, it becomes a partnership not only between those who are living, but between those who are living, those who are dead, and those who are to be born."[112]

Consequently, we who are now alive have obligations of stewardship for future generations. We do not have the right to pass on to our heirs a depleted and polluted planet. In the words of Burke once again, "one of the first and most leading principles on which the commonwealth and the laws are consecrated, is lest the temporary possessors and life-renters in it, unmindful of what they have received from their ancestors, or of what is due to their posterity, should act as if they were the entire masters; that they should not think it amongst their rights to cut off the entail, or commit waste on the inheritance . . . hazarding to leave to those who come after them, a ruin instead of an habitation."[113]

Margaret Thatcher explicitly applied this principle in her environment speech: "No generation has a freehold on this earth. All we have is a life tenancy—with a full repairing lease."[114] This is a principle fully shared with environmentalists. As Sierra Club executive director Carl Pope writes, "If there is anything that has distinguished the environmental movement during the past 100 years, it has been our insistence that we not plan for a one-generation society, that the future matters."[115]

Principled conservatives need to endorse the concept of sustainability or sustainable development for our economy. As the Brundtland report defines it, this means "development that meets the needs of the present without compromising the ability of future generations to meet their own needs."[116]

This concept of sustainability should be a natural one for conservatives. William Lind and William Marshner, in a work directed toward a cultural agenda for conservatives, write that "we also seek to conserve the environment and natural resources. Our responsibility to future generations includes leaving them clean air and water, good farmland, living forests, and wilderness areas for recreation. An attitude of 'trash it up and move on' is not consistent with culturally conservative beliefs about stewardship, personal responsibility and self-discipline."[117]

Here again, properly functioning markets can go a long way toward meeting this goal. Because of the many subsidies and negative externalities involved in extractive and manufacturing industries, the prices of many goods do not reflect their true social costs. Their prices are artificially low and they are therefore too often wasted. If consumers had to pay the full costs of the goods they consume, this would be considerable motivation to conserve and use them carefully. "Identifying—and ending—hidden subsidies for pollution would dramatically advance sustainable development."[118]

Moreover, some instances of depletion and unsustainability take the form of a "tragedy of the commons." A resource that is not privately owned but is exploited by many is a commons. There is a perverse motivation built into common ownership, for example, of a pasture. Each person can gain by putting one more cow in the pasture, but if all do that, the field is destroyed by overgrazing. No one has incentive to protect the resource because someone else will just take advantage of his or her restraint. The result is that resources in common ownership are often overexploited and destroyed. A conservative solution is to establish property rights and markets where they do not now exist. In this example, the carrying capacity of the pasture would be determined. Each farmer would be issued tradable quotas to graze specific numbers of animals on that field. The total number of quotas would not exceed the pasture's carrying capacity. (Policies of this type are discussed in Chapter 9.)

We will see later that there are some areas of environmental policy for which the market by itself would be insufficient to fulfill our obligations to future generations (e.g., in providing for wilderness or protecting rare plants and animals) and that some market institutions actually undermine sustainability (e.g., discounting the future). In protecting the interests of future generations, it is the task of principled conservatives to discriminate among these, to extend the market where it would help, and to design compensating policies where it is harmful.

PRUDENCE IS THE MOST IMPORTANT POLITICAL VIRTUE

Another important principle with major implications for the environment is prudence, which is the primary conservative virtue in politics. As William Harbour observes, "Prudence is the key word for the Conservative whenever it comes to dealing with specific political problems."[119] Edmund Burke ranked prudence as "the first of all virtues."[120] One of the last books by Russell Kirk is titled *The Politics of Prudence*. Under the aegis of that virtue, Kirk advocates acting with caution, deliberation, and moderation. Prudence also means "judging any public measure by its long-run consequences."[121]

Much of the traditionalist thought about prudence stems from Edmund Burke's reactions to the French Revolution. He severely criticized the revolutionaries for overthrowing a social structure that had been built up over the centuries and attempting to create a new one based on some new abstract theories. Prudence, Burke warned, is required for social reform, if it is to have any chance of being an improvement. "The individual is foolish, but the species is wise." Consequently, no person or group has the right "to risk the very existence of their nation and their civilization upon experiments in morals and politics; for each man's private capital of intelligence is petty; it is only when a man draws upon the bank and capital of the ages, the wisdom of our ancestors, that he can act wisely." Changes should only be made with great prudence. "If we feel inclined to depart from old ways, we ought to do so only after very sober consideration of ultimate consequences. Authority, prescription, and tradition undergo in every generation a certain filtering process, by which the really archaic is discarded; yet we ought to be sure that we actually are filtering, and not merely letting our heritage run down the drain."[122]

The traditionalist conservative analysis of society and its complexity, far beyond what any one person can understand, makes prudence the primary political virtue whenever we want to make social changes of any sort. That analysis applies equally well to an ecosystem and any changes humans want to make in it. Just as a society develops over hundreds of years, so have ecosystems over millions of years. Just as society is, in Burke's words, "of the greatest possible complexity," so are ecosystems.[123] As noted biologist Jack Ward Thomas, who capped his career as chief of the U.S. Forest Service, so often reminds us, "ecosystems are not only more complex than we think, they're more complex than we *can* think."[124]

Consequently, the conservative political virtue of prudence ought a fortiori to be the primary virtue whenever people want to make any changes in ecosystems. And Richard Weaver drew just this conclusion:

> we have before us a tremendous creation which is largely inscrutable. Some of the intermediate relationships of cause and effect we can grasp and manipulate, though with these our audacity often outruns good sense and we discover that in trying to achieve one balance we have upset two others. There are, accordingly, two propositions which are hard to deny: we live in a universe which was given to us, in the sense that we did not create it; and, we don't understand very much of it. . . . Therefore, make haste slowly. It is very easy to rush into conceit in thinking about man's relationship to the created universe.[125]

We now face several environmental problems that call above all for prudence in dealing with them. In several respects, we face a new situation in history: For the first time our actions have global effects in ways that can impair the entire planet's ability to support life.

We are exterminating countless species of plants and animals worldwide, causing a collapse of biodiversity equivalent to a handful of catastrophes in geological history from which it took life on earth millions of years to recover. And we are doing this from a position of massive ignorance. We do not know what possible use these vanishing species might be for us, we do not know what roles they play in the web of life, and we do not know at what point entire ecosystems will collapse as one after another of their vital parts are removed. This is not only impious in the extreme, it is surely the height of imprudence. Indeed, biologist Edward O. Wilson warns, "This is the folly our descendants are least likely to forgive us."[126] Prudence surely demands that we halt this impoverishment of the earth and do everything we can to preserve the full range of life on the planet. (Conservative policies to preserve biodiversity are considered in Chapter 8.)

Then, of course, there is the issue of global warming. We are altering the composition of the entire atmosphere in ways that are probably already causing the global climate to warm up much faster than the rate of natural climate cycles. The exact results are uncertain, especially at the regional level, but they could well be disastrous: coastal cities could be flooded as sea levels rise, the number of extraordinarily violent storms could increase, entire ecosystems could be devastated as climate shifts faster than they can adapt. It will take years of research to know for sure what the results will be. But prudence obviously indicates that we ought to act now to slow the rate of climate change toward its natural level. The key is to become more efficient in our use of fossil fuels. And since we are very inefficient in using them—grossly wasteful in the extreme would more accurately describe the United States—the task can be accomplished now with prudent actions that can even save us a lot of money in the long run. It would be the height of imprudence to do nothing, carrying on with our wasteful ways until drastic actions might be needed or until it is entirely too late. (This daunting problem is considered in Chapters 6–7.)

These two problems have received the most publicity, but they are far from unique. For example, we rely on all sorts of "ecosystem services" to an enormous extent, for everything from purifying our air and water to generating the soil to the final recycling and processing of the wastes we produce. But ecologists find that "human activities are already impairing the flow of ecosystem services on a large scale."[127] Since these services are seldom traded in markets, they are ignored by economists and economic analysis. Here, once again, our actions are highly imprudent. (See Chapter 9.)

Conservatives are in a good philosophical position to deal with these uncontrolled and largely irreversible global experiments we are so rashly conducting. It is the conservative virtue of prudence that would make these problems salient and motivate us to take corrective (and prudent) measures before it is too late.

CONSERVATIVES ARE NOT IDEOLOGUES

Finally, we must note that conservatism is not an ideology and conservatives are not ideologues. Conservatives insist on this frequently, and they mean something very specific by the terms. "Ideology" denotes an abstract, utopian idea of the perfect society. An "ideologue" is one who believes that people and society can be perfected, if only they can be forced to conform to that utopian notion. Communism is perhaps the clearest example; it was an attempt to create a new "socialist man" by imposing Marx's notions of ideal modes of production and property relations.

Conservatism and the conservative principles do not constitute any such thing. Gerhart Niemeyer insists that "no conservative ideology can be dreamed into existence."[128] According to Russell Kirk, "the conservative abhors all forms of ideology,"[129] and he devotes the first chapter of one of his last books to identifying and refuting "the errors of ideology."[130]

Conservatives do not believe that the world is perfectible, so we must act in it always with humility. Humans are fallen beings and consequently human institutions are always going to be imperfect. Conservatives, therefore, will be pragmatic in applying their principles to the solution of real-world problems. They will not automatically look to governmental regulation for ultimate solutions, as liberals are wont to do. But nor will they expect perfect solutions from the free market, as do some true believers on the farther extremes of libertarianism. As John Gray contends, a conservative is in the best position to confront this kind of dilemma. In selecting specific policies, the conservative "will make a choice between the imperfections of markets and those of governments, in the hope that the resultant mixture will best promote freedom and community."[131]

CONCLUSION

The contemporary conservative movement and its underlying philosophy were originally developed at a time when our environmental problems had not yet become acute.[132] When applied to our current problems, as we have seen, the relevant principles of conservatism clearly and unambiguously support protection of our environment, careful stewardship of our natural resources, and preservation of our natural heritage.

How far have so many conservative politicians and pundits fallen! There is absolutely nothing conservative about pandering to developers and polluters and extractive industries that want to maximize their profits by irresponsibly maximizing their negative externalities.

It is now the job of principled conservatives to condemn these politicians and pundits whenever they go astray, and to recall them to first principles. As Richard Weaver insists, conservatives must argue and act from principle or they will govern no better than the liberals have.[133]

In the rest of this book we will consider specific environmental problems and issues to see how conservative principles can be applied in analyzing policy options, both the policies we have now and alternatives for the future. Conservative principles often suggest policies that are rather different from the ones the liberals and bureaucrats have adopted. A nonideological conservative will want to determine which would work the best, and what the respective consequences might be. But in any event, the principled conservative will want environmental policies that will be effective in protecting our air and water and land, our health, our resources, and our natural heritage.

4

Pollution

WHEN THE SUBJECT OF ENVIRONMENTAL PROTECTION comes up, probably the first thing that most people think of is pollution control, so let us start with that. There are dozens of pollution issues, but in this chapter we will focus on major types of air pollution, water pollution, and control and disposal of toxic substances.

AIR POLLUTION

The federal government first got serious about cleaning up our air with the Clean Air Act of 1970. The Environmental Protection Agency (EPA) sets emission standards for major stationary sources of air pollution that specify the kinds of technologies they must install. These standards are usually implemented at the state level through state implementation plans (SIPs). The federal government also sets emission limits for cars and trucks. These regulations have done a great deal to clean up our air, but after three decades serious problems remain. Concern today centers primarily on sulfur dioxide (SO_2), ozone (O_3), particulates, and carbon monoxide (CO). Millions of Americans live in areas that still have dangerous levels of these pollutants.

Sulfur dioxide comes mostly from utilities and factories that burn coal. It first caused concern because it produces acid rain that damages forests in the Northeast and Canada and that acidifies many lakes. Emissions of SO_2 have been greatly reduced over the years, but the ecosystems have barely begun to recover.[1] Moreover, as it turns out, the most serious effects of SO_2 are on human health. Sulfur dioxide is harmful to children, to the elderly, and to people with respiratory diseases because it produces sulfate particulates.[2]

Ozone, a primary ingredient of smog, is a form of air pollution that is produced indirectly in a complex reaction from oxides of nitrogen (NO_x) and volatile organic compounds (VOCs) when they are affected by sunlight and heat. These ozone precursors come from the burning of fossil fuels by utility generating plants, factories, and cars and trucks. Since heat and sunlight trigger the production of ozone, the problem is much worse in the summer.

Ozone has serious impacts on human health. It causes difficult breathing and is especially bad for people who suffer from asthma. It causes eye irritation, headaches, sore throats, chest pain, and nausea. Ozone also damages materials and plants, including agricultural crops.[3] Although other pollutants have been reduced over time, there has been no improvement in ozone and almost every metropolitan area has reported violations of the standards.[4]

Particulates—tiny particles emitted from burning fossil fuels in power plants, factories, cars and trucks, and so on—are also dangerous. They reach the deepest levels of the lungs and cannot be expelled. They damage health in many ways, causing everything from runny noses and hay fever to bronchitis, aggravated asthma, and heart and lung diseases. Exposure to particulates can even cause death, shortening the human life span by years.[5] In 1997, the EPA concluded, after reviewing hundreds of studies, that particulates and ozone still cause some 64,000 premature deaths in the United States each year.[6]

Carbon monoxide comes mostly from cars and trucks. Concentrations can be especially bad at "hot spots" in cities where traffic congestion is heavy. Carbon monoxide is especially harmful for people with cardiovascular diseases.[7]

Improvements in air quality so far have produced tremendous benefits. The EPA estimates that between 1970 and 1990 implementation of air quality regulations cost approximately $523 billion and has avoided some $22.2 trillion worth of harm.[8] But air pollution is still a serious problem in the United States. Approximately 80 million people live in counties that still do not meet air quality standards for at least one major pollutant.[9] Further improvements would prevent further damage.

Conservative principles impel us to take action. Pollution is a negative externality that should be internalized for economic efficiency. More importantly, air pollution violates the property rights of its victims and infringes on their freedom by imposing costs and suffering that should not be theirs to bear. Polluters are evading a responsibility that is clearly theirs.

So, let us take a look at recent and current policies to control air pollution. Believe it or not, those policies are evolving in directions conservatives should approve of, at least in dealing with emissions from stationary sources, which we will consider first.

When the federal government first got involved in controlling air pollution, it adopted a command-and-control approach that set limits on emissions from each plant and specified the technologies to be used in meeting them. This regulatory approach had significant successes, especially in the early years, and air quality improved in many areas.

However, regulations that specify technologies have major limitations and a fairly standard list of them can be found in the literature. Command and control allows for little or no flexibility in meeting the requirements. It imposes a "one size fits all" solution, regardless of costs and regardless of the differences in costs from one plant to another. Command and control provides no incentives for innovation. Specific technologies are required and consequently there is nothing to be gained by looking for better ways to control emissions. For the same reason, there is no incentive to reduce emissions below the required level. Command and control also creates problems because it imposes tougher standards on new plants than on old ones. This creates a barrier to entry in the marketplace as well as incentives to keep old, dirty plants in operation instead of replacing them with new, cleaner ones. Command-and-control regulations typically do not place a limit on total pollution; they just limit emissions from each plant.[10]

On top of those inherent limitations, command-and-control regulation is now reaching its limits. It is relatively easy to regulate emissions from a limited number of large sources, such as power plants and large factories. That has mostly been done, and now the problems we face increasingly come from smaller sources—thousands and millions of them, which would be virtually impossible to regulate.[11]

Economists have long recommended turning away from command and control toward market-based incentives that would get people and companies to make their everyday decisions in ways that help the environment. Pollution is an externality, which means that the prices of products that cause it are too low. Consequently, consumers demand too many of them. For economic efficiency, the key is to get the prices right and include the full social costs of the damages that pollution causes.[12] Economists generally recommend two mechanisms for doing that: pollution taxes or tradable emission permits.

In the first case, the government would estimate the marginal cost of damages caused by pollution and then tax emissions at that rate. This typical recommendation by economists is based on cost-benefit analysis, but a pollution tax can also be adjusted to reduce emissions to a level based on human health and environmental amenity standards, without putting a price tag on them. After all, we try to reduce pollution not to confer benefits but to prevent harms that, according to conservative principles, the polluters have no right to inflict in the first place.[13] In either case, factories would reduce their emissions as long as doing so is cheaper than paying the tax.[14] (Conservatives would not want the size of government to increase, so the new taxes could be made revenue neutral by reducing income and payroll taxes.)

There are, however, some problems with pollution taxes. It is often difficult to place a price on the damages from pollution, so it is difficult to determine how high the tax should be. And a pollution tax does not place a limit on pollution; it merely raises its price. Tradable permits offer a better alternative.

The United States began experimenting with tradable emission quotas in the 1970s and 1980s, and they are now a major part of our air pollution control policies. Conservatives should applaud. In a tradable permit scheme, the government establishes a cap on total emissions. Quotas (allowances) are then given (or auctioned) to companies, totaling no more than the cap. When the goal is to reduce the level of emissions, the cap and quotas can be reduced over time until the goal is reached. A company can release emissions up to the level of its quota. It can reduce emissions below its quota and make a profit by selling the remaining allowances. An older plant, for which reducing emissions would be very expensive, could emit more than its quota by buying surplus allowances from others. (A cap-and-trade system thus does not "grandfather" old, dirty factories, which is one of its major advantages.)[15] To build a new plant, a company would have to buy surplus allowances to operate. This scheme creates maximum flexibility for each company. The regulatory agency does not specify how anyone has to go about limiting emissions, yet total emissions are capped. The emission quotas are not "rights to pollute"—they are temporary permits. They can be reduced year by year until air quality has improved to meet the standards.[16]

There are two keys that make a tradable quota system superior to command and control. First, it uses overall performance standards, not technology standards that have to be met by each plant. This gives every company flexibility to meet the standards in the best and cheapest way for its own operations.

Second, the costs of emission control vary greatly from one plant to another. This is what makes trading work. Costs for some can be as much as 100 times greater than the costs for others.[17] To reduce pollution at the least cost, the companies for which pollution control is cheapest should reduce emissions the most. Trading makes it profitable for them to do so because they can sell or lease their unused allowances to companies for which cleanup would be very expensive. (In economic terms, if the system works perfectly, the marginal costs of pollution reduction would end up being the same for all sources.) At the same time, the total amount of emissions is capped and can be reduced over time.

A cap-and-trade system can save huge amounts of money, compared to command-and-control regulation. Tietenberg summarizes the findings of a number of studies of various pollutants: command-and-control regulations can be as much as twenty-two times more expensive than the least costly means of control. Typical conclusions are that regulation is four to six times more expensive.[18] (These are theoretical savings. Real-world savings under cap-and-trade programs are less, but they are still quite significant.) Alternatively, for the same expenditures on pollution control, we could get much cleaner air by switching from command and control to cap and trade and tightening the standards even more.[19]

In the 1970s the EPA began experimenting with a few features of a marketable permit system. If a company wanted to build a new plant in an area that did not meet air-quality standards, it could do so by obtaining a reduction of emissions from other sources ("offsets"). Firms that had multiple sources of emissions could combine them under a "bubble" and use emission reductions from one source to offset increases from others in the total package. "Netting" allowed emissions reductions from a bubble to offset increases elsewhere. "Banking" allowed companies to reduce emissions and save the credits for use later.[20]

Even though they involved a lot of bureaucracy and red tape, these modest steps saved a lot of money without harming air quality. Hahn and Stavins estimate total savings at between $5 and $12 billion.[21]

The first real cap-and-trade program was instituted in the 1980s, when lead was being phased out of use as an additive in gasoline. Some refineries, especially smaller ones, would have difficulty meeting the new standards and would benefit from flexibility for a period of time. For the few years of the transition, refineries could reduce their use of lead more than required, earning credits that could be banked for use later or sold to other refineries that found it expensive to reduce their use of lead. An active market developed and the use of lead shifted among refineries, but it did not increase the total amount that could be used. The EPA estimated that banking alone saved refineries $226 million. Hahn and Hester concluded that total savings over those few years were much higher than that.[22]

In 1990 Congress passed amendments to the Clean Air Act that included a major shift of policy, away from command and control to a full marketable permit system for SO_2 from power plants. The system was implemented in two phases. Phase 1 began in 1995 and included 110 large utilities. Phase 2 began January 1, 2000, and includes virtually all power plants in the country. The law caps SO_2 emissions and reduces the cap so that by 2010 total annual emissions will be cut by 10 million tons from the level of the 1980s. Each plant was issued a quota of allowances based on its historic level of emissions. The quotas will be reduced over the years until the target is reached. Each plant must monitor its emissions continuously and must surrender allowances to cover its emissions. The allowances are dated and are fully marketable. They cannot be used until the year of their date, but they can be saved (banked) and used in later years. They can be bought and sold and leased by anyone on the open market. If an environmental group wants to help clean the air, it can buy allowances and retire them. No governmental approval is required for sales.[23]

Under the Clean Air Act Amendments each utility has complete flexibility to decide how to deal with its emissions. It can use "scrubbers" to remove sulfur; it can shift from eastern high-sulfur coal to western low-sulfur coal; it can blend fuels to meet its targets; it can retire older, dirty plants. Each utility can decide whether to reduce emissions just to the level of its quota, reduce emissions more than required and bank or sell the surplus allowances, or emit more than its quota and buy enough allowances on the market to make up the difference.

When the amendments were being considered, there was much speculation about what the effects would be—this was, after all, the very first large-scale experiment with a cap-and-trade system. Economic theory predicted that the marketable permit system would save a lot of money, compared with the regulatory alternative, which was, essentially, to require all plants to use scrubbers.

It was generally expected that the cost of reducing SO_2—and therefore the price of the allowances—would be as high as $500 to $700 per ton. But no one anticipated what actually happened. By 1994 an active market had developed and the price of emission allowances was not even close to the anticipated level. Early sales were in the $150 to $200 range and the price was continually coming down. By 1996 allowances were selling for as low as $63; they later stabilized at about $100 per ton.[24]

Several factors contribute to the low cost of controlling SO_2, but it is the flexibility allowed by the cap-and-trade system that lets the utilities take advantage of them. One major factor resulted from the deregulation of railroads: Their prices for shipping western low-sulfur coal came way down. Since transportation is a major part of the cost of coal, this made it economically feasible for plants farther and farther east to switch from eastern high-sulfur coal. Faced with this competition, the manufacturers of scrubbers developed new models and cut their prices drastically for utilities that burn high-sulfur coal. And the market for allowances lets utilities save on capital costs because they no longer have to invest in backup scrubbers in case one breaks down or has to be taken out of service for maintenance. Now they can buy allowances to cover the extra emissions while repairing a scrubber.[25]

The utilities included in phase 1 reduced their emissions far more than required: from 23 to 39 percent below requirements in 1995–1997.[26] Initially there were few trades, presumably because these utilities were banking their surpluses in anticipation of tighter standards under phase 2. Even with limited trading, the utilities were saving $1 billion per year.[27] The GAO estimates that by 2002 utilities could save $3.1 billion each year if they were to use trading to its full potential.[28] (These figures are comparisons with the costs of achieving the same emission reductions under command and control regulations.)

The success of the tradable SO_2 quota system confirms the economists' theory that it would be a major improvement over command and control regulation. Joskow and colleagues recommend that its success "supports further use of the tradable permit approach to environmental policy."[29] And that is precisely what is happening. The EPA is encouraging states and regions to adopt tradable permit programs, especially to control ozone and its precursors. This opens up a wider scope for market-based policies because NO_x comes from a variety of sources. Approximately one-third of NO_x emissions comes from power plants, but another third comes from cars and light trucks. The remainder is largely from semitrailers, buses, and off-road construction vehicles.[30]

Several trading systems are already in place. The RECLAIM program in southern California covers over 400 of the largest sources of NO_x and SO_2, from a wide diversity of industries. The plants included in the cap-and-trade system account for about 70 percent of the emissions from stationary sources. Smaller emitters can opt into the trading system or remain subject to command-and-control regulation. RECLAIM has succeeded in reducing NO_x by 35 percent and SO_2 by 25 percent, at a price 55 percent cheaper than regulations on each plant would cost.[31]

In the Chicago area ozone is a serious problem in the summertime. The Illinois Environmental Protection Agency developed a cap-and-trade program for ozone precursors for the region, and a similar program may be extended upwind.[32]

States in the Northeast have long faced a serious problem with ozone that is especially difficult to solve. They are downwind from the industrial regions of the Midwest, and thus much of their ozone and ozone precursors are blown in from elsewhere. In fact, the air coming into Pennsylvania already violates air-quality standards, and a study by the state of Connecticut concludes that it could completely shut down—turn off all factories and generating plants, park all trucks and cars—and it still would not meet air-quality standards.[33]

The Clean Air Act has a provision under which a downwind state can petition the EPA to require stricter pollution controls on upwind states. But until very recently, although several northeastern states have filed petitions over the years, the EPA had never granted a single one.[34] Of course, the upwind states have always done everything they can within the law to weasel out of taking responsibility for the damage they are doing to their neighbors.[35] At long last, the EPA finally granted a petition and ordered these utilities to reduce their NO_x emissions.[36]

The Northeast is also approaching the problem of ozone transport with a cap-and-trade policy. States have internal trading programs and there are several interstate trading agreements.[37] Estimates for the EPA indicate that trading will save 30 percent beyond the savings achievable under a cap with no trading allowed.[38]

Since ozone precursors come from many different sources, cap-and-trade policies can include more innovative provisions than SO_2 trading programs. One of these, used in both California and Chicago, recognizes that older cars are a significant source of pollution. Although the cap only applies to major stationary sources, these two programs let them go outside to earn emission reduction credits by buying and scrapping old vehicles. In a test program in Los Angeles, Unocal Corporation offered $700 each for old cars and removed over 8,000 of them.[39] (There may, however, be a more effective approach to scrappage programs than basing them on the age of the vehicles; see below.)

There are probably many other ways to reduce ozone and ozone precursors that could be incorporated as ways to earn emission reduction credits. For example, many businesses that are too small to be included in cap-and-trade systems emit pollutants. Add up all of these minor sources, and the total is a serious problem. Programs geared to the specific needs of small businesses that reduce their emissions could earn credits for major sources. Many cities have found that substantial reductions in air pollution from small businesses can be obtained simply by working with them. After being shown how to reduce emissions, many small businesses are eager to do their part.[40] Another possibility is to let a company earn NO_x credits by purchasing natural gas powered buses for the city to replace dirty diesel ones.[41] Fuel cell buses could be even better.

The kinds of market-based programs discussed here will probably be expanded in the future. The South, with a much hotter climate than the Midwest or Northeast, is suffering from air pollution problems that are getting worse. The Atlanta area, for example, faces serious smog problems that rapid growth is making ever worse. Recently Houston passed Los Angeles for the dubious honor of having the worst air quality in the country. These areas could probably improve their air considerably with cap-and-trade policies, at much less cost than direct command-and-control regulation. There are also proposals to develop multipollutant trading schemes. That would greatly increase the possibilities for market-based policies, if formulas could be developed to rate the relative risks that different pollutants pose to human health.[42]

Conservatives should applaud the expanding number of cap-and-trade pollution control programs. They hold industries responsible for cleaning up their acts, while giving them freedom to decide how to meet that responsibility. They even hold older plants responsible; command-and-control regulations typically undermine cleanup by grandfathering them. Cap-and-trade programs eliminate bureaucratic meddling in each plant's operations and replace it with market-based incentives and instruments. On top of all that, they let us meet our air-quality goals at much less cost than command-and-control regulations. These policies do not work everywhere, but wherever they can be employed, conservatives should actively support them.

Although tradable permit systems work well in many cases, they are not appropriate for all pollutants. Emissions that cause toxic "hot spots," cancers, or birth defects need to be controlled directly at each source. Currently, the EPA sets limits for each plant and prescribes the acceptable ways of meeting them. This has resulted in a heavy and costly regulatory hand on industry. And here conservatives should be interested in a fascinating but little-known study.

In 1989 the EPA and Amoco agreed to do a detailed study of the company's refinery in Yorktown, Virginia.[43] Refineries emit complex combinations of hydrocarbons that contribute to smog, and some of them are highly carcinogenic. In this study, they focused on hydrocarbons overall and on benzene in particular, which is known to cause cancer. They conducted a detailed analysis of every aspect of the plant to obtain a complete inventory of emissions from every possible source in the process of refining crude oil. With the help of outside experts in a workshop, they then identified the most cost-effective means of reducing emissions. They discovered that the refinery could meet its emission targets for about one-fourth of the cost of the actions the EPA regulations required.

This study deserves to be better known because it has some important lessons for policy. Every now and then you may run across a reference to it by an industry representative, but most misinterpret the study as a condemnation of ignorant and distant bureaucrats imposing costly, ineffective, and unnecessary regulations on a virtuous industry that could have done much better on its own. But that is most certainly *not* the lesson this study teaches.

The EPA regulations required the refinery to adopt end-of-the-pipe technology to control emissions from what were assumed to be the major sources. But it turned out that neither the EPA *nor the company* knew where most of the emissions were actually coming from. Both the EPA *and the company* assumed that what turned out to be a major source of emissions was only a minor one. *The company did not even know* about several options to reduce emissions that produced net savings. Moreover, the monitoring process to discover where emissions were really coming from—checking all of the seals, vents, loading mechanisms, and so on through the entire refining process—was very difficult and time-consuming. (In a small refinery there can be over 10,000 possible sources of emissions.) And it was something the company had never done before. None of the data the company had previously gathered (e.g., for toxic release reports) came anywhere close to disclosing the major sources of pollution from this plant. On top of all that, the study concludes that a key to its discoveries was the brainstorming workshop, in which the results of the monitoring were presented to outside experts who generated ideas for controlling emissions. This produced possible solutions much more effectively and rapidly than the agency or the company could have done on its own.

The bottom line of the study is most impressive. Using the best ideas generated in the workshop, the Yorktown refinery could cut its emissions by 95 percent of the amount required for 20–25 percent of the cost of the regulatory requirements. For just a bit more, total emissions could actually be cut by more than required.

The amounts of money involved are substantial. To meet the regulatory requirements and reduce emissions by 7,300 tons per year would involve a capital cost of $53.6 million and annual costs of $17.5 million. But using the most cost-effective options discovered, the Yorktown refinery could reduce emissions by 7,500 tons at a capital cost of $10.7 million and annual costs of $3.8 million.[44] (Emissions at that time were 15,500 tons per year, 90 percent of which were released into the air.)[45]

The conclusions of this study include numerous suggestions for reducing emissions more cost-effectively that could be applied widely. For example, at many plants, the cost of controlling emissions is charged to a separate budget, not to the departments that generate the pollution. The departments, therefore, have no incentives to manage their emissions more effectively.

But for environmental policy, the most important conclusion is that "Congress, EPA and much of industry have become used to command-and-control, end-of-pipe treatment approaches based on twenty years of experience. These well established problem solving approaches are difficult to change. . . . [However], many of today's problems are sufficiently different than those of the early 1970s that they can benefit from alternative approaches."[46]

Pollutants that cause smog and cancer and birth defects clearly need to be controlled. It seems to me, on the basis of the Amoco study, that a viable conservative policy would be to give companies the option of alternative approaches. They could stay with the status quo, obey governmental regulations, and adopt the technologies they specify; or they could propose alternatives that would be at least as effective in controlling emissions. If a company is willing to devote the effort and expense to detailed monitoring of its plants in order to discover more cost-effective methods of pollution control, it should be encouraged to do so. Perhaps a regulatory agency has to operate initially on the assumption that one size fits all; it cannot possibly study every facility that emits pollution. But individual companies should be given flexibility to prove that different approaches would be at least as effective in their plants. The regulatory agency, that is, may have to operate initially on technology-based standards but individual plants should have the option of meeting equivalent performance standards. The Yorktown refinery study indicates that this will not be easy for companies to do, but if they invest the time and effort to monitor their entire plant, the end result could be substantial savings in addition to protection of the public from pollution. Giving them this option looks like sound conservative policy.[47]

The policies considered so far concern stationary sources. Even the old car scrappage programs are used to generate emission reduction credits for utilities and refineries and factories. But mobile sources are also significant. On-road vehicles spew out 64 percent of total carbon monoxide, 35 percent of NO_x, and 27 percent of VOCs.[48] Fully half of all smog in major cities comes from vehicles.[49]

Cars and light trucks have to meet emission standards. Trucks get away with more pollution than cars do, but that is being corrected. And the EPA is now beginning the process of reducing emissions from semitrailers and buses.[50] Off-road construction vehicles, however, still have no emission controls, so an obvious next step is to include them with semis and buses. There is no good reason why these machines should not have pollution control standards applied to them, especially if they are to be used in areas that do not meet air-quality standards.

Personal transportation vehicles have pollution-control devices installed at the factory, but there is no guarantee that they work very well. Some do not and here we find a significant opportunity to reduce air pollution in our cities. In fact, most pollution from cars and light trucks comes from a small percentage of them and not just older ones. Studies have consistently shown that half of all carbon monoxide from vehicles is emitted by just 7 percent of all vehicles, and half of all hydrocarbons (VOCs) from vehicles is emitted by just 10 percent of them. Although emissions do tend to increase with the age of the vehicle, that is not the most important factor. Any well-maintained car or truck will be fairly clean. But the dirtiest 20 percent of vehicles from any model year, even very recent years, emit a hugely disproportionate amount of pollutants. They are superemitters because they are poorly maintained or their pollution controls have been tampered with.[51]

A commonly proposed solution is to require cars and trucks to be inspected regularly, with ever more expensive and time-consuming procedures, and to require their owners to maintain them properly. These enhanced inspection programs are very unpopular and always face major political barriers.

Conservatives believe that all of us should take responsibility for the consequences of our actions. All of us, therefore, are responsible to maintain our vehicles and limit our pollution. But there seems to be a way of enforcing that responsibility that would be less costly, less intrusive, and less demanding on the average driver than frequent mandatory inspections for all, and that is "on-road monitoring." Remote sensing devices on the side of the road can measure emissions of cars as they go by, even at very high speeds. The monitors can be installed in vans, so they can be both inconspicuous and relocated frequently.[52]

Since most pollution comes from relatively few vehicles, these monitors can be used to identify the superemitters. Once they are identified, there are several options policy could take. A successful experiment in Denver was entirely voluntary. A monitor measured emissions as vehicles passed and a camera photographed their license plates. An automated sign told each driver the level of his or her emissions—poor, fair, or good. Drivers whose cars got poor and fair ratings were sent letters reminding them to repair their vehicles. Just one monitor and sign on a busy freeway exit ramp got 16,000 drivers to repair their vehicles voluntarily, lowering CO levels in the area by thirteen tons per day.[53]

A proposal by Daniel Dudek of the Environmental Defense Fund would let private individuals or companies use remote sensing to identify superemitters. They would then offer to buy and scrap those vehicles to earn emission reduction credits that could be marketed in a cap-and-trade system.[54] This would be more effective in cleaning the air than scrapping cars based solely on their age. If a reasonable formula could be devised, they could repair high emitters to earn credits even more cost-effectively. Scrappage programs pay $700 to $1,000 per vehicle, but it only costs about $200 on average to repair a high emitter.[55]And, of course, high emitters could simply be required to have their vehicles fixed, or be given the choice of repairing them or accepting a scrappage bounty.[56]

With remote sensing, full inspections that are required of all cars and trucks could be spaced farther apart—say every four years instead of every one or two. Thus remote sensing can be used to make a mandatory inspection and maintenance program less intrusive and less burdensome on the average driver. Alternatively, remote sensing could be used to identify high polluters and full inspections could be required only for those vehicles. (Remote sensing is not as accurate as full inspection, so complete tests are still important before requiring the owners to make repairs.) Major emission reductions would be achievable that way at much less cost and much less hassle than annual, universal inspections.[57]

Over the last three decades, emission limits on vehicles have become ever more stringent. But much of the gain in lower emissions has been offset by greater numbers of cars and trucks, and by an increase in the number of miles they are typically driven. From 1969 to 1990, vehicle miles traveled doubled. The number of vehicles rose, and the average car was driven 22 percent more miles in 1990 than in 1969.[58] By 2005 these factors will overwhelm tighter emission controls and total pollution from cars and light trucks will increase once again.[59] So, for the future, travel demand management—policies to reduce miles traveled in personal vehicles—will become increasingly important to improve air quality.

The federal government recently tried a command-and-control policy to reduce miles driven, requiring employers to institute travel reduction programs for their employees. The goal was to reduce the number of people driving alone to work, since some 73 percent of us drive to work by ourselves and trips to work account for one-third of our total mileage.[60] Companies had to try to reduce the number of personal vehicles brought to work each day by organizing car pools, encouraging use of mass transit, providing van service, and the like. Naturally, these regulations were extremely unpopular, a political disaster—as any conservative could have predicted—and Congress soon repealed them.[61]

But the problem of pollution from vehicles remains and gets worse. In the future, states' transportation policies and planning will have to conform to the Clean Air Act air-quality requirements.[62] Merely tightening pollution controls on factories and utilities no longer produces sufficient gains. We need to come up with innovative policies that change the incentives for the use of personal vehicles. Several proposals should be attractive to conservatives.

One intriguing proposal would tackle the problem of people driving alone to work, but in a way that is totally different from the federal regulatory disaster. Almost all employers provide free parking for their employees and some 93 percent of all auto commuters park for free.[63] Free employee parking is also subsidized, since employers can deduct it as a business expense. This proposal would require employers who provide free parking to offer their employees a choice: they could take the parking space or they could take its cash equivalent. A significant number of people, it turns out, place less value on the parking space than it costs their employers. These people would opt for the cash and find alternative ways to get to work. Where it has been tried, solo driving to work has declined by 20 percent. If all employers adopted it, total household vehicle miles traveled would be cut by 5 percent and gasoline consumption would drop by 4.5 billion gallons per year.[64] Expanding choices is surely something conservatives should support.

A broader and more general approach to reducing driving, which conservative principles would support, is to move toward full-cost pricing in transportation by removing subsidies. Drivers pay only a fraction of the costs of driving. As we will see in greater detail in Chapter 7, all sorts of those costs are passed on to the general taxpayers—everything from a substantial portion of the costs of building and maintaining the roads to emergency and support services to a large part of the costs of accidents. And all of the environmental externalities are passed on to their victims: all of the health impacts of pollution from cars and trucks.

The direct costs of driving, according to market principles, are seriously underpriced. Consequently, we do too much of it.[65] All of us who drive should face up to the fact that we too are welfare bums. We should get off the dole and pay our own way.

Kessler and Schroeder cite a survey of English drivers who said that 30 percent of their car mileage was "not at all" or "not very" important.[66] Results would likely be quite similar (or even higher) for U.S. drivers. If we had to pay the full costs, many of these trips would be eliminated. If gasoline taxes covered all costs, demand for driving would go down, economic efficiency would improve, and pollution would be cut substantially.[67] And the general taxes that now subsidize driving could be cut.

As a complement to offering cash for employee parking and full-cost pricing of driving, there is another policy that should appeal to conservatives—opening the market to paratransit services. Regulatory barriers in our cities and states currently prevent private operators from offering consumers a variety of services, such as "dial a rides" and fixed-route jitney services. "Local and state governments can improve travel alternatives at no cost by removing restrictions on entry and exit, fares, and scheduling of private transit services and concentrating instead on ensuring driver fitness and safety. . . . Expanded transit alternatives would strengthen pricing mechanisms as tools for reducing the amount of [single occupant vehicle] driving and thus emissions."[68] This would expand consumer choices by making available convenient alternatives to driving solo.

Other incentive-based proposals have been made, such as feebates for cleaner cars (buyers of higher-emission cars pay extra fees, which are used to give rebates to buyers of cleaner cars), congestion pricing (charging more to use roads during rush hours than at other times—which may or may not reduce pollution), and taxing vehicles on the basis of miles traveled each year, adjusted for emission levels.[69] Longer-term proposals include zoning and urban planning changes so that more people could live, work, and shop within walking distance.

The challenge for conservatives is to evaluate proposals such as these and to design new ones that would change incentives so people would drive—and pollute—less. Travel-demand management is clearly the next major challenge in solving air pollution. Conservatives should be actively involved in developing market-based policies to avoid the imposition of command-and-control restrictions on our choices of transportation.

WATER POLLUTION

The federal government first got serious about reversing the pollution of our rivers and lakes with the Clean Water Act of 1972. This act divides sources of water pollution into two types, point and nonpoint. The CWA focuses almost entirely on the former, in a classic command-and-control system.

Point sources of pollution are just what the name indicates: dischargers of effluent that goes into a river or lake from a readily identifiable point. This includes factories that discharge wastewater from a pipe or ditch, and it includes city sewage treatment plants. These facilities must obtain permits that require them to use specific technologies to treat the water before discharging it. There are also pretreatment requirements for factories that send their effluents into municipal sewer systems. As with the Clean Air Act, the permits are typically administered by the states, according to EPA standards.

Nonpoint pollution is primarily rainwater runoff from farms and cities. The important differences from point sources of pollution are that nonpoint pollution cannot be monitored or measured directly at its source; it is episodic and random, depending on rainstorms, snowmelt, and the like; and there is only imperfect knowledge about the relationship between inputs (e.g., fertilizers and pesticides) and resulting pollution of rivers and lakes.[70]

Polluted water from point sources can be treated before it is sent into a river. Control of nonpoint pollution, on the other hand, depends largely on management practices, such as reduced use of fertilizer and pesticides, conservation tillage to prevent erosion that sends sediment and chemicals into the water, buffer strips between cropped fields and water bodies to filter out sediment and chemicals, animal waste treatment lagoons for feedlots, and the like. (The Clean Water Act specifically excludes small feedlots and irrigation return flow from the point source category, even though they are often like point sources. Factory farms, however, are considered point sources.)[71]

Nonpoint pollution is basically uncontrolled under the CWA. The act leaves it entirely up to the states to deal with, since controls are site specific and depend on land use management, which is traditionally a state and local concern.[72] Although the act encourages states to develop best management practices (BMPs) programs for farms, they are entirely voluntary.[73] The end result is that, until very recently, nonpoint pollution has been basically ignored.

Strictly controlling point sources has been fairly successful and over the years, water quality in the United States has improved. Unfortunately, the job of cleaning up our waters is far from done. Fish consumption advisories have increased 72 percent in the last few years.[74] Pollution coming down the Mississippi River creates a huge dead zone in the Gulf of Mexico every summer some 5,000 to 7,000 square miles in extent.[75] The EPA reports that 38 percent of all rivers, 44 percent of all lakes, and 32 percent of all estuaries are still polluted.[76] And the agency has identified 18,000 water bodies that will not attain water quality standards even if all point sources meet their technology-based requirements.[77]

To make further progress on cleaning up our rivers and lakes, we must tackle nonpoint sources as well. "Nonpoint source pollution, primarily agricultural runoff of nutrients and sediment, is the chief cause of impaired surface water quality today."[78] Almost half of the current load of pollutants in our waters comes from nonpoint sources. Nearly half of all estuaries in our country—the most productive of all water bodies—are susceptible to eutrophication from agricultural pollutants.[79] (The nutrients stimulate algae blooms, which use up all of the oxygen in the water and block sunlight. At that point, nothing can live in that water.) According to the EPA, of all polluted water bodies, agricultural runoff affects 57 percent of the lake acres and 60 percent of the river and stream miles.[80] Clearly, controlling only point sources is insufficient to clean up our nation's waters. In the future, nonpoint sources will also have to accept responsibility for the damage they cause.

Recently, a long-ignored provision of the CWA has come into play that changes the system considerably. According to the law, if a water body cannot be brought up to its quality standards by technology-based regulation of point sources, the government must take a different approach. It must determine the total maximum daily load (TMDL) of pollutants that the water can receive and still meet its standard. That total budget then must be allocated among point sources, nonpoint sources, natural causes, and a safety margin that must be included.

As so often in the history of environmental law, the TMDL provision has long been on the books but the EPA simply ignored it. It took a series of citizen lawsuits to force the agency to enforce the law.[81] Now the EPA refers to TMDLs as the "technical backbone" of its water quality program.[82]

TMDLs force the EPA and the states to look at each watershed as a whole, rather than just at individual point sources. They have to consider water quality and everything that affects it.[83] If the technology standards in discharge permits are insufficient to assure clean water, other measures must be taken. TMDLs also force communities to face possible limits on growth—which should get their attention and concentrate their minds mightily.[84] They must confront the possibility that discharge permits may have to be denied, even for good facilities, if the water quality is still impaired. That means that they can no longer ignore nonpoint sources. "Local governments will no longer be able to avoid curbing livestock and urban runoff when rivers and lakes don't meet water-quality standards, since permits for other businesses and for sewage plants will be at risk."[85] But, as we will soon see, this new TMDL approach may also open up some new opportunities for cost-effective pollution control through trading schemes that are similar to the ones now used to control air pollution.

A few trading programs are now in place. However, they are very limited and largely ineffective. Two that have received much attention are for the Dillon Reservoir in Colorado and the Tar-Pamlico River basin in North Carolina. Dillon Reservoir faced a potential problem of eutrophication from nutrient pollution. Local officials developed a phosphorus budget for the reservoir and allocated the load to all sources. They determined that it would be cheapest to control nonpoint pollution. So, in order to accommodate anticipated growth, they allowed sewage treatment plants to increase their discharges if they would finance nonpoint source control projects on a 2 to 1 ratio (i.e., to increase their discharges by one unit, they would have to reduce nonpoint pollution by two units). However, the sewage plants upgraded their operations so that even with complete buildout they will be below their allotment and they will never need to trade.[86]

The Tar-Pamlico program was considerably more ambitious. Municipal sewage treatment plants and a few factories on the river were facing more stringent standards from the state government, which they estimated would be very expensive to meet. Several of them formed an association that proposed an alternative plan. The association funded a computer model of the watershed, to determine where the sources of pollution were and how the pollutants were transported through the river basin. They then proposed a scheme in which the members of the association could trade pollution allowances among themselves, with a cap on their collective discharges. Since it would be much cheaper to reduce nonpoint pollution than to upgrade their plants, they proposed that if they exceeded that cap, they would contribute to a state fund that financed BMP runoff controls on farms in the area. Trades of that type would also allow for future growth of the cities. The new state standards would be met, but at much less cost than if the sewage plants had to bear all of the burden. (The association calculated that the state plan would cost them between $50 and $100 million; their proposal would meet the same water quality standards for $11.7 million.)

The association proposal was accepted and is now in place. However, the members hired an engineer to help them optimize their current facilities—and he showed them how to reduce their discharges so much, with no additional capital investment, that none of them has yet had any reason to engage in trading.[87]

Although these trading programs have not been active, the EPA is encouraging other watersheds to develop similar policies. The agency has even published detailed guidelines for watershed-based trading.[88] Just as cap-and-trade policies for air pollution have much to recommend them to conservatives, so do these. However, they face many more barriers than air pollution policies face, and these difficulties come from both law and nature.

Although the Clean Water Act permits cap-and-trade programs, several of its provisions pose major barriers to them. Discharge permits are based on specific technologies, and so are pretreatment requirements for plants that send their wastewater into city sewers. But for trades to work, each facility needs to be able to decide for itself how to meet its limits. The act also has "antibacksliding" and "antidegradation" provisions for each factory and treatment plant. These also remove flexibility from individual facilities. Once a waste treatment system is in place, a company cannot increase emissions (backslide) and make up for it elsewhere by buying surplus allowances in a trade.[89] These provisions severely limit the possibility of any trades, between point sources or between point and nonpoint sources. (The Tar-Pamlico association was able to devise its trading system because it was dealing with nutrients—nitrogen and phosphorus—that were not governed by CWA permits.)

Moreover, nonpoint sources have not been required to do anything to reduce their pollution of our rivers and lakes. Consequently, the kinds of trading policies for Dillon Reservoir and the Tar-Pamlico basin—which are just the kinds of schemes the EPA is now advocating—are not like trades under the Clean Air Act, where utilities that clean up more than is required can profit by it. They are more like bribes or subsidies paid to farmers to get them to do something they do not have to do, so that a factory or city can avoid having to do something very expensive.[90]

The legal situation is now changing, and the new structure, with TMDLs, may well open up more and better opportunities for market-based control of water pollution. The most important change being made is, at long last, to include nonpoint sources in pollution-control policies. This is important for two reasons. First, as we have seen, nonpoint sources are major polluters, so if we do not tackle them we will not be able to clean up our rivers and lakes. Second, it will be much cheaper to clean up pollution from runoff than to force point sources to do even more than they are now required to do.[91]

States must now include nonpoint sources in their pollution-control efforts because the TMDLs set limits on discharges from all types of sources. To meet TMDLs, the states will have to set pollution budgets for runoff from farms and cities and take steps to make sure they do not exceed them. Some states are already adopting land-use regulations to control agricultural runoff.[92] And under the Coastal Zone Management Act, states on the coasts are required to enforce BMP runoff controls.[93] Once nonpoint sources are required to control their discharges, they should be interested in participating in trading programs for maximum cost-effectiveness.[94] There is also a provision in the CWA that may allow point sources to avoid the antibacksliding and technology-based provisions for water bodies subject to TMDLs.[95]

The second barrier to increased use of cap-and-trade programs for water pollution is the nature of nonpoint pollution. It has long been assumed that since it is diffuse, it cannot be measured. And since runoff is not continuous like a factory's discharges but random due to rainstorms and snowmelt, it would be very difficult to control. These assumptions are now being challenged. If they can be overcome and methods can be developed to monitor and control runoff, trades can be facilitated. (You have to be able to measure and control amounts of pollution in order to trade them.)

Some agricultural sources that are exempt from point source regulations, such as small feedlots, are actually like point sources and could be treated as such.[96] In some places, diffuse runoff could be collected and channeled, allowing it to be monitored and treated like a point source.[97] Other agricultural sources are actually a lot of little point sources combined. For example, this is what the Environmental Defense Fund found when it studied the problem of selenium leaching from fields into irrigation water, with the return flow going into the San Joaquin River. It developed a market-based proposal, including trading pollution allowances among irrigation districts, to control this very serious water problem.[98] In some places irrigation inflows and outflows are controlled by conveyance canals, which could be monitored.[99]

Where direct monitoring of nonpoint pollution cannot be done, the alternative is indirect monitoring by computer modeling. Computer simulation models can be used to estimate relationships between pollutant loadings and water quality, to assess the assimilative capacity of water bodies, and to determine the fate and transport of nonpoint pollutants. Models are also used to estimate the effectiveness of management practices.[100] "More elaborate and detailed computer simulation modeling now in the refinement stages of development translates site-specific land use conditions . . . into estimates of effluent runoff. These estimation models are capable of estimating baseline nitrogen, phosphorus, and sediment runoff from an individual parcel as well as changes in load that occur from land use changes."[101]

In other words, in many cases nonpoint pollution can be measured and monitored, at least indirectly, and the effectiveness of BMP pollution controls can also be measured, or at least estimated, and the management practices themselves can be monitored. Thus "quantifying nonpoint source discharge with sufficient certainty for trading is not an insurmountable technical obstacle to trading."[102]

There will, of course, almost always be more uncertainty in monitoring runoff than in monitoring a sewage plant's discharge pipe. In a trading scheme this is compensated for with a trading ratio that must be determined for each program.[103] For example, at Dillon Reservoir the ratio was 2 to 1; for a point source to increase its discharge by one unit, it had to reduce nonpoint pollution by two units.

A study commissioned by the EPA contends that point-nonpoint trading is the wave of the future.[104] But once farms and cities are required to clean up their act, the way should be open to trading between nonpoint sources as well. If the barriers in the Clean Water Act could be removed, at least for point sources that join a trading association, the way would also be open for trading between point sources in more watersheds.[105] This could apply to factories with pretreatment requirements as well. For example, the jewelry industry in Rhode Island must limit discharges of cyanide and several metals in wastewater that enters city sewage treatment plants. One study concluded that placing them under a bubble, with a cap on total discharges and trading among factories within the bubble, could save 50 percent of the pretreatment costs without increasing toxic wastes.[106] Overall, the EPA estimates that if states adopted effluent trading among point sources, nonpoint sources, and pretreatment dischargers, total savings would be at least $700 million and could reach $7.5 billion.[107]

Not only can trading save money, it can also accommodate economic growth, even in areas where the TMDLs are fully allocated. "Trading allows new sources to develop and attain permits. If trading was not allowed in areas where the TMDL was completely allocated, a new industry would have to wait until permit renewals came up to attempt to receive any allocation. . . . Thus effluent trading can help the economy of a locality by allowing industrial growth in areas where the TMDL is completely allocated."[108]

Trading of capped water pollution allowances will not work everywhere. Trading programs would obviously have to be limited to individual river basins, and many watersheds may not be suitable, depending on the nature, number, and locations of dischargers and the different wastes they are discharging.[109] One study of an EPA database of water bodies identified 943 with impaired water quality that could potentially benefit from nutrient trading programs.[110] An EPA study of thousands of water bodies found that in 10 percent of them both point and nonpoint sources contribute the same kinds of pollutants, so there are potential benefits from trading. That is still over 4,000 water bodies.[111]

Where trading programs may not be feasible, another incentive-based policy to consider is effluent charges or taxes. They have played virtually no role in U.S. pollution control policy but are widely used in Europe. They are not used for incentive effects but rather to raise money for treatment plants. But in one country—the Netherlands—the charges resulted in a 27 percent reduction in organic pollution and a 50 percent reduction in industrial heavy metal pollution.[112]

But for our pollution-control policy, a charge might work better for nonpoint sources. This would involve a tax on inputs (e.g., fertilizer and pesticides) that would internalize their social costs and motivate farmers to use less of them.[113] Following the European example, perhaps the revenue could be earmarked to fund BMPs. Many states already have cost-sharing programs for BMPs on farms, and this would give all states a continuous source of revenue for that purpose.[114]

Overall, however, water pollution control involves many more obstacles than air pollution control that may impede market-based programs. Consequently, even libertarians concede that direct regulation may be needed.[115] However, market-based options would be preferable wherever they could be made to work. As with air pollution, these policies embody conservative principles, placing responsibility for cleanup where it belongs but allowing maximum flexibility in deciding how to fulfill that responsibility and removing bureaucratic meddling in the operations of each and every plant.

TOXIC SUBSTANCES AND HAZARDOUS WASTES

One significant feature of our economy is the explosion in the use of chemicals in the years since World War II. Although we gain many benefits from living in a "chemical age," we also face increased risks from many of these substances. In this section I will consider two areas of environmental policy that try to protect the public from danger: controlling substances that cause cancer and birth defects, and regulating the disposal of hazardous wastes.

Toxic Substances

The federal government, under a number of statutes such as the Clean Air Act, Clean Water Act, and Toxic Substances Control Act, has for many years attempted to determine for specific chemicals exposure levels that are dangerous to human health. The EPA then regulates release of those chemicals to protect the public. These efforts have been, by and large, a failure. After many years, the EPA has set standards for only a few chemicals.

The main reasons for this failure are the incentives and procedures built into the process. It is up to the government to determine the safe exposure limits for each chemical. The burden of proof is on the EPA. Until a standard is set, the chemical is *presumed* to be safe at any level of exposure and its use is unregulated. Industries, therefore, have every incentive to delay the process as long as possible, and to litigate any standard once it is proposed.[116]

Although federal regulatory attempts to protect the public against carcinogens and substances that cause birth defects have been largely unsuccessful, another federal policy has had considerable success—and it does not involve regulation of chemical use at all. Since 1988, the federal government has required all businesses and industries that release toxic substances into the environment above a minimal amount to file a report. The EPA compiles these and publishes its annual *Toxics Release Inventory.* In the TRI you can find out how much and what kinds of toxic substances were released in each state and locality, by each industry, by each company, and so on.

The results have been very interesting—and just the sort of thing conservatives should approve. If a company releases toxics into the environment, surely it has an obligation to tell the public what it is doing, and surely the public has a right to know what it is being exposed to. But *merely having to admit in public* what it is doing is a powerful incentive to quit doing it, or to reduce it as much as possible. And that is precisely what has happened. Although regulatory standards have succeeded in reducing our exposure to relatively few dangerous substances, merely shining the light of day on industries' releases has resulted in substantial reductions and continuing motivation to reduce them even more in the future. As a result of the TRI reports, "dozens of Fortune 500 companies have announced voluntary reductions in toxic releases."[117] According to the EPA, "the 1998 TRI shows a 45 percent decrease—1.5 billion pounds—among manufacturing industries monitored over 11 years (1988–98)."[118] A 1991 report by the National Resources Defense Council concludes that requiring TRI reports has done more to reduce toxic air pollution than twenty years of regulation under the Clean Air Act.[119]

California carried this process even farther, with a policy that conservatives should seriously consider advocating nationally. In 1986 the voters approved Proposition 65. It passed by a 2 to 1 margin in spite of massive opposition by the business community, which claimed it would eliminate farming in the state, cripple the economy, create chaos in the courts, and so on—all nonsense.[120]

Proposition 65 is really very simple in principle. It says that the state government must make a list of all chemicals that are known to cause cancer or birth defects. This list is to be updated annually. (As of 1999 it included some 660 chemicals.)[121] There are two requirements placed on companies using those chemicals: they are prohibited from dumping them into the drinking water supply, and if they expose people to those substances they must warn them, unless the level of exposure constitutes no significant risk. The law, notice, does not ban any substances. It does not even prohibit exposing employees or customers or the public to significant risks. It just says that if you do so, you must warn the people you are affecting. And here is the key: the law goes into effect whether or not standards have been set to determine the "no significant risk" level. If the state has adopted a standard, then businesses know precisely where their level of safety lies. If they expose people at levels below the standard, no warning is required. But if the state has not yet set a level of no significant risk and a business exposes people to that chemical, the burden of proof is placed on the *business* to prove the exposure is safe. Otherwise, it must warn the people.

Thus Proposition 65 completely reverses the incentives from those in the federal policies. At the federal level, safety from liability for a company lies in ignorance and absence of exposure standards. But in California, safety for a company lies in having a standard. At the federal level, companies try to delay standard setting as long as possible. In California, companies want standards set as soon as possible.

The results of Proposition 65 have been most impressive. In the first twelve months, California set twice as many safety standards as the EPA had set in twelve years.[122] After five years, the state had set standards for 282 chemicals. It reported that it would take the EPA a full century at its current rate to establish that many exposure standards.[123] Equally impressive, not a single one of California's standards has been challenged in court.[124]

The purpose of Proposition 65 was much more than setting standards and attaching warning labels to dangerous products. Its goal was to get companies to reduce using dangerous chemicals and reduce exposing the public to them. The idea was that companies would fear that consumers would reject their products if they had warning labels and competitors' products did not. Even more, companies would fear the bad publicity the press and citizen organizations would give them if they had to admit they were exposing their customers or workers or the public to dangerous chemicals. The hope was that Proposition 65 warning requirements would get industries to reformulate their products and processes to reduce or eliminate their use of toxic chemicals.

That is precisely what has happened. Litigation has produced several high-profile results. For example, the ceramic ware (china) that we use on our dinner tables can contain large amounts of lead, which can leach into our food. In 1991 the Environmental Defense Fund conducted a major testing program and found lead leaching well over the standard from a broad range of ceramic ware. As a result of Proposition 65, many manufacturers removed or greatly reduced their use of lead. Before the proposition passed, only 650 lines of china met the standards for safety; now 8,000 lines do.[125]

Another noteworthy example that required litigation concerned the lead content in brass faucets. The lead leaches into tap water, often at levels far over the standard of safety. Due to Proposition 65, the manufacturers have greatly reduced or eliminated use of lead, especially in kitchen faucets.[126]

There are numerous such instances of Proposition 65 lawsuits resulting in negotiated reformulation of products.[127] (Note that Proposition 65 cannot force a company to reformulate its products. The changes are made simply to avoid having to warn people.) But most such changes in products have been done behind the scenes, with no lawsuits and no publicity—precisely to avoid lawsuits and publicity. As business columnist Daniel Akst says, "Most of the good arising from Proposition 65 remains hidden, in the form of companies that quietly assess what they're doing, and presumably, clean up their act to avoid the brouhaha that might arise from having to tell all the neighbors they're being poisoned."[128] Moreover, many of these safer products are now sold nationally—few manufacturers want to make separate lines for California and for the rest of their markets—so all of us are benefiting from Proposition 65 in countless ways that we will never know.

It is impossible to ascertain just how many products and manufacturing processes have been changed because of Proposition 65. But there are some overall statistics that demonstrate just how effective it has been. As we have seen, the federal TRI, which is only an annual report to the government, motivated many companies to reduce their emissions drastically. Proposition 65 has been even more effective. Between 1988 and 1996, air emissions of lead in California dropped by 99 percent, compared to a 42 percent drop in the rest of the United States.[129] During that time, air emissions of 147 chemicals subject to both Proposition 65 and the TRI declined by 73 percent in California, compared to a drop of 48 percent in the rest of the country. At the same time, there was no difference in rates of decline of chemicals subject to TRI but not to Proposition 65.[130] "The reductions were also accomplished without apparent industrial strain. (Indeed, the industries responsible for the reductions always had the option, under both TRI and Proposition 65, of simply disclosing their emissions and giving warnings, instead of reducing them.) This demonstrated potential for large-scale reductions, without industrial disruption and without mandates, should come as extremely good news for anyone concerned about the failure of conventional risk-based policy."[131]

Proposition 65 has a great deal to recommend it to conservatives as a national policy. It has demonstrated that changing incentives can be more effective than regulating. Federal regulations have accomplished little because the incentives for industry are to delay setting standards as long as possible, and to tie the process up in litigation for years on end. "There is a clear lesson in the poor results produced by the current federal regulatory system: more of the same will not work."[132] As David Roe, principal author of Proposition 65, explains, "the idea behind Proposition 65 . . . is to give industry a compelling incentive to remove nonessential carcinogens and reproductive toxins from its products and processes. And it does so with very little arm twisting. If the choice is warning or finding a new chemical . . . most companies will opt for the latter."[133]

Proposition 65 is a market-based alternative that has proven to be more successful than regulation, and conservatives should surely appreciate that. For markets to work properly, all parties must be informed. As Tom McGarrity says of Proposition 65, "the free market economy is predicated on the informed consumer. It is the fundamental assumption about how markets work. The most conservative of theoreticians should love this approach."[134] A 1997 report to the Presidential/Congressional Commission on Risk Assessment and Risk Management refers to Proposition 65 "as an example of how government can 'aggressively' seek alternatives to command-and-control regulation."[135] William Pease explains that "with its incentive-conscious approach to toxic chemical control, Proposition 65 attempts to apply a market-based strategy to the control of toxic chemicals rather than to rely on the traditional command-and-control approach."[136]

Proposition 65 has, in fact, been approved by several conservative politicians. Governor Deukmejian opposed the initiative, but once it passed Tom Warriner, who administered the law, became a convert. He concluded that "there is nothing inherently wrong in telling people if you expose them to chemicals at high levels."[137] William Reilly, head of the EPA in the Bush administration, praised Proposition 65 as an important shift in the philosophy of environmental protection. He approved it as "consistent with [President Bush's] call for more economically-oriented regulatory programs."[138] The Proposition 65 concept should be seriously considered by conservatives as a national approach to protecting the public from toxic chemicals.[139]

Hazardous Wastes

When we think of hazardous wastes, probably the first things that come to mind are Love Canal and Superfund. The Superfund law was passed to clean up abandoned chemical waste sites, but over the years it has been much criticized, from practically all sides.

James Strock, writing in 1994, gives a good summary of the earlier problems with Superfund. The law makes practically anyone who had any connection with an abandoned waste dump or contaminated site liable for the entire costs of cleaning it up (strict, joint, several, and retroactive liability). The EPA could go after one or a few parties, require them to pay for the cleanup, and let them sue all of the other potential parties to recover portions of the costs. Consequently, "transaction costs" were enormous—nearly a third of the costs were being spent on litigation rather than on cleanup. Singling out one or a few parties was seen as unfair. And frequently many companies that contributed to the problem have long since gone out of business. The costs of cleaning up their "orphan shares" was imposed on the parties that were still around. Small contributors to a hazardous waste site were often sued for large portions of cleanup costs by large contributors. Holding all parties liable for cleanup delayed urban redevelopment. Developers were reluctant to buy old industrial ("brownfield") sites for fear of being held liable for messes they did nothing to make. And retroactivity was also criticized: companies were held liable for actions that were perfectly legal at the time they were done.[140]

Many (but not all) of these problems have been solved or greatly reduced by new regulations. The EPA now contributes a portion of orphan share costs, and this has greatly reduced litigation. Alternative dispute resolution through mediation is now standard operating procedure, which also reduces litigation. Transaction costs have been cut by 30–50 percent. Companies that made only minor contributions to a contaminated site are now divided into two groups: *de minimis* and *de micromis* parties. Companies that dumped only a small portion of waste at a site (*de minimis* parties) are now offered cash-out settlements early in the negotiations about assigning cleanup costs. Once they have paid the negotiated amount, their liability is satisfied. More than 15,000 parties have accepted these terms. Companies whose wastes contributed only tiny portions at a site (*de micromis* parties) are now excluded by the EPA from any potential liability. If a large contributor sues a *de micromis* party, the EPA will intervene and settle with the large contributor. The importance of these new rules can be seen in a toxic site in North Carolina: the EPA reached settlements with over 200 *de minimis* parties. The agency and the larger contributors also agreed to waive rights to sue over 1,000 *de micromis* parties. And that is at just one site. The new rules also remove liability for companies that want to clean up brownfields and build on them.[141]

Although there are still some problems with the law, as EPA administrator Carol Browner says, "the Superfund program is fundamentally different and better. It is faster, fairer, and more efficient."[142]

Desposit/Refund Policies

Although old toxic dumps are subject to Superfund, current disposal of industrial hazardous waste is controlled by the EPA under the Resource Conservation and Recovery Act. The system is a "cradle-to-grave" tracking system under which all significant quantities of hazardous wastes generated by industry must be accounted for from the time they leave the plants until they reach approved disposal facilities. The EPA also sets design and performance standards for storing and treating the wastes at each step. "It is in hazardous waste management, if anywhere in environmental policy, that the infamous 'command and control' approach is to be found."[143]

However, some types of hazardous wastes are not included in this system and are often dumped improperly. Small volumes of industrial wastes are not covered and some common things used by individuals are excluded. Perhaps most prominent are car batteries and motor oil, which pose serious disposal problems.

Car batteries use most of the lead produced in the United States. Over the years, on average, about 70 percent of defunct batteries have been recycled. But a lot of them—more than 20 million per year—are disposed of improperly. Quite a few end up in landfills, where the cases eventually break and the liquid tainted with lead can get into groundwater. If they are incinerated, lead can be released into the atmosphere. The lead in dead batteries can be recycled many times, so it is a valuable resource.[144]

Motor oil is recycled now by garages that handle large quantities. But there are a good many "'do-it-yourselfers' whose recycling rate is only 5% and who are collectively responsible for nearly 50% of illegal dumping."[145]

Market-based policies would be more successful in controlling these wastes than regulation would. With small quantities of hazardous waste, and with things used by many individuals, the problem of monitoring disposal is impossible. So Clifford Russell asks, "why not pay for proper disposal? . . . If the amount of the payment is tuned correctly, the source should have an incentive not to try to conceal its waste . . . but rather to work to collect the reward. Presto! The terribly difficult monitoring problems seem to be solved. Toxic wastes all end up in the right places."[146] Since these rewards should not be a drain on the taxpayers, a deposit and refund system could be used. This would be similar to the "bottle bills" some states have adopted. When you buy soda pop, beer, wine, or liquor, you pay a deposit on the can or bottle. To get it back, you return the empty to a store. (I lived in one of those states for several years and the system clearly works: the reduction in littering is obvious.)

Under an extension of this idea, when people buy car batteries or oil, they would pay a deposit and get a receipt for proof of purchase. When they replace them, they could take the used batteries or oil to approved disposal facilities, along with the proof of purchase, and get their deposits back. The keys are (1) to require a deposit large enough to motivate people to get it back, but not so large that they would go out and steal batteries, and/or (2) to design a proof of purchase that would be difficult to fake (to discourage theft). Several states already have deposits and refunds for batteries.[147]

A deposit and refund scheme might also work for some kinds of hazardous materials for industrial use, especially ones that can be containerized and are used in small amounts. With a deposit and refund system, motives for illegal dumping are reversed. The company that uses the toxic substance would want to dispose of it at an approved facility. Currently, the burden of proof is on the government to prove charges of illegal dumping—nearly impossible for small quantities of waste. Under a deposit and refund policy the burden of proof is on the company to prove it is returning what it claims. (That is not difficult with a battery but could be a problem with things such as industrial solvents.) A deposit/refund scheme for industrial materials would also provide incentives for companies to reduce or capture losses of the materials in the production process.[148]

Clearly, there are many policies to control pollution that conservatives should find attractive. Many of them could replace at least part of the current regulatory regime. But it must be stressed once again that conservatives should search for policies that would be *more effective* than we have now. *Proposals for mere deregulation, for weakening controls to let polluters get away with fouling our air and water and land, are violations of conservative principles.* We should find the most effective, and most cost-effective, policies that place responsibility for cleanup precisely where it should be: on the polluters.

5

Public Lands

FOR A NATION WITH AN ECONOMY based on capitalism and private property, governments in the United States own a surprisingly large amount of land. Although states and local governments have extensive holdings, it is the federal lands, concentrated in the western part of the country, that are caught up in the most contentious environmental issues.

Most federal land is administered by four agencies. The Forest Service in the Department of Agriculture has 191 million acres of national forests and grasslands. The other three agencies are in the Department of the Interior. The Bureau of Land Management controls 264 million acres, most of which is grazing land. The Fish and Wildlife Service administers some 92 million acres of wildlife refuges. And the National Park Service—probably the only one of the four that the average American has ever heard much about—runs the 77 million acres of our national park system.

Public lands have been a bone of contention for a very long time, and probably will be for a long time to come. They have often been abused, mostly as a result of iron triangle politics—the cooperative coalition of business interests that want cheap access to public resources, a bureaucratic agency that provides those resources, and the politicians beholden to the special interests who appropriate the taxpayer subsidies that finance the cozy scheme and give the agency appropriate rewards. Because of the accumulation of abuses to the land, environmentalists became concerned, and now some of our most contentious environmental issues focus on the protection and rehabilitation of these lands.

It must be emphasized at the outset that conservatives *should* have become concerned as well. There is absolutely nothing conservative about iron triangle politics. Quite the contrary; it is always a violation of conservative principles. The public land iron triangles were largely formed during the days of Democratic domination in Washington. But whenever Republicans gained power, to their considerable shame, they played the same game for all it is worth.

There are now many conflicts over the use and protection of public lands. In this chapter we will look at four issues—logging (primarily on national forests), grazing (primarily on BLM land), mining, and federal water projects—and consider what kinds of policies might come from applying conservative principles.

These issues are concentrated in the public land states of the West, but they are of increasing concern to Americans all over the country because of the explosion in recreational use of public lands. Although recreation creates problems of its own, it is also a key to solving many of the land use and abuse issues.

THE FOREST SERVICE

The national forest system was first established in the late nineteenth century, when several million acres were set aside from the public land to be kept as forest reserves. The Forest Service took its current administrative form during the administration of Theodore Roosevelt, who greatly expanded the national forests. Later presidents added to the system. Since these first reserves were created from public land, they were all in the western part of the country. The national forests in the East were mostly acquired during the Depression, as the federal government bought up heavily logged land that was in need of rehabilitation and reforestation.

National forests were created as a reaction against the destructive practices of private logging companies. In the nineteenth century they had gone across the United States, from east to west, "mining" the forests and moving on, leaving a trail of devastation in their wake. "Cut and get out" was the name of the game. The national forests were meant to be managed, by contrast, as examples of excellent forestry.

For many years the Forest Service managed its land largely on a custodial basis. Timber companies had plenty of trees of their own to meet production needs. During World War II logging increased as part of the war effort, but it was only after the war that the national forests became an important source of timber. The pent-up demand for housing that had developed during the Depression and the war created a housing construction boom in the late 1940s and 1950s. By this time, much of the private forests had been logged out and the industry turned increasingly to the national forests.

And here is where our current problems began. A classic example of the iron triangle developed, in which timber companies got access to enormous amounts of timber at subsidized prices, the senators and representatives from public land states got control of the mostly obscure committees that appropriated the subsidies, and the Forest Service got its budgetary and bureaucratic rewards for serving the industry.

Timber production became the dominant purpose of the Forest Service, regardless of what its "multiple use" legislative mandates said, regardless of how much damage was done to the forests, and regardless of how little economic sense it all made. Two major problems resulted: the land was devastated and the taxpayers took enormous losses. For nearly half a century the Forest Service practiced what the late Senator Talmadge used to call "idiot forestry." We taxpayers have been systematically fleeced for the privilege of having our forests destroyed for the benefit of a few in the timber industry. And those forests are far more valuable for other purposes: recreation, watershed protection, habitat for wildlife—values that logging degrades and often destroys.

It needs to be recognized that the Forest Service has undergone profound changes in recent years. It now has far more biologists, ecologists, hydrologists, and other scientists than it had in earlier years, when the agency hired only road engineers and silviculturists. And these "ologists" have far more influence within the agency than they previously had. In fact, as I write, the current chief of the Forest Service, Michael Dombeck, and his predecessor, Jack Ward Thomas, are both biologists, not foresters.

The agency now claims that it has given up the bad old ways and is going in new directions. No longer, they say, do they want to turn the national forests into tree farms for the industry. "Ecosystem management" is the new slogan. Biodiversity and recreation are the new priorities.[1] The House Committee on Natural Resources in 1994 found that of all Forest Service activities "timber is the one area in which agency policies are currently most in flux."[2] We can hope.

But our hope must be cautious, and we have to be vigilant. The new rhetoric sounds right, but has it made much difference on the ground? Years ago Forest Service managers said clearcuts would be limited in size to around forty acres and now they say they are not clearcutting at all. Yet in 1995 the Wilderness Society found very large clearcuts, up to 344 acres, being sold and in 1997 the Forest Water Alliance found clearcuts of old growth exceeding the maximum being offered for sale.[3] Forest managers say that they are protecting streams with buffers these days. Yet in 1997 the Forest Water Alliance found logging in streamside reserves and a timber sale in Oregon that would clearcut straight across several streams.[4]

Moreover, the history of the Forest Service is not encouraging. For years, "multiple use" and "sustained yield" were operative slogans. Yet on the ground, it turned out that any possible use of a national forest always required logging. And sustained yield promises never prevented the agency from cutting trees at rates even it admitted would result in serious declines in the future. Moreover, in the spotted owl case, Judge William Dwyer characterized the Forest Service's actions as "a remarkable series of violations of the environmental laws. . . . The most recent violation . . . exemplifies a deliberate and systematic refusal by the Forest Service . . . to comply with the laws protecting wildlife."[5] If the agency did not even comply with the laws in the late 1980s, to what extent can we expect it to comply today with what could be merely a passing fad in its own rhetoric? Will the current slogans be anything more than a smokescreen for "business as usual"?

A recent study compared new and old forest plans on three national forests, drafted before and after the Forest Service leadership adopted ecosystem management. It found that two of the forests had converted to the new way of thinking but the third had not.[6] Moreover, in 1999 the inspector general studied twelve timber sales and found that all of them were flawed, with illegal tree cutting, failure to prevent environmental damage, and an absence of surveys for endangered species.[7]

To a great extent, of course, the old days are gone because federal courts have been willing to force the Forest Service to obey the laws. To that extent, environmentalists have been able to weaken if not break the old iron triangle.

Yet the Forest Service budget structure is unchanged and, as we will see, it is the budget structure that gives the Forest Service greater incentive to cut down its trees than to preserve them. The old iron triangle still raises its ugly head from time to time. For example, in the infamous "salvage rider" of 1995 Congress exempted many timber sales from the normal evaluation procedures, gave the industry a lot of perfectly healthy trees that had absolutely no need to be "salvaged," and let them avoid complying with environmental standards in the process. While that rider was in effect in 1995 and 1996, the Forest Service's subservience to the industry was clearly demonstrated once again. It offered numerous timber sales that violated environmental laws, even though the secretary of agriculture had ordered the service to obey them.[8] Moreover, the subsidies are still in place—the Forest Service still sells timber and still loses money doing so.

In sum, there is reason to hope that the future of the Forest Service will be better than its last fifty years. But there are a number of reasons to fear that the changes it has made are far from enough.[9] And that gives us good cause to look at forest destruction and losses to the taxpayers to see what conservative policies might help ensure that the Forest Service will live up to its current rhetoric on protecting the forests and will finally put a stop to the drain on the treasury.

DESTRUCTION OF THE NATIONAL FORESTS

As the Forest Service got into the timber business in a serious way after World War II, and as iron triangle politics took over, the national forests were subjected to increasingly intensive logging. Within a couple of decades, timber production had become virtually the sole goal of the Forest Service, and the national forests were "hammered" by logging at the expense of all other values the forests could provide. The agency set out to eliminate natural forests wherever possible and replace them with monoculture tree farms for the wood products industry. This, of course, required extensive clearcuts to liquidate the forests. The result now is widespread destruction on the land, seriously degraded watersheds, loss of biodiversity, and enormous costs passed on to the taxpayers.

The damage to the forests happened when the Forest Service reversed its original ideas that logging should be done carefully and selectively. It adopted the industry's abusive clearcutting as the dominant procedure for harvesting. Clearcutting is just what the words imply: loggers simply chop down everything in the area to be harvested, remove the logs, burn the "slash," and leave the place looking like a World War I battlefield. Cecil Andrus, then governor of Idaho and himself a former logger, describes the results in his state: the Forest Service "has just devastated the area for the past 30 to 40 years."[10]

The irony of this should not escape us. The national forests were originally set aside as a reaction against precisely this destructive type of logging. The Forest Service long insisted that the best logging is selective cutting, in which foresters carefully identify specific trees to be harvested, leaving the forest basically intact.

The change was gradual. At first the Forest Service started using clearcutting on Douglas fir in the Pacific Northwest, which it justified by claiming that this species is "shade intolerant" and thus needs the open space and unobstructed light of a clearcut to regenerate. To transform natural forests into large Douglas fir plantations, it made large clearcuts.

The Forest Service then expanded clearcutting to other species and to forests far from the Douglas fir region. In the 1950s it began clearcutting in the Rocky Mountains and the intermountain West. In the 1960s clearcutting replaced selection cutting in the East, and in California the change came in the 1970s.[11] By then, clearcutting was the dominant harvesting method everywhere and remained dominant well into the 1990s.[12]

Moreover, to give the industry access to more and more timber in ever more remote locations and to turn the forests into intensively managed tree farms required the construction of an extensive system of roads. On our national forests there are now more than 380,000 miles of roads, eight times as long as the entire interstate highway system.[13] The Forest Service has been called "the world's largest socialized road building company."[14] Many of these roads are in poor condition. Only some 40 percent of them are maintained to proper environmental and safety standards, and there is a backlog of $10.5 billion in road reconstruction and maintenance needs.[15] These roads are important here because, as we will see, they do even more damage than clearcutting does.

Moreover, much of the land in national forests is fragile, often on steep slopes. Except for the Pacific Northwest, most of the national forests are not very good for timber production. They are, after all, lands that the government could not give away in earlier years. Most are in arid regions or at high altitudes with short growing seasons and harsh climates. These lands produce mostly poor-quality trees of little commercial value. That has not, however, kept the Forest Service from logging them to meet timber goals.

The Forest Service is supposed to determine which of its lands are suitable for timber production and which are not. The unsuitable lands are to be removed from the timberland base. Unfortunately, the Forest Service does not decide whether land is suitable or not on the basis of its ability to grow trees profitably. The agency decides that particular land is suitable for production if it needs it to meet its timber goals![16]

As timber production goals went up over the years, more and more marginal and poor land in ever more remote areas was reclassified as suitable for logging. Of course, no private company would do that on its own land. But a company has to make a profit. The Forest Service only needed to meet production goals.

Combine poor-quality land on steep slopes with clearcutting and miles and miles of roads and the result is massive destruction from erosion. The clearcutting by itself allows the surface soil to wash away because there is no longer any vegetation to hold it. The specifics, of course, will depend on the particular soil and terrain. One study found erosion on a clearcut in Idaho was 60 percent greater than before logging.[17] In an unstable soil zone in Oregon, erosion was 2.8 times greater on clearcuts than on undisturbed forest.[18]

When the roots of the trees that have been removed finally decay several years after logging, the slopes become vulnerable to slipping and the result is "mass wasting" from landslides. A study on the H. J. Andrews Experimental Forest found that with heavy rains there were 0.4 landslides per 1,000 acres on undisturbed forest, but 3.9 landslides per 1,000 acres on logged areas.[19]

The roads are even more damaging than the clearcuts.[20] The study in Idaho found that roads increased the rate of erosion by 750 times over the natural rate.[21] In the Oregon study, landslides were thirty times greater along roads than on forested land.[22] On the H. J. Andrews Experimental Forest, heavy rains one winter produced 126 landslides from road construction per 1,000 acres.[23]

The most notorious example happened in Idaho on the South Fork of the Salmon River. The Forest Service had built over 600 miles of roads in the watershed and over 8,500 acres had been clearcut, most of it using the most destructive mechanical processes of all. In 1965 after two weeks of rain on snow, entire hillsides over a distance of twenty-five miles simply gave way. Some 500 landslides dumped 400,000 cubic yards of sediment into the river.[24]More recently, the heavy rains in the Northwest in the winter of 1996–1997 produced massive mudslides on clearcuts that destroyed many homes and killed a dozen people.[25]

Clearcutting does much harm to the soil beyond the loss from erosion and landslides. The use of heavy machinery for logging can result in severe compaction of the surface soil.[26] Clearcutting exposes the forest soil to intense sunlight, which raises its temperature and increases evaporation, destroying or at least damaging the normal life of all of the microorganisms that are important components of the soil.[27] Clearcutting also results in leaching of nutrients that are essential to tree growth.[28]

All of these effects combined can reduce the productivity of the soil, which reduces the sustainability of the forest. "Of all the structural components of the ecosystem, soil tends to be the least renewable, and examples of loss of ecosystem sustainability have generally, though not exclusively, been related to loss or reduction of soil function." Such examples are "unacceptably frequent" in our managed forests.[29]

One of the most valuable functions of a forest is that it provides clean water to cities in its watershed. Heavy clearcutting destroys its ability to filter water and release it slowly, and adds all the sediment eroding from the cuts and roads. In February 1996 heavy rains in Oregon washed mud and debris from clearcuts and logging roads into the North Santiam River, from which Salem gets its water supply. The dirty water overwhelmed the city's filtration equipment, forcing industrial users to scale back or stop production for over a week. Water for residents was rationed.[30]

This function that forests provide for free is so important to Portland that it got Congress to declare its watershed off-limits for logging.[31] Small towns do not have that kind of political clout and have to suffer the consequences.[32]

Erosion from logging also degrades the streams and rivers as habitat for fish and other aquatic creatures.[33] When the sediment settles, it ruins the streambed for spawning, damages fish eggs, and reduces the survival rates of fry after they emerge. Sediment reduces the food supply for fish as well.[34]

The national forests have been so heavily cut that many fish are now threatened or endangered. This is especially a problem with the several species of salmon in the Pacific Northwest. At least 106 populations are now extinct and another 214 are at risk of extinction.[35] Although there are several factors that contribute to their peril, including the many dams on the rivers, logging is a major cause. The American Fisheries Society calculates that for at least ninety of the 214 endangered salmon stocks, clearcut logging in national forests is the primary problem.[36]

When fish populations are hurt by logging, so is the fishing industry, both commercial and recreational. The destruction of habitat in the South Fork of the Salmon River, described above, destroyed the most productive chinook salmon stream in Idaho. The Forest Service estimates that it caused $100 million in damages to the fishery.[37] The value of the timber that was cut in that watershed was about $14 million.[38]

The fishing industry is an important part of the economy of the Northwest. Overall, its fisheries are a billion dollar industry affecting some 60,000 people.[39] A staff report of a House committee concluded that "the loss of fisheries damages an important local industry, impairs recreational activity, and creates a conflict with Indian tribes that possess treaty fishing rights."[40]

The Forest Service has not only done much damage to our national forests and their watersheds by massive overcutting, it has also failed at reforesting much of this land. The most productive timberlands—relatively flat and at low elevations, that will easily regenerate after cutting—are almost all privately owned.[41] Most national forestland in the timber base requires replanting, and this has long been a major problem for the Forest Service.

In the first place, replanting costs money. The Forest Service has for years and years reported to Congress that it has a backlog of several million acres in need of reforestation. Congress almost never appropriates more than a fraction of the money required to do the job—although it approves almost all of the timber-cutting budget requested and sometimes more.[42]

Even if the money were available, the Forest Service would still have serious problems. As Gordon Robinson explains, "Much of the forest being logged today is very difficult to reforest and does not respond well to clearcutting and planting. . . . Often the Forest Service will repeatedly replant open areas, and in some situations success may never be achieved."[43]

The Forest Service *claims* that its reforestation program is successful. But Catherine Caufield found that "it is difficult to know for certain. The agency's records on reforestation and growth rates are incomplete and, in some cases, unreliable. The Forest Service defines reforestation so narrowly that it can claim very high rates of restocking, but these claims may be misleading . . . [because] the agency sets standards only for the number of trees planted, not for the number that survive."[44] She found an example in the Siskiyou National Forest, where officials claimed that more than 99 percent of the logged areas had been replanted. But one area had been replanted to Forest Service standards six times since 1961, and there was every indication that the latest planting would fail as well. Moreover, "the agency keeps no records of how fast its plantations are growing, or how robust they are."[45] As Julie Norman, president of Headwaters, an Oregon environmental organization, says, "The agency doesn't want to know what's happening out there."[46]

That is not just an environmentalist's hyperbole. That is precisely what the staff of the House Committee on Interior and Insular Affairs found when it investigated this issue. They published a scathing criticism of the Forest Service timber management program, calling it "the loss of accountability." The committee staff looked at sixteen national forests for several problems, including reforestation failures, which they found on fully half of those forests.[47] In calculating its timber base to determine how much can be cut, the Forest Service does not take into account "reforestation failures and slow regrowth rates on tree farms" and makes "over-optimistic projections of future tree farm growth." "Vague claims" of growth on replanted tree farms "were not supported by documentation." Reinforcing precisely what Julie Norman claims, the committee staff observe that "the agency has a distinct disincentive to monitor implementation because degenerating conditions reflect poorly on the agency and its managers."[48]

The Forest Service justifies its timber production goals on the basis of growth projections and reforestation assumptions. But when reality turns out to be very different, they do not make corresponding reductions in the cut. For example, the Forest Service calculated that using herbicides would let it increase its cut by up to 37 percent. Herbicide spraying was later banned in Oregon and Washington, but the Forest Service did not reduce its harvest goals to compensate.[49]

Another serious problem is that clearcutting and replanting with a single species, even when it is successful, destroys the complexity of forest ecology. As Chris Maser says, to have sustainable forestry you first have to have sustainable forests.[50] Jerry Franklin of the University of Washington is one of the leading experts on forest ecology and much of his work, and that of his colleagues, has for some years now stressed the importance of complexity in forest ecosystems. Natural forests, he observes, are characterized by "high levels of structural complexity" that offer an "abundance of habitats or niches that, in turn, support a large array of specialized species." Much of this complexity survives natural disturbances, such as fires and windstorms, and is "incorporated into the recovering ecosystem" as "biological legacies."[51]

The complexity of natural forests makes them "resilient," that is, able "to absorb stress or change without significant loss of function." Resilience is necessary for a forest to be sustainable. The principle for proper forest management is, therefore, "maintenance or enhancement of complexity—genetic, structural, and spatial" and the maxim to be adopted is: "simplification is rarely beneficial."[52]

The clearcutting regime that the Forest Service (and private industry) have long practiced is based entirely on simplification. Single species of trees are planted in place of natural diversity. All trees are of the same age. Successional stages, both early and late, are eliminated. The result is a tree farm, biologically sterile by comparison with a forest, and with much of its resiliency removed. Long-term sustainability of such a system is questionable at best.

Conservatives should be able to appreciate the basis of the recommendations by Franklin and his colleagues. "It is not even known what all the parts of a forest ecosystem are, let alone what they do. And past experience suggests that they are important."[53] "Our woefully inadequate knowledge of forest ecosystems should humble us. . . . Already we are learning that parts of forests that we never considered seriously are proving significant, even essential, to ecosystem functioning. And through this new information, we are finally recognizing the many fallacies and weaknesses in our traditional forestry dogma . . . which moves us toward a homogenized, simplified forest. . . . Have we been proceeding prudently? The answer, inescapably, is no."[54]

Clearcutting destroys old-growth ecosystems that are now rare and increasingly important.[55] When old growth is harvested, obviously, it is gone. But logging has severely fragmented the old growth that remains, which often destroys its ecological value. Although the northern spotted owl has gotten all the publicity, there are many species that depend on old-growth forests for all or part of their life cycle. In the Pacific Northwest, seventy-six other species of wildlife use old growth as their primary breeding habitat, and sixty-five other species use it as their primary feeding habitat.[56]

Clearcutting also results in increased flooding, especially from rain-on-snow events, because of greater snowpack in the clearings.[57] In the Idaho Panhandle Forests, the *Spokane Spokesman-Review* found that "holes in the forest canopy and the 10,000 mile network of roads cause 100- and 500-year floods almost every winter."[58]

Excessive logging increases the risk of forest fires. There are two sources of these fires. In the first place, after a clearcut the slash is normally burned and there is a danger that these fires can get out of control. A study on the Mount Hood National Forest found that all major damage from fires had been caused by escaped slash burns.[59]

The second source of fire danger comes from the logging roads that open previously unroaded forests to increased access by people. And careless people cause a lot of fires. "Roads are in fact the primary reason that humans now rival lightning as the most important cause of fires on the national forests."[60]

Finally, clearcutting seriously degrades the forests for recreation.[61] Clearcuts obviously do major aesthetic damage to the area, which reduces its attractiveness for hikers and campers. Very few people choose to spend their vacation time and dollars looking at stumps and debris in clearcuts. On top of the aesthetic devastation, the damage caused to streams and rivers reduces their value for anglers. And building ever more logging roads into roadless areas destroys their value for primitive recreation, the demand for which is increasing explosively.

The result is a clear conflict between logging and recreation on the national forests. The overall significance of this point will become clear later because it is the enormous and ever increasing demand for recreation on our national forests that offers a key to solving many of their problems.

As well as practicing the destructive logging it formerly condemned, the Forest Service has long harvested timber at an unsustainable rate, despite its legal mandate to manage its forests on a "sustained yield" basis. After criticizing private companies for "cut and run" logging, following World War II the Forest Service reversed itself and did the same thing. The cut from the national forests increased rapidly after the war and continued to increase, except during recessions, until it reached a very high plateau from the 1960s through the 1980s.

As early as 1969, the Forest Service publicly acknowledged that its timber sales were seriously overextended. Its *Douglas-Fir Supply Study* in that year offered eight alternatives and every one of them indicated that there would be a sharp drop in production near the end of the century.[62]

In 1989, after he had retired as chief of the Forest Service, Max Peterson told a meeting of agency employees: "Anyone—on the back of an envelope—could have figured out that the rate of harvest cannot be sustained."[63] But whenever the political corner of the iron triangle thought the agency was becoming insufficiently subservient to the industry, the politicians intervened and legislated higher timber production targets.[64]

As I indicated at the beginning of this look at the Forest Service timber program, there are some reasons for cautious optimism today. The Forest Service claims to be using "ecosystem management," the goal of which is to preserve ecosystems from abuse and keep them intact by using them with care. "New Forestry" practices are being adopted which harvest timber in ways that preserve the complexity and structural diversity of the forest.[65] The concept of sustainable forest management is being developed, with detailed performance criteria and indicators to measure success and failure.[66] And the total cut from national forests is far below the level of the late 1980s. (National forest production averaged between 10 and 12 billion board feet during the 1960s, 1970s, and 1980s. It is now down to between 3 and 4 billion board feet per year.)[67]

Nevertheless, as I noted earlier, there are also reasons for skepticism. The Forest Service has long attempted to obscure abuse of the land in a rhetorical fog of "multiple use" and "sustained yield" slogans. And the political pressures from Congress to keep the national forests in the pocket of the timber industry are still there—witness the disastrous salvage rider of 1995. On the other hand, when the political corner of the iron triangle tried in 1998 to make the provisions of that rider permanent, they were defeated in the House.[68] But the abuse of the forests has been so bad in the past that conservatives—and all others, for that matter—need to be aware of what happened. Iron triangles do not easily dissolve, and the institutional structure that led to the abuses is still intact. Conservative policies for reform will have to be addressed to that system.

BELOW-COST TIMBER SALES: THE TAXPAYERS LOSE

It is not just the type of personnel but the *institutional structure* of the Forest Service—specifically its Byzantine budget—that produced both the abuses of the forests and the second major problem: the loss of massive amounts of taxpayer money year after year after year. The Forest Service sells timber at prices that do not even come close to covering the costs of the sales, and this problem continues to this very day. As Representative Jim Leach says, "The U.S. government is the only property owner I know of that pays private companies to deplete its own resources."[69] The authors of *Project 88* concluded that "if the Forest Service were a private firm, the value of its assets would place it among the top five of the *Fortune* 500 list of largest corporations, while in net income terms it would be classified as bankrupt."[70]

The authors of *Project 88* found that in the late 1980s Forest Service timber sales lost more than $400 million every year.[71] A General Accounting Office report combines Forest Service timber sales for fiscal years 1992–1994. In those three years, the Forest Service lost $1 billion of taxpayer money on timber sales.[72] In fiscal 1995, according to the Council of Economic Advisers, the Forest Service lost $234 million on timber sales.[73] Another GAO report combines fiscal years 1995–1997. In those three years, the Forest Service lost another $1 billion selling timber.[74]

These deficits are from timber sales. But the Forest Service has other resources as well, such as grazing lands and, most important of all, some of the most popular recreational lands in the country. And it loses money on all of them. Overall, each year the Forest Service loses about $2 billion.[75]

Before we can consider solutions to the problems with the Forest Service, it is important to take a look at its budget structure to see just exactly why the agency incurs these losses. Basically, the Forest Service continually loses money selling timber because its *own* budget *increases* even when the taxpayers lose. This stems from a number of interesting and entirely perverse incentives built into the Byzantine structure of the Forest Service budget. A detailed analysis of all the nooks and crannies of the budget is beyond the scope of this discussion. A special issue of *Forest Watch*, entitled *The Citizens' Guide to the Forest Service Budget* (April 1992), gives a good summary in some thirty pages. For a book-length analysis, Randal O'Toole's *Reforming the Forest Service* is excellent. For our purposes here it will suffice to look at the major budget provisions that produce losses for the taxpayers. And it is there that we will find the keys to designing reforms.

As we look at these provisions, we need to keep in mind that the budget structure contains the incentives and disincentives for an organization that direct its activities. Former Forest Service chief Dale Robertson says, "We can talk a lot of philosophy about how the national forests ought to be managed, but let me tell you it's the budget that energizes the Forest Service."[76] Randal O'Toole makes a persuasive case for the claim that Forest Service actions can be explained by only one thing: it attempts to maximize its budget.[77] Even if this is going a bit too far, it is clear that budget incentives are powerful motivators in any organization. And at the heart of our problem is a budget structure that *rewards* the agency even when it *loses* money.

The Forest Service gets annual appropriations from Congress, including funding for timber sales, reforestation, recreation, research, road maintenance, and so on. At the legislative end, Congress has long been willing to finance timber sales to the fullest, but seldom appropriates anywhere close to the needs for the other items. That is why the Forest Service has long had a backlog of millions of acres in need of reforestation and its backlog of needed road maintenance now totals $10.5 billion.

The Forest Service also gets revenue from some of its commodities and services. The original idea, of course, was that the income would be returned to the Treasury and would exceed the appropriations. But there are many quirks in the Forest Service fee structure that virtually guarantee that will never happen.

The major provision relating specifically to timber sales dates all the way back to 1930 in the Knutson-Vandenberg Act. The costs of arranging timber sales are paid by the taxpayers. The K-V Act allows the Forest Service to keep money from timber receipts to be spent in the vicinity of the sales. The original purpose was to provide reliable funding for reforestation, but no limit was set on the amount the agency can keep. The National Forest Management Act in 1976 extended the use of K-V funds to wildlife habitat management and general sale area improvements. Moreover, the portion of K-V funds that goes to overhead has expanded, and now it is shared by all levels of the Forest Service bureaucracy. Consequently, the Forest Service relies on K-V funds for all sorts of things, which means that it relies on timber sales to pay for all sorts of things. And, since the *costs* of the sales are paid by *appropriations*, the Forest Service *increases its own budget even if the sales lose money*.[78] There are some other similar funds in the budget, such as grazing fees that can be retained, but the K-V fund is by far the largest.

All sorts of strange things result from this provision. For example, although logging harms water quality, the Bighorn Forest sold timber to raise money for watershed mitigation. It also identified some historic sites that might be eligible for listing on the National Register of Historic Places. To fund an archaeological survey, the forest sold timber, but the logging destroyed one of the sites.[79] In the Panhandle Forests of Idaho, the Forest Service had virtually no money to repair logging damage. To restore some areas, it had to barter with timber companies, trading trees for environmental repair. But many of the trees to be cut were in the same drainages that logging had damaged in the first place.[80] The Gallatin National Forest needed funds to close roads to protect grizzly bears. To get the money, it had to sell timber and build roads in other grizzly habitat.[81]

As these instances illustrate, because Congress fully funds timber sales but seldom appropriates more than a fraction needed for other activities, the Forest Service often turns to K-V funds to pay for them. "In effect, nontimber resource specialists have become addicted to timber cutting."[82]

Another perverse provision of the budget is county payments. Since national forests pay no property taxes to the counties where they are located, Congress decided to give 25 percent of national forest revenues (from most of their accounts) to the counties.[83] Since revenues come almost exclusively from timber sales, the counties apply political pressure to maximize the amount of timber sold. Just as the Forest Service increases its budget even when the taxpayers lose money on a sale, so do the counties.

One more major aspect of timber sales that contributes to deficits is the way in which the Forest Service appraises the trees it wants to sell. In this procedure, the Forest Service does not even *consider* any costs to the taxpayers of offering the timber for sale, including costs of building roads that are to be paid out of appropriations.[84] The Forest Service sets a price that will be low enough for the timber company to make a profit. But "it makes no attempt to ensure that the government will make a profit or at least break even on the sale."[85]With all these perverse incentives, plus a number of minor ones, Forest Service managers "never even compare costs with receipts."[86] So they have lost hundreds of millions of dollars every year from selling timber.

But the Forest Service also subsidizes other activities, and the most important is recreation. The national forest system is the country's largest provider of quality outdoor recreation, with more visitors per year than the national parks. The total contribution to our economy made by recreation on the national forests simply dwarfs the contribution from all extractive industries combined. Assistant Secretary of Agriculture Jim Lyons provides some interesting statistics. Total contribution to the economy from all uses of the national forests is just over $130 billion annually. Of that, recreation generates $98 billion, to which hunting and fishing and other wildlife activities add another $13 billion. Timber from national forests generates all of $3.5 billion per year.[87] The Forest Service estimates that 74 percent of all jobs that rely on the use of its lands are in recreation; only 3 percent of the jobs are in logging.[88] And most of the recreational values of the forests, of course, are jeopardized and degraded by logging. As Jim Lyons told the Outdoor Retailers Summer Market in 1996, the economic facts are relatively unknown in Washington because the outdoor recreation industry does not have an outspoken lobby and is not well respected by Congress.[89]

Recreation has also not been well respected by the Forest Service for a very specific reason: its budget. The agency provides excellent recreational opportunities on its trails, rivers, lakes, and mountains. But it is prohibited by law from charging for most of these uses. It can collect fees from a very few things—boat landings, commercial developments, and developed campgrounds—but most of the receipts have to be deposited in the Land and Water Conservation Fund. Of course, any costs of collecting them would come out of the Forest Service budget. As one district ranger says, "Since I don't get to keep the receipts, I don't even try to collect them."[90]

Congressional appropriations for recreation have not come anywhere close to keeping up with the explosion in recreational use of the national forests.[91] So we have a situation in which recreation, the economically most important (and rapidly growing) use of the national forests, is seriously underfunded, and the Forest Service has no way to "profit" from it. One of the least economically important uses of the national forests, logging, generates much revenue that increases the agency's budget no matter how much the taxpayers lose. When you also consider that the lesser-value use destroys the forests for their greater-value use, the conclusion seems clear that the Forest Service timber program is more than just a loser; it is actually antieconomic. But the Forest Service budget structure gives the agency every incentive to chop down all the trees and gives it little or no incentive to preserve them for their highest value.

Even in narrowly economic terms, the Forest Service is in considerable need of reform. Neither timber companies nor hikers and hunters pay their own way. And reform, to be lasting, needs to be structural. Even though today we have biologists as chiefs and timber beasts have been replaced in many offices, the old budget structure is still there. The old perverse incentives are still in place and the old badly skewed funding is still taking its toll on the forests, albeit more slowly. Forest Service chief Michael Dombeck recently told a meeting of foresters that timber production still drives the priorities and the reward system, and they still need to be changed.[92]

REFORMS

So what possible solutions would conservative principles suggest? The national forests have been massively abused—a clear violation of piety toward nature and an abdication of responsibility for damage caused. They have been logged for decades at a rate that even the Forest Service knew was unsustainable—a clear violation of our duty to future generations. Taxpayers have lost billions—a clear violation of the economic principle against subsidies. Several solutions have been proposed.

Out of sheer frustration with past attempts to reform the Forest Service, Sierra Club members voted (against the recommendations of their national leaders) to support a policy of banning all logging on national forests. A bill to do just that was introduced in the 105th Congress by conservative Representative Jim Leach. So little of the nation's wood supply now comes from national forests—under 4 percent—that eliminating all logging on them would have little impact on the country's economy. But this would still be only a partial reform. It would leave the other subsidies in place, eliminate revenues that are important to some counties, and would not by itself introduce new incentives for Forest Service managers.

A recent development that seems to be spreading is the formation of local "consensus groups" or "collaboration groups." Divergent local interests that have been fighting each other in the forest planning process, in the political realm, and in the courts get together to see what common ground they can find. Some have had notable success in cooperating and developing, along with Forest Service representatives, management plans for the forests they previously fought over. The most famous is the Quincy Library Group in California. The Forest Service has, to a limited extent, endorsed various levels of "collaborative planning."[93]

Conservatives should like the idea of local stakeholders getting together to try to work through local problems, and certainly their input should be highly valued. However, the forests are not just the property of the locals, and the stakeholders are not limited to those who live in the vicinity. Moreover, these consensus groups cannot change the perverse incentives in the Forest Service budget and they cannot eliminate the subsidies for those who use the national forests. At best they offer only a small step toward true reform.

One suggestion that periodically raises its head is to turn the federal lands over to the states. This has little or nothing to recommend it to conservatives. In many states in the West iron triangles appear to be even more firmly entrenched than in Washington—which, of course, is presumably the real motive behind most of these proposals. And it cannot be stressed too often that iron triangles violate important conservative principles. A detailed study by the Thoreau Institute of the lands now owned by states concludes that they do no better job administering their lands than the federal agencies do.[94] Moreover, recent proposals would let the states choose which lands to take over, leaving many millions of acres in federal ownership.

Because of the unique historical developments by which the national forests are still owned by the American public, traditionalist conservatives would probably be comfortable retaining them as public lands. Libertarians, however, usually favor privatizing them. But privatizing the national forests would create additional problems, social, environmental, and political.

Timber companies would only want the economically productive forests of the Pacific Northwest, plus perhaps a few areas in the South. In the Northwest, they would want above all to liquidate the last fragments of old growth—unless constrained by the Endangered Species Act, in which case they would probably not be interested in owning the land.

Moreover, privatizing the productive forests would probably compound the environmental destruction from logging. On their own land, timber companies clearcut *much* larger areas than the Forest Service allows, so the damage they do to the land and watersheds is likewise greater. They have long been dominated by a "cut and get out" mentality that creates devastation and leaves it for others to clean up.[95]

"Responsibility" has not been a strong point of the industry, to say the least. For instance, look at what happened in Oregon in the extremely wet winter of 1996–1997. The timber companies had clearcut vast swaths of private lands on steep slopes, even though they had been warned of the consequences. Heavy rains that winter caused massive landslides that devastated the streams and rivers, destroyed many homes, and killed a dozen people.[96] "The Oregon timber industry, fearing lawsuits from homeowners wiped out by mudslides, quietly won indemnification through special state legislation."[97] So much for stewardship of the land and assuming responsibility for the consequences of their own actions.

In sum, although there are some notable exceptions, the history of the private timber companies is in many ways even worse than that of the Forest Service. Thus there is reason to believe that privatizing the productive national timberlands might simply exacerbate the problems from poor forestry.

National forestland that has scenic value would be attractive to developers, who would close it to public access and subdivide it for trophy homes, golf courses, and resorts. The extent to which this is being done in the West, as ranchers sell out to subdividers, is already a major social problem in the area. (See practically any issue of *High Country News.*) Privatizing national forestlands for subdividing would merely make it worse.

From a conservative point of view, in no way is it desirable to turn all beautiful places in America into suburbs and shopping malls and resort complexes. Russell Kirk addressed this very problem some years ago, with reference primarily to the East and Midwest: "Farmlands, woodlands and wetlands retreat before suburban sprawl, gigantic highways, 'resort' and tourist development and industrial expansion. . . . It has gone so far that restraining action has become necessary—unless we have lost interest in the survival of a tolerable society. . . . Public policy ought to encourage landowners to refrain from 'development' that would be baneful to posterity. . . . Isn't it high time to help, rather than to crush, the people and institutions that have conserved something of America's countryside?" He praised Vermont's legislature for adopting policies to "restrain thoughtless development by bulldozer and speculator."[98]

The subdividing of the West is also creating ecological problems. John Sawhill, president of The Nature Conservancy, writes that sprawl is "the most serious threat facing the biodiversity of the Greater Yellowstone ecosystem. . . . The development going on around Yellowstone—and in other rapidly growing rural areas across the country—is threatening to undermine the ecological integrity of many of North America's last great wide open landscapes."[99] From an ecological point of view, it is ever more important to keep the large tracts of public land intact. The last thing we should do is let suburban sprawl onto the national forests.

In sum, privatizing the national forests would likely exacerbate environmental problems from logging and create additional problems from "development." Moreover, any conceivable attempt to privatize the national forests would still leave many millions of acres in public ownership. Besides, any effort to privatize the national forests would call up such a firestorm of political protest that it would certainly fail.

Philosophically, only the "true believer" libertarians want everything possible to be in private ownership. Most conservatives accept that many public goods and public amenities are properly owned and provided by the government. (This does not, of course, mean that all such public amenities must be given away "for free.") So, let conservatives accept as given that, for better or for worse, our history has left these lands in public ownership for any foreseeable distance into the future.

For all these reasons, conservatives should consider something short of privatizing the national forests. It would be much better to make major structural reforms in the Forest Service, including new incentives to reinforce and institutionalize the apparent shift from maximizing timber production to ecosystem management.

The first thing we certainly should do is establish reserves to preserve *all* remaining old growth on federal land that is still in large enough blocks to be ecologically viable. (There is almost no old growth left on private lands.) These reserves should be surrounded wherever possible by forests that will be managed so that they can eventually become old growth and replace the inevitable losses in the core preserves from natural disturbances. These core preserves could be used for non-abusive recreation, for scientific study, as genetic banks, and the like.

Another thing that should be done, obviously, is to eliminate the subsidies for all destructive uses of the forests. If a timber company is permitted to purchase trees in a national forest, it should have to pay all of the costs of preparing the sale, building the logging roads, harvesting the timber, reforesting the land, and removing the roads, plus a "profit" for the taxpayers. If timber companies had to pay all of those costs, many thousands of cut over acres would never have been logged at all. Without the subsidies, most of our national forests would still be intact, their watersheds would still be protected, and their streams and rivers would still be the superb salmon and trout habitat they used to be.

If the timber subsidies were eliminated, timber sales on all but a handful of forests would cease. To ease the transition in local economies, we could take the tax money now being lost and for several years use it to hire the loggers and mill-workers for reforestation, stream rehabilitation, road removal, and the like—as is now being done in a small way in the Pacific Northwest as part of the management plan.

However, eliminating subsidies for extractive industries, important as it is, would only be a start. I believe that conservatives should carefully consider the kinds of proposals for "marketizing" the Forest Service suggested by the Wilderness Society, the authors of *Project 88*, the recent report by the Committee of Scientists to the secretary of agriculture, and developed in considerable detail by Randal O'Toole and his Thoreau Institute.[100]

The national forests should remain in federal ownership, but fair-market user fees for all uses of the forests should be charged. All subsidies should be eliminated—for logging and all other activities. For recreation, the national forests would presumably charge a day-use entry fee, which could be collected on every visit or could take the form of an annual pass good for one or more forests, as the national parks do now. Additional fees would be charged for backcountry camping permits and for use of developed facilities such as campgrounds and boat docks.

Strict standards would have to be enacted for appropriate recreational uses to ensure protection of the resources. Off-road vehicles, for example, do so much damage and interfere so badly with all other users of the land, both human and nonhuman, that they need to be strictly regulated, regardless of how much their owners would be willing to pay.[101]

Congressional appropriations for the Forest Service should be phased out. The national forests should be funded entirely from a portion of net revenues. Congress should not be able to set timber production targets (or any others).

A portion of all user fees should be placed in a "biodiversity trust fund" that can be used to finance protection of endangered species, which would not itself generate revenue for the forest managers. A portion of user fees could also be set aside in a "wilderness trust fund" to finance wilderness preservation, since back-country hiking and camping in wilderness areas will probably not generate enough revenue by themselves.

Counties should receive a portion of all revenues, not just timber income. Local governments would then benefit from all uses of the forests, not just from chopping them down. The counties would also benefit because their income would be more predictable than the highly cyclical timber income is.[102]

All roadless areas should remain roadless.

All relevant environmental laws, such as the Endangered Species Act, Clean Water Act, and Clean Air Act, should remain.

If the national forests were funded out of revenues charged to all users, recreation would clearly be the dominant source of income. Recreation fees will simply "swamp commodity receipts in the vast majority of national forests if the average fee collected is only $3 per visitor day."[103] According to Forest Service estimates, the market value of recreation on the national forests is over $6.6 billion per year. A "willingness to pay" survey produced an even larger figure for the value of recreation in national forests: $11.2 billion per year.[104] These figures are far larger than the total Forest Service appropriations, which in the mid-1990s were about $2.4 billion per year and are currently about $3 billion. Add in its receipts of some $1 billion (mid-1990s), and the potential income from recreation still dwarfs the total Forest Service budget.

Within strict constraints to protect natural resources, the key to this proposal is that "recreation fees can act as a proxy for numerous environmental values, giving managers incentives to protect wildlife habitat, clean water, and scenic beauty."[105] The Forest Service would then have predominant incentives to preserve the forests, whereas now their incentives are to liquidate them. Clearcutting degrades the recreational values of a forest. Under this proposal, the Forest Service's income would be decreased. Thus there would be a budgetary incentive to use careful selection logging, if the forest sells timber at all.[106] Recreation fees, clearly, would institutionalize an economic reinforcement for ecosystem management practices.

Charging user fees for recreation would quite likely be generally acceptable to the public. Hikers and campers would know that their money is actually going to protect the resources they enjoy. As I write this, a pilot program is in place that allows several national parks, national forests, wildlife refuges, and BLM districts to charge extra fees, which are then retained where they are collected. Public acceptance has been generally good. And many states have long had entry fees for state parks and additional user fees for campgrounds and other developed facilities.[107]

Overall, there is good reason for conservatives to give careful consideration to proposals to "marketize" the commodities and services our national forests provide, while retaining them in federal ownership. They would embody important conservative principles that are now being violated. They would adopt important market mechanisms that would make users pay their own way, and in a manner that would protect the resources for the long-term future. By changing the incentives within the Forest Service, they would go a long way toward solving both the environmental problems and the fiscal drain inherent in our current system.

LIVESTOCK ON THE PUBLIC RANGE

Grazing on public land is another prime example of iron triangle politics. A relative handful of western ranchers are permitted to graze cows and sheep on federal land. Some have permits for Forest Service land but most are involved with the Bureau of Land Management (BLM). These ranchers overgraze some of the least productive pasture in the country and are kept going only with huge subsidies from the taxpayers. The politicians beholden to them have long kept the BLM subservient to the ranchers. Just how subservient is well illustrated by a case that the General Accounting Office uncovered. An area manager caught a rancher illegally cutting trees in a sensitive zone by a stream and ordered him to stop. The manager was soon ordered by his superiors to apologize and to deliver the wood to the rancher's house.[108] The result of this iron triangle has been the continued destruction of public land and huge losses to the taxpayers. The system has been well described as "cowboy socialism" and the beneficiaries as "welfare ranchers."

One major difference from the destruction of our national forests is that the rangeland in the West had been devastated before the iron triangle was formed. In the late nineteenth and early twentieth centuries, the public rangeland was open to anyone who wanted to graze livestock on it. The result was a classic tragedy of the commons. Everyone tried to get his cattle to the grass first, before someone else got it, and no one had any incentive to take care of the land. The Taylor Grazing Act finally put a stop to this in 1934. The open range was closed and a permit system was created for access to the public land. To get a permit, a rancher has to own a base property in the area, typically for hay and winter forage. The permit to use public land is attached to this base property. Although grazing on the public land is a privilege and not a right,[109] the value of the permit contributes to the value of the ranch. Unfortunately for the taxpayers, the permits have value only because of an extensive system of subsidies.[110]

Several aspects of the system are astonishing: how much land is involved, how few permittees there are, how little the public range produces, how bad the destruction is, and how much the taxpayers lose to produce practically nothing. In the western United States the BLM manages 167 million acres for grazing and the Forest Service has 91 million acres, so the total area is enormous. But it is used by a relative handful of ranchers: only some 27,000 individuals, partnerships, and corporations.[111] And the grazing allotments are actually concentrated in the hands of a few: 27 percent of those holding BLM permits use 87 percent of the BLM's forage. The vast majority of ranchers in the West do not use public land at all. Only 22 percent of western beef producers and 19 percent of sheep producers have federal grazing permits.[112]

And what do they get from that vast land? Only 2 percent of the total feed consumed by cattle in the United States![113] This total is so insignificant, of course, because the federal land is very unproductive. Much of it is desert, some of it is at high altitude with very short growing seasons, and almost all of it is arid. These are, after all, the lands the government could not give away in the days of homesteading.

Unfortunately, for that miniscule portion of our meat production, the destruction of the land has been massive and it continues to this day. The cow is an animal adapted to a wet climate that ranchers are trying to raise in an arid one. We need to look at the kinds of damage this mismatch can create and then see what condition the western land is in as a result.

Much of the western range is divided into two ecological systems: dry uplands and riparian zones along streams. The latter are less than 1 percent of the land but they are vital. They are the "thin green lines" that are biologically the richest habitats in that arid land.[114] Overgrazing of federal lands has severely damaged both systems.

Cows do not graze like native animals—elk or bison—which keep moving. Cows tend to stay in one place, which is fine in wet climates with productive soils. But in the arid West, over the years they can turn large areas into wastelands.[115] Where they overgraze the uplands, the grasses are sometimes simply wiped out. Often they are replaced with vegetation that is less palatable to cows and is less able to stabilize the soil against erosion. Trampling by the cows compacts the soil and adds to surface runoff and erosion during rainstorms and spring snowmelt. Once the soil has eroded away, uplands may never recover.

Moreover, in many arid places the land is covered with fragile cryptogamic crusts, which are made up of various algae, lichens, and mosses. They are critical for protecting the soil from erosion and for fixing nitrogen in these ecosystems. Cattle trample and destroy the crusts, severely reducing the productivity of the land. Even low levels of use maintain high levels of damage.[116] Even after cattle are removed, the crusts can take more than a decade to recover from trampling.[117]

As a consequence of overgrazing, much of the western range is degraded. Some 10 percent of it has been severely "desertified"—an ugly word, to be sure, that describes an ugly result: the loss of topsoil and vegetation that eliminates the ability of arid lands to support life.[118] A study for the EPA reports that desertification has reduced the productivity of 225 million acres in the West.[119]

The worst damage from cattle is in riparian zones. Unless forced to move, cows will congregate near and in streams and stay there. They trample the streambed and banks and pollute the water, which degrades it as habitat for fish and for users downstream. They wipe out the grasses by the streams and compact the soil by trampling. This reduces the ability of the soil to absorb water, which runs off and increases both erosion and flooding. Since the rains and snowmelt run off, they no longer recharge the ground water and the water table is lowered. As a result, many streams that used to run year round have become dry washes and arroyos. As the water table is lowered, wetlands beside the streams dry up. Sediment in the runoff, which used to get trapped by the plants, no longer builds up the stream banks. It simply runs into the streams, adding to the pollution. The streams become wider and shallower, with no protective cover along the banks, which often become mere muddy quagmires. Damage to the land by cattle, especially the destruction of riparian areas, endangers many native species.

For many years, range policy treated riparian systems as "sacrifice areas."[120] The BLM estimates that 70–90 percent of the original riparian systems on its lands have been destroyed, and up to 80 percent of the remaining ones are in unsatisfactory condition.[121]

Cattle have had devastating impacts on the ecosystems in the arid West. This raises an important question about the overall condition of the public range, which is a very contentious issue. For many years, critics of federal grazing policy have claimed that over half of the public range is in less than good condition. Supporters of the status quo responded that the range is in the best condition of this century.[122]

The supporters' claim, obviously, does nothing to refute the critics. Range condition may be the best of the twentieth century, but that is primarily because the land was in such terrible shape before the closing of the range.[123] Moreover, the standards and categories for those earlier evaluations were narrowly focused on forage production and not on the overall condition of the ecosystems.[124]

In 1995 the Interior Department adopted new criteria for evaluating the condition of the land, and new regulations for grazing. The new criteria look at ecosystem functioning as a whole, not just at forage for cows and sheep. And they differentiate between uplands and riparian areas, whereas the old standards did not. The new categories classify rangeland ecosystems as "properly functioning," "functioning at risk," or "nonfunctioning." According to these standards, uplands are in much better shape than riparian habitats. Of the BLM's upland range, 57 percent is properly functioning, 30 percent is at risk, and 13 percent is so degraded that it is nonfunctioning. In riparian zones, however, only 34 percent is properly functioning, 46 percent is at risk, and 20 percent is nonfunctioning.[125] As Wayne Elmore concludes, the changes in grazing practices since the 1930s have improved uplands, but riparian conditions continue to decline in most areas.[126] In fact, a study for the EPA concludes that riparian conditions are the worst in history.[127]

So why are conditions not better? The tragedy of the commons on the open range was ended some sixty-five years ago. Why is so much of the public land still in such bad shape? To answer that, we must look at the political system that governs grazing on public land, a system that makes it "profitable . . . to mine the inherent productivity of the fragile . . . rangeland as if [it were a] nonrenewable resource."[128] The system also leaves the taxpayers in pretty bad shape.

Ranchers with base properties near federal land obtain permits to graze their livestock on the public range. These permits are issued in terms of animal unit months, or AUMs. An AUM is the amount of forage that will feed one cow and a calf or five sheep for one month. The agency also specifies the areas that can be grazed under the permit and the time period the livestock can be on the federal land.

For the permit, the ranchers pay the BLM or Forest Service a fee that is far below the going market rate for grazing on similar private land. Historically, the rates have been one-fourth (or less) of fair market prices. And half of the fees are spent for "range improvements."[129] Consequently, BLM and Forest Service revenue from grazing does not even cover their administrative costs.[130] This is the subsidy that has generated the most heated debate, but, as we will see, it is only a part of the total.

Not only do the permits *allow* grazing on federal land, they *require* it—and here is another problem with the system. A rancher has very limited ability to reduce stocking to allow the land to recover. If a rancher does not use her permit, the agency finds someone else who will and transfers it to that person. Nonranchers cannot buy out grazing permits to remove cows and restore the land. The Nature Conservancy tried to do this, but its attempt was ruled illegal.[131] The 1995 BLM grazing regulations included a provision for "conservation use," but the federal courts threw it out.[132]

The permit system institutionalizes overgrazing in several ways. When the system was first established, the land was being overgrazed. Rather than reduce the stock level to carrying capacity, the government allowed too many AUMs from the very beginning.[133] And charging below-market rates is an economic incentive to place more cattle on the land.

The Taylor Grazing Act established local grazing advisory boards to make recommendations to the range managers. These boards were entirely made up of ranchers who, of course, wanted maximum use of the public range, no matter what its condition. In a classic manifestation of an iron triangle, the ranchers dominated grazing policy. The GAO concluded that "the BLM is not managing the permittees, rather, permittees are managing the BLM."[134]

These boards have now been replaced by resource advisory councils with wider representation, including environmentalists. The BLM is required to consult with the new councils, and the councils can appeal to the secretary of the interior if they believe their advice is arbitrarily ignored.[135] It remains to be seen how much this will help the land.

According to the laws, the agencies have the right to reduce the number of cattle grazing on their land. In fact, they are supposed to protect the land and reducing grazing is one clear way of doing so. But because of political pressure, the agencies are very reluctant to do it. The BLM has long had an implicit "numbers maintenance policy."[136] And for a good many years, the process it used to justify reducing a permit's grazing allowance severely limited its ability to do so. There was a monitoring requirement that was expensive and time-consuming. With limited budgets and personnel, the BLM seldom had the ability to gather enough data to build a case for reducing a permit that would stand up on appeal.[137] Consequently, stocking numbers were rarely reduced, even where the land was clearly being badly abused.

Under the 1995 regulations, BLM managers are *required* to modify grazing in their districts if it is a significant cause of degradation of the range. With new standards for making that judgment, it should now be easier for managers to reduce the number of cattle on an overgrazed allotment.[138] The final result is difficult to predict because the subsidies and other perverse incentives remain intact.

Pressure to maintain high stocking levels comes from another source as well: the banks. The value of the permits that attach to a base property increases the value of that property. Where a private land rancher maintains the long-term value of his or her ranch by sustaining its productivity, a public land rancher maintains the base property value by keeping the permit levels high.[139] The collateral for mortgages on ranches and for loans from the banks includes the value of the permits. The bankers, as well as the permittees, get very upset if the BLM tries to reduce permit levels.[140] Thus the system includes pressure from several sources that keeps the public land overstocked and overgrazed. But the perverse incentives do not stop there. Other subsidies also result in overgrazing.

One of them is the emergency feed program. The Agriculture Department pays half of a rancher's cost of buying feed during a drought. This gives ranchers incentives to keep stock rates far too high in the arid West: if the land does not produce enough feed, they can get the emergency subsidy. And it should not surprise anyone that the subsidy is often available regardless of the amount of rainfall.[141]

Moreover, instead of reducing the numbers of animals to let the land recover, the BLM tries to compensate for overgrazing by expensive "range improvements." These include water development, brush control, seeding, and the like. Some of these programs have been nothing but expensive boondoggles, all billed to the taxpayers, of course.[142]

Then, of course, there is water, very cheap water. The federal government has built an extensive system of dams and reservoirs and canals throughout the West. Most of this water—85 percent or so—is used for agriculture, and half of that irrigates hay and alfalfa for cows.[143] Ranchers pay only a tiny fraction of the cost. (More on subsidized water projects below.)

So how much are we paying for the privilege of having our lands degraded? Karl Hess, with the Cato Institute, estimates that the total cost to the taxpayer to keep cows and sheep on the public land is about $500 million per year.[144] A similar estimate is given by economist Robert Nelson with the Competitive Enterprise Institute.[145] (And these figures do not include subsidized water.)

The absurdity of this system is apparent when you look at the bottom line for the ranchers. The BLM's *administrative costs alone* are greater than the total profits the ranchers make from grazing on agency lands. For every $4 the BLM spends managing rangelands, ranchers earn only $3.[146] We could *quit* managing BLM lands for grazing, hand the ranchers the cash they would make, have them keep their cows off the land, and the taxpayers would come out ahead. And that does not even count the savings we would make from other programs.

Who benefits from this fleecing of the taxpayers? So little meat is produced from the public land that it certainly is not the consumer. As we saw earlier, AUMs are highly concentrated in the hands of a few permittees, and these include some multibillion dollar companies and several of the wealthiest individuals in America.[147] They certainly do not need, or deserve, these handouts.

But at the other end, the majority of permits are held by individuals whose operations are so small they cannot be economically viable.[148] These are hobby ranches or marginal operations whose owners depend mostly on outside income. For these people, we are subsidizing a way of life—questionable public policy at best.[149]

Not only are taxpayers losing a lot of money on public land grazing, cows on the public land can be harmful to local economies. In many places there may be economically better uses of the land. "The lands managed by the BLM . . . contain tens of millions of acres of spectacular desert, mountain, and canyon scenery that constitutes a major recreational resource for millions of Americans. BLM lands support an estimated 72 million annual visits for sightseeing, hiking, backpacking, wildlife watching, hunting, fishing, off-road vehicle use, and other recreational pursuits. The economic value of the recreational use of BLM land alone far exceeds the value of the livestock forage that the same land provides."[150] Using BLM statistics and cost-benefit model values, Joseph Feller calculates that the forage on BLM lands is worth about $80 million per year. Recreational use of BLM lands is worth nine times as much, over $720 million.[151]

In many places, cattle grazing degrades the recreational value of the land. What we saw with the national forests applies here as well: dust and flies and cow pies are even less photogenic than stumps. A BLM management plan draft for the Elko, Nevada, Resource Area concludes that "gains in recreation-related jobs and income would exceed losses in livestock-related jobs and income if livestock grazing were eliminated in the . . . Area."[152]

Consider the case of some spectacular canyons in the Comb Wash grazing allotment in Utah. Five redrock canyons drain into the wash. They contain thousands of archeological sites and they used to provide outstanding recreational opportunities. However, the grazing permittee grazes cattle not only in the wash itself but in the canyons as well, keeping part of the herd in each one for a month at a time. There is little forage there (and there is more unused forage elsewhere on the allotment), but the damage is severe. The streams and riparian areas have been devastated. The cattle trample the archeological sites and threaten to topple the standing ruins. Hikers and campers now find cow-bombed canyons, polluted streams, dust and flies, and manure so thick that campsites have to be shoveled out.

A local outfitter had to discontinue pack trips into the canyons because his customers found conditions completely unacceptable. His losses are some $15,000 per year. He estimates that if cattle were kept out of the canyons, they could support two or three other outfitters and generate $20,000 to $30,000 per year for the local economy. The annual value of the forage in the canyons is about $2,500.[153]

Why does this happen? Because "the BLM simply does not take into account the types of environmental impacts occurring in the Comb Wash canyons, nor does it consider whether the public interest would be better served by eliminating grazing in such places."[154] And the law prohibits outfitters, or anyone else, from buying out permits or parts of permits in order to exclude cattle. As the

Thoreau Institute contends, "the entire system runs increasingly counter to what Americans want and expect from their public lands."[155] Times have changed but policies have not.[156]

Conservative reform of this bankrupt system should have two goals. As stewards of the earth, we should adopt policies to restore the degraded rangeland. Fortunately, much of it can be returned to good health. We should also move grazing policies toward the free market.

To restore the land, we should first require the agencies to take a serious look at their land and determine what areas are ecologically unsuitable for grazing. Livestock should not be permitted on those lands. So far, the agencies have simply assumed that virtually all of their land is just fine for cows.[157]

Second, the agencies need to determine the carrying capacity of the land that is suitable for grazing and reduce the overstocking to conform to the ability of the land. Karl Hess estimates that the federal range is overstocked by some 3.5 million AUMs and another 2.5 million AUMs should be retired from ecologically fragile land. At market value, he calculates that $300 million is all it would take to buy out the permits and accomplish both of these steps.[158] A small portion of the current federal budget surplus would do the job in one year.

But the reduction of stocking to carrying capacity will not by itself be sufficient. Fewer cows will still trash the streams. What is required is good management—and numerous examples demonstrate that with good management many uplands can eventually recover and riparian areas can make very dramatic improvements.

When riparian areas are rested from grazing, grasses grow again. Cottonwood and willow saplings return and thrive. The new vegetation stabilizes the soil, preventing further erosion. It catches sediment that is washed down and builds up the stream banks. It slows rain and snowmelt runoff and allows it to sink into the soil and recharge the groundwater. The water table rises and supports a wider, lush meadow. Streams that merely used to channel runoff or run only intermittently begin to run year round. Trees and shrubs shade the water and trout and other aquatic creatures thrive once again. A "thin green line" can often return to life.

But for this to happen on a badly degraded stream, cows have to be fenced out for a few years and then allowed back only on short duration. This requires ranchers to give up the "Columbus method of ranching"—turning the cows out to pasture in the spring (or when the permit begins) and then going out in the fall to discover them.[159] The cattle must be moved from pasture to pasture to avoid overgrazing, and they must be kept away from streams until they recover. In many places, it will not even be necessary to reduce the number of animals, just to manage them properly.[160]

Surely, we should require ranchers who are using the public land to use excellent management practices. At a bare minimum, we should adopt a policy similar to the "sod buster" and "swamp buster" provisions in the old farm program: if the rancher does not practice good management on federal land, all subsidies are forfeited.[161] The new 1995 grazing regulations are based on ecosystem management to improve rangeland health and are a clear step in the right direction. They may be useful in improving cattle management. Unfortunately, they still leave most of the old perverse incentives and rules in place.

The second step in reform is to move toward a free market. The proposal that has gotten the most publicity—and the most heated debate—is to raise the price of the permits to free market levels. Certainly, over a period of years, this should be done. But by itself this is far from enough. The other subsidies should be phased out as well. Since there are different ways of doing that, conservatives should consider several options.

The subsidies now are enormous. Taxpayer losses are so large that they could buy out all public land grazing permits at their fair market value every four years.[162] And that is an option conservatives should consider. The ranchers would get the market value of their permits, and the taxpayers could soon quit throwing money into this bottomless pit.

Another option, if grazing is to be allowed on public land, is to lease it at fair market rates but offer the option of buyout by the agencies. Many of the smaller, marginal operations may depend on cheap allotments and would take the chance to get cash instead. "Many ranchers on these lands would welcome the option of selling of their grazing privileges. They recognize grazing is not financially effective but lack opportunities to sell their permits at a reasonable price."[163]

More than that, I believe conservatives should consider a wider package of the following policies.[164] First, marketize grazing permits, and let their current holders sell or lease them at market prices. Eliminate the requirements that they must be used for grazing and that to acquire a permit you have to own a base property. That way, anyone could obtain a permit (or a portion of one). Conservation groups could buy permits from willing sellers for the purpose of retiring them and restoring the land. So could others, such as the outfitter in Utah. Some groups are now doing this on state lands, and the Conservation Fund recently bought out the grazing permits in Great Basin National Park in order to remove cows permanently from the park.[165] This should be allowed on BLM and Forest Service land as well. The government could also buy out permits to reduce overstocking and protect ecologically fragile areas.

Second, charge user fees for all uses of the BLM land, just as proposed earlier for all uses of the national forests.

Third, phase out all subsidies, for cows and sheep but also for hiking and mountain biking and everything else as well. Just as with the national forests, however, restrictions would have to be enacted to protect the resources from damage by recreation (e.g., off-road vehicles) as well as by cows.

Fourth, set aside a portion of user fees for a biodiversity trust fund, to be used to protect habitat for endangered and rare species. Another portion could be set aside for restoration of the land, which could be used for a multitude of projects, including buying permits and retiring them.

We could even go one step farther and, as the Thoreau Institute suggests, fund the BLM out of net receipts, not out of tax money.[166] The iron triangle would be broken because income from recreation on BLM land would swamp revenue from grazing. Moreover, just as with the national forests, recreation fees would give the BLM greater incentive to protect its land.

MINING

Mining is another industrial use of public land that needs reform. The major problems today are from "hardrock" mining—minerals such as gold, copper, silver, lead, and so on. On federal land, hardrock mining is still governed by the General Mining Law of 1872. It originally applied to extraction of all minerals, but over the years Congress has removed oil, gas, coal, sand, and gravel from this antiquated law. The law was written to encourage the lone prospector with his burro and shovel, but it is still on the books in the days when mining is dominated by multinational corporations using trucks that carry 200 tons of ore at a time. It has left us with a sorry legacy. We have pollution from both active and abandoned mines. Taxpayers get nothing from mining valuable minerals on public land. Mining projects are proposed in places that may be more valuable for other things, in which case the only recourse we have is for the government to buy them out. Worst of all, in far too many cases a mining company extracts millions of dollars worth of metal, pollutes the land and streams with its wastes, declares bankruptcy, and leaves the taxpayers stuck with a very expensive mess to clean up. Moreover, as a House subcommittee staff says, the law provides "the most generous range of subsidies and benefits" of all of the public land-use programs.[167] Much of the current mining on federal land is for gold, over 80 percent of which is used for jewelry.[168] Why in the world, a conservative must surely wonder, are taxpayers subsidizing jewelry? We need to take a brief look at this law, at the environmental problems hardrock mining creates, and at some solutions conservatives should consider.

The 1872 General Mining Law contains several provisions that may have made sense a century ago but simply have no place in today's society and economy, and with today's mining technology. In the first place, the 1872 law establishes mining as the single most important use of federal land. Anyone can look for gold (or other hardrock minerals) on federal land. Wherever it is found, it can be claimed and mined—no matter how environmentally risky the site and no matter how much more valuable that location is for other things. The Bureau of Land Management, which oversees mining on federal land, has no authority to stop the gold from being claimed and mined. There is, however, no requirement that a claim must actually be mined, only that a nominal annual fee be paid to keep the claim active. There are about 330,000 active claims on federal land.[169]

Second, when a mining company finds a valuable hardrock mineral deposit, the company has the right to the mineral—for free. Even though it is on public land, the public gets nothing from the company—no royalties, no rent, nothing. If a company mines on state land or tribal land or on private land it does not own, it pays royalties. A coal mine on federal land pays royalties to the U.S. Treasury, and so does an oil well. But a gold mine or a copper mine does not. The American people simply give away $2–$4 billion per year[170] in what is, as Taxpayers for Common Sense says, a clear case of "corporate welfare."[171]

Third, when a claim has been made, the claimant has the right to buy the land (to "patent" it, in the legal jargon) for $2.50 or $5.00 per acre, depending on the kind of claim. Those, of course, are 1872 market prices that have never been adjusted. A mining company does not have to buy the land in order to mine it, but it if wants to, it cannot be denied.

Besides giving away public property, patenting creates other problems. The patented claim becomes a private inholding, which can create additional management problems for the federal agencies. But the major problem is that once the company owns the land, it is no longer subject to the weak federal mining regulations but operates under even weaker state rules.[172] When mining reform was being debated in Congress a few years ago, there was a rush to patent mining claims in order to evade any stricter rules that might emerge.[173]

Fourth, the 1872 Mining Law contains absolutely no regulations for environmental protection and no requirements for reclamation of mines once the ore is gone. Mining is inherently destructive—obviously. Hardrock mining in particular has been well described as "digging into a sealed toxic waste dump."[174] If it is not done properly, and if a mine is abandoned at the end, the environmental destruction can be widespread and can continue for hundreds or thousands of years. Coal mining is subject to strict environmental rules, but not hardrock mining. Coal mines must be reclaimed and the land restored once the coal is gone, but not hardrock mines. And the federal law does not require mining companies to post bonds sufficient to cover the remedial costs if the company leaves a mess behind. The taxpayers pick up the bill.

If mining is not done properly, it causes severe environmental problems. As practiced today, hardrock mining—particularly for gold and copper, two of the most important metals—uses enormous machinery to scoop out and move truly incredible amounts of earth and rock. Modern technology makes it profitable to mine ores containing only minute amounts of metal. Using a very efficient process that extracts almost all of the gold in the ore, on the average it takes thirty-five tons of ore to produce a single ounce of gold.[175] A mine in Montana with very low grade ore processed 60–100 hundred tons of ore per ounce of gold.[176] And that does not count the huge amounts of earth and rock ("overburden") that have to be removed in order to reach the ore. As a result, modern mining creates enormous open pits, equally enormous piles of rocks ("waste"), and huge amounts of depleted ore ("tailings"), which are often left in ponds behind retaining dams. The pits are often below the water table, and when mining is done they fill up and become highly toxic lakes.[177]

Many of today's mines use a "heap leach" process of extracting metal. Waste rock is removed to get to the ore, and this waste is piled up wherever convenient. The ore is removed and placed in a heap on a pad with a liner (usually thick plastic sheets). These heaps can be 300 feet high and can be spread over hundreds of acres.[178] In a gold mine, a cyanide solution is sprayed on top. (Copper mines use sulfuric acid.) As the liquid percolates down through the ore, the metal bonds to it. The solution is drained out of the bottom of the heap and channeled into a retaining pond. The liquid then goes through a processing facility where the gold is removed. In these mines, the depleted ore is simply left in the heaps.

Mines present potentially severe pollution problems from four sources.[179] First, the waste rock and tailings often contain sulfides. When they are exposed to air and water they produce sulfuric acid, which then drains into streams and lakes—the notorious acid mine drainage. If they are not sealed from exposure, the acid drainage will continue until the sulfides are gone. In many cases this means, essentially, forever.

Second, the waste and tailings often contain heavy metals as well—cadmium, zinc, copper, lead, aluminum, arsenic, and the like—which are highly toxic in more than minute concentrations. Rain and snowmelt dissolve these metals and they run off into the streams and lakes. Acid drainage increases the leaching of these metals when it runs over the rocks. Tailings ponds can overflow from rainstorms and their dams can break, sending floods of toxic soup into streams and rivers.

Third, the chemicals that are used to extract metal are toxic. Cyanide and sulfuric acid can leak through tears in the pad liners and can be accidentally spilled, contaminating both surface water and groundwater.

Fourth, sediment can be eroded from the waste piles by wind and water and get washed into streams. Even if it is not acidic or toxic, the sediment itself is a pollutant and can harm fish and downstream users of the water.

Mining presents a number of serious hazards to the environment. According to the Bureau of Mines, pollution from mines has contaminated 12,000 miles of streams and rivers and 180,000 acres of lakes.[180] According to the Western Governors Association, mines are the largest source of water pollution in some western states.[181]

Threats from mining pollution can be severe, both to wildlife and to people. Retaining ponds attract birds, with deadly results.[182] Acid drainage and toxic metals have killed many thousands of fish and severely degraded waters far downstream from their sources. The Bingham Canyon mine in Utah has contaminated 200 square miles outside Salt Lake City.[183] A groundwater pollution plume now extends for seventy square miles and threatens the water supply of a suburb of 70,000 people.[184] In Arizona a toxic plume is headed for Lake Roosevelt, source of Phoenix's water supply. The cost of stopping the threat is some $100 million.[185] In the area around Butte, Montana, site of one of the worst Superfund sites of all, mortality from several serious diseases and rates of several kinds of cancer are much higher than average.[186] These examples could easily be multiplied many times.

It is entirely possible to mine in responsible ways that prevent these problems from happening. Modern technology and knowledge of geology and hydrology can be used to extract minerals while preventing pollution of streams and aquifers.[187]

Far too often, however, even modern mines have had devastating effects. The Summitville mine in Colorado is one of the worst cases. This was a heap leach gold mine that operated from 1986 to 1992. The mine had problems from the very beginning. Cyanide leaks from the heap contaminated both surface and groundwater, and acid drainage and toxic metals were carried into the Alamosa River. Aquatic life in seventeen miles of the river below the mine was eradicated. (Even today, this section of the river shows only small signs of life.) The company had posted a $6.7 million reclamation bond. It negotiated with state regulators about mitigation measures and they returned $2 million. Two weeks later the company declared bankruptcy and the taxpayers are now stuck with a Superfund cleanup job that will cost $170 million.[188]

Summitville is one of the most expensive cases, but it is hardly unique. In 1998 the Brohm Mining Corporation abandoned the Gilt Edge mine in South Dakota. The state holds $6 million in reclamation bonds but the costs will be over three times that amount.[189] Also in 1998, the company that owned the Zortman-Landusky gold mine in Montana declared bankruptcy and abandoned the mine. Throughout its history the mine had numerous cyanide leaks and spills, acid drainage problems, and wildlife fatalities. Half of the streams in the area are now seriously polluted with acid and heavy metals. The bonds held by the state will not cover water treatment and reclamation costs. Taxpayers are likely to be billed $60 million.[190] (The citizens of Montana got so fed up with the problems caused by heap leach mining and the weak protection provided by state and federal regulations that in 1998 they voted to prohibit any future mines from using that process.)[191]

There are now sixty-seven mines on the Superfund priority list.[192] The EPA estimates it will cost at least $7.5 billion and perhaps as much as $50 billion to clean them up.[193] Many of these are the legacy of the distant past, but some are recent mines, such as Summitville. Our current laws do not do an adequate job of making mining companies act responsibly to avoid new Superfund expenses in the future. Taxpayers get stuck paying the cleanup bills, which is, of course, another enormous subsidy, and a completely unjustifiable and unconscionable one.

Environmental standards for mining are generally weak. The 1872 law contains no environmental protection provisions whatsoever. The Clean Water Act could be used to protect surface water, but it allows the states to set standards for nonpoint sources of pollution, which is what most polluting mines are. And western states are extremely lax in enforcing any standards on mining.[194] Moreover, the act does nothing to protect groundwater.

Although mine wastes and tailings produce acids and toxic metals, most of them are exempted from EPA regulations under the Resource Conservation and Recovery Act. They are subject only to state solid waste programs, with no independent enforcement by the federal government.[195]

The Forest Service and BLM now require that operational plans be submitted and approved before mining can begin. But Forest Service regulations are weak and the BLM ones are, as Charles Wilkinson puts it, "more dismal still."[196] There are no provisions for fines or penalties for noncomplying operations. Mines that disturb less than five acres are exempt, even though they too can present serious pollution problems.[197] And there is no requirement for a hardrock mining site to be reclaimed and restored.[198] Moreover, a mining company can patent the land as a means to escape from federal regulations. It is then subject only to state mining regulations, which are even weaker and are very poorly enforced.[199]

Thus current laws and weak enforcement leave pollution from mines largely unregulated. The mining industry has clearly demonstrated that it cannot be trusted to act responsibly on its own. The EPA found that poor practices of the sort that resulted in many of the Superfund listings are still used by the mining industry today.[200]

The current laws confront us with other problems as well. Since mining takes priority over any other use of federal land, the BLM has no authority to veto mining projects in places where mining simply should not be done. The proposed New World gold mine just outside Yellowstone National Park is one example. A mine in that location would have had a terrible impact on one of America's "crown jewels" and threatened to pollute the pristine rivers in the Yellowstone ecosystem. But the government had no legal authority to prevent the company from mining there and had to pay $65 million to buy it out.[201]

Congress has permanently withdrawn some land from prospecting, such as national parks and wilderness areas. Federal agencies have the authority to withdraw other areas as well. Recently, for example, 429,000 acres of national forestland in Montana were withdrawn.[202] Existing claims, however, would have to be honored or purchased.[203] Moreover, agency withdrawals last for only twenty years and can be reversed.[204]

With no provision for royalties to be paid on minerals taken from public land, the companies get all of the profits and the taxpayers get nothing out of a valuable resource that they own. Other property owners—state, private, tribal—charge royalties for mining on their land. As Thomas Power points out, royalties are a normal cost of doing business in the mining industry. For one major source of minerals not to collect them creates an economic distortion that artificially increases the demand for mining on federal land and artificially decreases demand for mining elsewhere.[205]

The patenting provision is also an unjustifiable giveaway of public property. The prices—$2.50 or $5.00 per acre—are 1872 prices that have never been updated. And the claim and patent system is open for abuse that can cost the taxpayers extra. For example, when it was becoming clear that the federal government would choose Yucca Mountain in Nevada for the permanent storage of nuclear wastes, one enterprising chap staked some claims on the mountain. It cost the American taxpayers $250,000 to buy him out.[206]

The weak and antiquated laws have left us another legacy as well: hundreds of thousands of abandoned mines, many of which pose significant threats to the environment and to human safety. According to a study by the Mineral Policy Center, there are over 557,000 abandoned mines in the United States. Many do not present significant problems, but 231,900 of them need landscaping and revegetation to prevent off-site impacts. Some 116,300 are safety hazards that need to be eliminated to protect the public. Surface water is being contaminated by 14,400 abandoned mines, 500 are contaminating groundwater, and many (now sixty-seven) are on the Superfund priority list. The total cost to clean up these mines and protect the public and our water resources is at least $32.7 billion and may be over $70 billion. There is no program, federal or state, to reclaim these abandoned mines, except for the ones that are bad enough to make the Superfund list.[207]

In past years, problems such as the above also came from mining coal, sand, gravel, and other minerals, and from drilling for oil and gas. Congress has long since reformed the legal system for extracting these resources, but not for hardrock minerals. It is long past time for the last remaining part of this antiquated system to be eliminated.

What policies should conservatives consider? Clearly, reform should have two main goals. The first is to eliminate the subsidies. There is no reason for taxpayers to support the mining industry—certainly not for something so trivial as jewelry. The second is to require mining companies to take responsibility for the consequences of their actions (to internalize their negative externalities). It is entirely possible to extract minerals and process them in ways that protect both surface water and groundwater. When a company does not act responsibly, the taxpayers should not get stuck with the bill. With these two goals in mind, several options should be considered.

- Mines on federal land should have to pay fair market royalties, just as the companies pay for mining on private or other nonfederal land. Requiring royalties would provide the public with income from its minerals and remove an economic distortion.[208]
- The patenting option has to go. Congress in recent years has passed an annual moratorium on patenting.[209] It is time to repeal it for good.
- There is a $200 annual fee to keep a claim active (originally it was $100). This should, at the very least, be adjusted for price changes since 1872. Ac-

cording to the Congressional Budget Office, that would be about $1,000 per year now.[210] At that rate, many claims would be abandoned, which would simplify management of the land and prevent potential problems in the future. (When Congress decided to get serious about enforcing the $100 annual cost, over half of all outstanding claims were dropped.)[211] Another option to consider is to add a requirement that claims must be mined within ten years or they lapse, as is currently the case with coal leases.

To protect the environment and prevent new Superfund expenses, several other measures should be considered.

- The federal government should determine where mining is an appropriate use of its land and where other values are more important. It should have the authority to veto mining projects where they simply do not belong—such as the proposed mine just outside Yellowstone National Park.
- Before mining begins, the company should have to post a bond that would be sufficient to cover *all* potential cleanup expenses in case the company abandons the mine. This should be required of all mines, regardless of whose property they are on, because acid drainage and toxic metals will not stay on the property but will run into streams, rivers, and aquifers.
- Clean Water Act standards should be established for streams in all mining areas, which means that if the states continue to be lax in setting water quality standards (TMDLs for the pollutants mines produce), the EPA should do so. These should be performance standards, not technology requirements, and the mining companies could then determine the best ways to meet them.
- Reclamation and restoration of the site should be required for all mines on federal land once the ore is gone. The mandatory bonds should also be sufficient to cover these costs, in case the company abandons the site.
- The royalties and holding fees could be earmarked to pay for reclaiming abandoned mines. This would provide a continuing source of income for solving these problems.[212]

Policies such as these would place responsibility precisely where it should be: on the mining companies. Such policies are hardly draconian or unusual; these proposals are very close to current requirements for coal mining. There is no reason to expect any less from hardrock mining companies.

The mining industry has opposed all attempts in Congress to make it behave responsibly, and it will surely do so in the future. It will claim that it cannot afford to pay royalties to the American public. This is, as the *Economist* puts it, "bunkum."[213] The industry already pays royalties on 65 percent of all gold that it produces.[214] If the industry is not willing to pay a normal cost of doing business on federal land, then it should not do business there at all.

If a proposed mine is carefully designed, constructed, and operated, full-cost bonding requirements should not be an obstacle. If a company designs a mine poorly or has a poor record and cannot post the bonds, then it should not be mining anyhow, anywhere. To forgo improperly designed mines by irresponsible companies is far better than sticking the taxpayers with reclamation expenses later. The industry will complain that adopting such reforms will "waste" valuable resources. But the minerals in the ground do not turn into pumpkins; some competent and responsible mining company may come along later.[215]

Conservative principles clearly support mining reform. The huge subsidies the industry now gets should stop. We should not pass on to the next generation what we received from the industry: abandoned unreclaimed mines, polluted lakes and rivers and aquifers. As stewards for the future, we should do what we can to remedy the legacy we got stuck with and make sure we do not create new Superfund sites. Mining companies must be made to take responsibility for the consequences of their actions. They must no longer be able to walk away with the gold and leave the taxpayers with an expensive mess.

FEDERAL WATER PROJECTS IN THE WEST

From federal land we now turn to federal water. Almost all of the United States west of the 100th meridian is arid, yet westerners waste water at a much greater rate than do people in any other part of our country. "The arid American West has some of the worst water conservation practices in the world."[216] A lot of that waste is at taxpayer expense, and at the expense of the environment as well.

The problems connected with the western water system are extremely complex. The legal system governing water use is complicated and varies from place to place. It consists of at least four levels of laws and regulations: (1) some general principles that were created by the early miners and pioneers, (2) state laws that embody those principles but vary from state to state, and (3) the extensive system of federal dams, reservoirs, and canals that have their own agency regulations governing water distribution and use. Then there are (4) the many irrigation districts (subject to state laws but with their own governing structures, rules, and motives) that distribute water at the local level. In the following discussion I can do no more than touch on some generalizations.[217]

The basic principles of western water law were created by miners and settlers who quickly discovered that eastern principles did not work in the arid West. In the East rights to use water are attached to riparian properties, with provisions for protecting downstream users. But in the West, water is usually needed far from rivers and streams and has to be diverted. So the principle of prior appropriation was established: first in time, first in right. The first one to divert water from a stream had the right to take as much as he was using. His right was "senior." People who came later could take whatever was left, and in the order and amount in which they first did so. But their rights were "junior," in chronological order. In a drought, those with senior rights could continue to take all the water they were entitled to, and those with junior rights far down the list might well be left with nothing.

A second principle is that the water must be put to "beneficial use." A short list of approved uses was soon established: mining, agricultural, industrial, municipal, domestic, stock raising, and hydropower. Leaving water in the river for the fish, to preserve a viable aquatic ecosystem, was not beneficial; the water had to be taken out of the river. That was the law throughout the West for a long time, and it is a major source of environmental degradation in the region. Only in very recent years have some states accepted the preservation of in-stream flow as a beneficial use.[218]

A corollary to beneficial use was that the holder of a water right could not waste the water. However, the notion of waste had little or nothing to do with efficiency of use. If the water was taken out of the river and used, however extravagantly, for irrigation or another recognized use, it was not being wasted. But leaving water in the river was wasting it; the rights to it were forfeited and could be appropriated by someone else.

Such were the general principles of water use in the West. As territorial and state governments were established, they enacted water codes that embodied those principles, but with variations from state to state. They also severely restricted or even prohibited the right to buy and sell water rights, although many of those restrictions have since been reduced.

As the West was being settled, it quickly became clear that rivers and streams offered only a limited supply of water, and one that varied tremendously from year to year and season to season. Most of the water in western rivers comes from melting snowpack in the mountains, so the rivers run full in the spring and are very low by late summer and fall. The solution was to capture the spring runoff and store it for use later. Many small-scale dams and reservoirs were built, but to get serious about water storage would require damming the larger rivers. Only the federal government had the resources for projects of that scale. Beginning in 1902 with the Reclamation Act, the federal government began a massive construction effort that lasted well into the 1970s and even continued into the 1990s, at a reduced level. It included some of the world's most impressive engineering feats: Grand Coulee Dam, Hoover Dam, Glen Canyon Dam, the "taming" of the Colorado River. It turned most of the rivers of the West into complex plumbing systems.

The massive federal water development program provided reliable supplies of water, hydroelectric power, and lakes for recreation. But it also created many environmental problems, and fiscal ones as well. Very quickly an iron triangle formed in which a few powerful people saw ways to profit from subsidized water projects. Their senators and representatives got control of the relevant subcommittees and appropriated the funds. The Bureau of Reclamation and the Army Corps of Engineers carried out the engineering and construction projects and extended their own bureaucratic power over the water and electricity they provided.

These projects made the West bloom, enriched a few, fostered some political careers, and aggrandized a couple of agencies. But very few of them made any economic sense. Robert Gottlieb describes the iron triangle in action:

> This three-way interlocking network of interests—Congress, the water agencies, and the local water industry groups—held sway over several generations of water projects. It was difficult to buck the system: annual appropriations were rarely established on the basis of project merit, let alone evaluated on anything but the narrowest and frequently manipulated definition of cost-benefit analysis. Instead, projects were most often selected on the basis of who had the clout to make it happen. . . . This was "pork barrel" politics in its most blatant form, a system so effective that the Iron Triangle became, in effect, a "subgovernment."[219]

The iron triangle system also did everything it could to make sure that Indian tribes got little or none of the water to which they were entitled—but that is a subject for another day.[220]

Almost all of the water used in the West is for agriculture. Irrigation accounts for 90 percent of total consumptive uses in the region.[221] The federal government supplies about one-third of all irrigation water used in the West.[222]

The West now has a water system that is highly subsidized, encourages gross waste of a supposedly scarce resource, and has caused tremendous damage to river ecosystems and their inhabitants. We need to look briefly at each facet of the situation.

The subsidies are pervasive and exist in so many different forms that it is virtually impossible to account for them all. The staff of a House committee reported in 1994 that they could discover no comprehensive study of water subsidies that included all of the possible factors.[223]

The basic subsidies are found in the repayment provisions of federal water projects. Irrigators are supposed to repay the construction and operating costs of the portion of water projects that are allocated for irrigation. (Parts of the costs can be allocated to other uses, such as flood control, hydropower, and municipal and industrial use.) But they pay no interest at all on forty- to fifty-year loans. Repayment does not begin until all major parts of a project are officially declared complete, which can take decades. Those who benefit from the project early on get water for many years before repayment begins. The irrigation part of costs can also be reduced, based on an "ability to pay" calculation. And the operating costs of water delivery are subsidized because the power to pump water to irrigators is set at below-cost rates.[224]

As a result, irrigators pay only a tiny fraction of the cost of the water they get from federal projects. For example, in the Central Valley Project in California, the full cost of an acre-foot of water in the Westlands district is over $45, but farmers pay only $8. (An acre-foot is the amount of water it takes to cover one acre one foot deep.) In the Central Arizona district of the Central Arizona Project, the cost of an acre-foot of water is $209 but irrigators pay only $2.[225] In the Truckee-Carson district in Nevada, an acre-foot of water costs $33 but farmers pay only $2.19. The Central Utah Project is probably the last major federal water development that will ever be built. It will deliver water at a cost of $400 per acre-foot, but farmers will pay only $8.[226] (By contrast with federal prices, in California irrigators who get their water from the State Water Project pay up to $200 per acre-foot.)[227] With such low prices, farmers and ranchers will never repay the costs they are supposed to. Central Valley Project irrigators, after using its water for forty years, had repaid only 1 percent.[228]

How much are taxpayers losing every year selling water below cost? The Bureau of Reclamation estimates annual irrigation subsidies in the West as $2.2 billion.[229] A recent study by David Pimentel and colleagues calculates the total annual subsidy for irrigation in the West at twice that much: $4.4 billion.[230]

But those figures only begin to tell how little economic sense federal water projects make. Half of the water is used to grow low-value hay and alfalfa for cows.[231] Much of it is also used to grow surplus crops that were subsidized under the old farm program. Federal water was being used to expand irrigated acres of cropland at the very time that Department of Agriculture programs were trying to reduce acreage in production.[232]

All these subsidies may have increased the income of farmers in the West, but at least some of it was at the expense of other farmers. Western reclamation projects reduced the income of non-reclamation farmers and forced out of use 5-18 million acres of farmland in the East.[233] Economically, it surely makes more sense to grow thirsty crops where the rainfall will support them and not in the arid West. But the purpose of iron triangles is to get things done that do not make good sense.

With all these subsidies, and with water so cheap, it should surprise no one that lots and lots of water is simply wasted, that water in the West is used very inefficiently. "With low prices, users have no incentive to consider alternative technologies and lifestyle changes that would save water."[234] Water leaks from unlined canals and ditches and evaporates from canals and storage reservoirs. Less than half of the irrigation water ever gets to the plants. And western farmers tend to use flood irrigation, one of the most wasteful methods of all.[235] According to one study, agricultural water use could decline by 15–20 percent through conservation without decreasing production.[236]

Of course, the basic legal principles also discourage efficient irrigation. "Use it or lose it"—so farmers and ranchers use far more than the crops need in order to preserve their water rights.[237] If a farmer saves water by adopting a more efficient system of delivering water to crops, he or she risks losing the right to the conserved water. If it can be saved, it must have been wasted in the past and not really put to beneficial use. It could be forfeited and made available for appropriation (without compensation) by someone else.

The environmental impact from turning natural rivers into closely regulated plumbing systems and diverting much of their water has been severe. The consequence that has gotten the most publicity is the plight of salmon in the Pacific Northwest. Salmon spend their adult lives in the ocean, but they hatch in freshwater streams inland and return there to spawn. Some dams completely blocked rivers and eliminated any possibility of those runs of salmon getting to their spawning streams. Even some dams with fish ladders have severely reduced salmon populations. Although the adult fish may be able to swim upstream, the juvenile fish have a very difficult time making it downstream. Instead of a swiftly flowing river that washes them down to the sea, they face a series of slack water lakes in which predators abound. And a significant portion of them get sucked into the turbines that generate electricity, where they perish. Many salmon runs are now extinct and many others are on the endangered species list.[238]

But ecological problems from dams are not limited to the Northwest. All major dams drastically alter the river environments. Flood regimes and flow rates are changed. Sediment that used to flow downstream drops out in the reservoirs. The temperature of water released from dams is often very different from the natural temperature of the river. As a consequence, many native fish are endangered.[239]

Smaller streams, and even some rivers, can be completely dried up by diversions for irrigation, especially in years of drought, with drastic consequences for fish and other aquatic creatures. In many cases, this could be avoided. For example, in 1987 irrigators in Montana took all of the water out of the Ruby River, causing a major fish kill. But while the fish were dying, farmers had water standing in their fields.[240] "Beneficial use" and "use it or lose it" doctrines caused this gratuitous disaster.

The soils in the West contain various kinds of salts, and when the fields are irrigated the water dissolves them. Irrigators apply enough water to remove these salts in the runoff. Some of the irrigation water evaporates and some is taken up by the crops. Consequently, the concentration of salts in the remaining water that drains away increases. The runoff returns to the rivers and is used again and again by irrigators as it goes downstream, picking up more salts with each use. The farther down a river basin you go, the more saline the water becomes. As a result, at times during the summer the Red River in Texas and Oklahoma is more saline than the ocean.[241] The Colorado River accumulates an enormous burden of salts. The Bureau of Reclamation estimates that every additional part per million of salinity in the Colorado River costs urban users several hundred thousand dollars per year because of its corrosive effects on infrastructure.[242] The return flow often contains pesticides, herbicides, and fertilizers from the fields. Drainage water is a leading source of pollution wherever it returns to the rivers.[243]

Other problems from irrigation projects come from toxic minerals that leach from the soil. The most notorious instance occurred in the Kesterton Wildlife Refuge in California. A drainage system was built for local irrigators to collect runoff water and keep the soil from becoming waterlogged. The water was channeled into the wetlands of the refuge. But the irrigation water leached selenium from the fields, and selenium concentrations in the refuge increased. Soon birds were dying and many were found with grotesque deformities. The government closed the refuge and filled the ponds, but now selenium is being deposited in private ponds and basins and is building up there. Kesterton is hardly unique. Selenium poisoning has been found in other states, and at least fifteen other water projects contain concentrations as high as those at Kesterton.[244] Selenium poses a serious threat to human health as well.[245] Overall, what Harrison Dunning says of the Central Valley Project applies to many of the water projects in the west: "Whatever CVP's benefits . . . [it] has unquestionably been a debacle for the environment."[246]

The West has changed from what it was when most of the large dams and water projects were built. Population has increased dramatically and is concentrated in a very few places—the West is the most highly urbanized section of the country. Demand for water in these cities, for domestic and industrial use, has also increased dramatically. Moreover, we now know a lot more about ecology, so we can see how much damage the western plumbing system has caused. We now know how important it is to maintain sufficient in-stream flows and to replicate as closely as possible the natural cycles of those flows. And today, even in narrow economic terms, the value of water in a stream—for tourism, fishing, rafting—can be greater than its economic value for irrigation.[247]

Although surface water in most western states is fully appropriated, the era of large water project construction is over.[248] Environmental concerns, combined with more realistic cost-benefit analyses than were ever used in the past, would block practically any new major proposal. Besides, almost all valleys where really big dams are feasible already have them. The challenge for the future will be to reallocate the water supplies we now have.

The West does not need any more big water projects. There is plenty of water to meet all needs—agricultural, domestic, industrial, and environmental. A mere 7 percent reduction in agricultural use would support a 100 percent increase in all other uses.[249] Transferring to cities a mere 10–12 percent of the water now used on fields would meet all needs for decades of population growth.[250] So much water is wasted, and so much is used for low-value crops, that the opportunities for reallocation are enormous.

What policies should conservatives consider? If we were starting from scratch, we would never create anything like what we have. But many farmers and many communities have come to depend on the huge infrastructure now in place and on cheap water, so some of the needed changes will have to be made gradually.

Some have proposed regulations to reallocate water, requiring farmers to use it more efficiently and taking conserved water for in-stream flow and cities. But from a conservative perspective, a better way to reallocate the water would be to reduce subsidies and expand water markets. As Zach Willey of the Environmental Defense Fund says, "you're not going to do it by wholesale taking away of resources from . . . farmers. . . . You're going to do it through a system of incentives." His approach is to "go out and make some deals."[251] But some fundamental changes in water law and policy will be required for this to work on any significant scale.

First, the federal government should begin raising water prices for its irrigation water. There is no way to recapture past losses, but in the future subsidies should be gradually reduced and eventually eliminated.[252] This will induce significant changes in the way farmers use the water. Low-value crops will be replaced with higher-value ones. Thirsty crops will be replaced with ones that require less water. Farmers will adopt more efficient irrigation systems and marginal land will be fallowed or turned into pasture. The result will be to make a lot of water available for reallocation. In the agricultural sector, studies have found that a 10 percent increase in price will reduce consumption by 4–7 percent.[253]

As water prices go up, some of the increased income will go to the federal government and should be earmarked for environmental purposes, to purchase water to maintain flows and to restore river ecosystems.[254] (A few steps in this direction were made for one water project in the Central Valley Project Improvement Act of 1992.)[255]

Second, the laws must be changed so that farmers who save water retain full rights to it. "The single change in the definition of a water right that would most increase water use efficiency is 'salvage legislation' that clarifies the rights to water saved through reductions in consumptive uses."[256] That is, property rights to water must become firm and clear. The "use it or lose it" principle has to be abandoned. The "beneficial use" requirement must permit the holder of a water right to conserve water and to sell or lease it to others.

Third, in-stream flow must be recognized as a beneficial use. Fourth, markets for water should be expanded.[257] With active markets, farmers could sell or lease conserved water for other uses: to cities for municipal and industrial use, and to conservation organizations and governmental agencies for in-stream flow and wetland restoration. This would allow farmers to cover the higher prices that the federal government should charge and to profit by conserving water.[258]

The potential for further expansion of water markets is enormous.[259] But today there are only limited possibilities to market water. Some irrigation districts have water "banks" in which farmers can "deposit" surplus water and "withdraw" water when needed. But these are not true markets, and the prices are usually fixed at a low level that prevents anyone from making a profit.[260]

A few major water transfers have been made. The most widely known is the agreement the Metropolitan Water District of Southern California made with the Imperial Irrigation District. MWD paid to line IID canals to prevent water from leaking away, and then transferred the water that was saved to the cities, without taking any agricultural land out of production.[261] But these large market deals are rare, and even MWD has failed in some of its attempted purchases.[262]

Some cities have bought farmland in order to acquire the water rights. Environmental organizations in some states are buying water rights in order to leave the water in the streams. The Nature Conservancy, for example, has negotiated several such deals.[263]

But western water is still not easily marketed. For water markets to do the job they are capable of doing—encouraging efficient use and conservation, reallocating water from low-value to high-value uses, and expediting purchases for in-stream flow to protect and enhance river ecosystems—numerous barriers will have to be removed or at least lowered.[264] In many places, water rights have not been finally adjudicated, so they are not clear and firm. For markets to work well, property rights must be clear and enforceable.[265] Most states do not let agricultural water be freely traded for in-stream flow. Bureau of Reclamation water is difficult to transfer. Irrigation districts often have the responsibility to contract for water, so it is uncertain who would have the right to sell or lease water: the users individually or only the district. Major water projects often prohibit trades outside of their areas, and irrigation districts are reluctant to allow such trades even if they are legally permissible. There is considerable opposition to transferring water out of a basin, since local communities fear that the economic consequences will be severe.[266] There also may be problems with monopolistic owners of canals inhibiting transfers of water.[267]

There are, in sum, many barriers—legal, institutional, and political—that stand in the way of marketing water. Although markets are developing slowly, "active water markets remain a relatively local and often sporadic phenomenon."[268] As Terry Anderson and Pamela Snyder contend, "the challenge for the next few years is to develop policies—national, regional and local—that allow trades to take place."[269] Conservatives should support policies that remove or lower these barriers.

Measures such as those suggested here would go far toward reallocating federal water to higher-value uses. More realistic prices will encourage conservation, which will make water available for other uses.

Markets can efficiently reallocate water, even in a drought. A case in California clearly shows how much water really is available. In 1991, California faced a fifth consecutive year of drought. Reservoirs were so low that many water users would receive only a small fraction of their normal supply, and some might get no water at all. The state Department of Water Resources established a drought water bank, which offered to buy water for $125 per acre-foot and sell it for $175. In just three months, it was able to buy over 800,000 acre-feet. It sold nearly half of it, mostly to cities, and stored much of the rest for use the next year.[270] Even under the worst conditions for supply, markets can make a lot of water available for reallocation.

Measures such as those suggested here would also help the environment. Changing the concept of beneficial use to include in-stream flow would give environmental uses of water a firm legal status. Enhanced ability to market water would make it easier for conservation organizations and government agencies to acquire water rights to maintain wetlands and streams.

Indian tribes often want to use their water rights to maintain fisheries, which requires that the water be left in the rivers. Some states allow this but some do not.[271] With in-stream flow established as a legitimate beneficial use, tribes in all states would have the right to maintain their traditional fisheries, and the environment would be a winner as well.

With higher prices encouraging water conservation, and with the extra federal revenues dedicated to environmental purposes, a lot of ecological restoration could be done. A steady source of income would be available to acquire water rights for environmental purposes. One model that could be used widely is the Yakima River Basin Water Enhancement Project. Congress provided funds for projects that would result in more efficient use of irrigation water. Two-thirds of the conserved water will be left in the river for improved aquatic habitat, and one-third will be available to the irrigators.[272]

Other uses for the income could be to modify dams so they can release water from different levels of the reservoir, as was recently done on the Shasta Dam in California.[273] This lets the operators control the temperature of released water to keep it close to natural in the rivers below the dams. Some dams could simply be removed to restore rivers to their natural state. "For many older U.S. dams, environmental damage outweighs economic value and makes them ripe for removal."[274] Some of the money could be used to restore wetlands in riparian areas. The possibilities are nearly limitless. The missions of both the Bureau of Reclamation and Army Corps of Engineers have been changed, making environmental protection and restoration an important part of their jobs.[275] A steady source of earmarked income could help them carry out these new missions.

Reducing subsidies, removing restrictions on the use of water, clarifying property rights, expanding markets, conserving water, and purchasing water rights to enhance or restore river ecosystems should constitute a natural policy package for conservatives.

WILDERNESS

Finally, before leaving the subject of public lands, I want to address the subject of wilderness. There is currently a proposal to preserve up to 60 million acres of national forests that are still roadless. Debate also continues about preserving several million acres of roadless desert in Utah. It seems to me that conservatives should support protecting all these areas, and any others like them. As stewards of the earth, we should preserve the few natural areas that remain. As our population increases and "development" relentlessly turns all desirable private land into parking lots and malls, subdivisions and trophy homes, we should be thankful that we still have some natural areas on public land, pretty much as God created them.

However, when environmentalists propose expanding the wilderness system, anti-environmentalists often accuse them of valuing nature over people. This is absurd. Environmentalists are not advocating turning Manhattan back into pristine forest or Los Angeles into coastal sage scrub. Environmentalists want to preserve natural areas simply because *there are so few of them left.*

Opponents of wilderness often accuse environmentalists of wanting to "lock up" valuable resources. But this demonstrates a complete misunderstanding of the issue, even in the accusers' own narrow terms. Wilderness areas *are* valuable resources—far more valuable as wilderness than for anything that could be extracted from them. There is increasing demand for wilderness—although it is not part of the market economy—and the supply is scarce. What little is left is being destroyed, usually for lower-value uses. As we saw earlier, recreation on the national forests makes a vastly larger contribution to our economy than do all extractive uses of those forests combined. Recreational demand on our wilderness areas is growing rapidly, and use will probably have to be limited in the future to protect the land. But turning those areas over to the extractive industries would almost certainly be antieconomic.

The traditionalist contribution to conservative thought would support wilderness preservation for a cultural reason as well. Wilderness is an inherent part of the American past and played an important role "in forming and maintaining a set of distinctive national character traits."[276] Wilderness thus has an important "heritage value" for us. Just as it is important to preserve historical buildings and battlegrounds and monuments that were involved in the shaping of our nation, so it is important to preserve wilderness, for it too is a vital and distinctive part of our nation's history. You do not have to be a backpacker to appreciate this "existence value" of wilderness. As Wallace Stegner says, "We were in subtle ways subdued by what we conquered. We simply need that wild country available to us, even if we never do more than drive to its edge and look in."[277] Our wilderness heritage has left us with "the nagging knowledge that when it's all gone, when we've beaten back and paved over what pioneers have fought and cussed for three centuries, something basic in the American grain will have gone with it."[278] We should make sure that it is never gone.

We have so far protected a mere 5 percent of our land as wilderness. All the rest that is intact should be preserved as well.

6

Global Warming: The Problem

THE ENVIRONMENTAL PROBLEMS we have examined so far have been largely local or regional, such as the kinds of pollution discussed in Chapter 4: urban smog, polluted rivers, toxic waste disposal. But we now face several daunting problems that are global in scope and have potentially catastrophic effects that will last for centuries or longer—problems that are unprecedented in human history. Robert Solo explains:

> Throughout his habitat on earth, [man's] technologies have been formed on the assumption that the autonomous system that produces the environment needed for life cannot be reached by what we do nor destroyed by us. And here, I think, a crucial change has come. The [earth's] life system itself is no longer beyond the reach of man's technology nor beyond his power to disarrange, degrade, and destroy. This is a danger no age has ever faced before. The volume of activity and the magnitude of consumption increase with terrifying rapidity, and technology, following an ancient momentum, takes no account of the limited capacity of the biosphere to rearrange what man disarranges.[1]

The world we inherited was well-adjusted for human life, but through our actions we are altering it in fundamental ways. We are conducting uncontrolled experiments on the entire planet, with consequences that cannot be predicted with any certainty but could very well be devastating. We will consider global warming in this chapter and the next, the collapse of biodiversity in Chapter 8, and the loss of "nature's services" in Chapter 9. Global warming requires so much attention because the issues are very complex and are widely misunderstood. Besides being an extremely important problem on its own, it will exacerbate other critical problems if we do not take effective measures to solve it.

Of all the environmental issues, global warming is surrounded by the most extensive confusion in the public debate. If you get your information about global warming from television news, you will know approximately nothing. If you get your information about it from the news and science sections of the quality press, you will know something about the issue—they give limited coverage to this complex problem. But if you rely on the op-ed page and letters to the editor, you will know less than nothing because you will just be confused by all of the contradictory assertions.

There have been many attempts to mislead the public about the overwhelming weight of scientific evidence. Typical tactics have been to pick a few studies with findings that differ (or *seem* to differ) from the overwhelming weight of evidence, or to find some maverick scientist, often speaking outside his or her own field of expertise, who rejects the general scientific consensus. They then present these minority views as the gospel truth (without telling people that they are minority views), and often accuse the rest of the scientific and political community of lying.[2] There is absolutely nothing conservative about such attempts to deny the weight of evidence and to confuse the public about the general conclusions of contemporary science.

When global warming first became a headline issue, industries with vested interests in the status quo got together and acted in the same way that the tobacco industry had for decades: they ran ads claiming that there was no hard proof that they were actually causing a problem. However, as evidence accumulated that altering the atmosphere really is a serious concern, the industrial front broke down. Many major companies have admitted that the evidence is sufficient for great concern, and they have established a new organization to deal responsibly with the problem. Even BP Amoco, Shell, and a few others have broken away from the original oil industry party line and admit that the scientific evidence on global warming presents the world with a serious problem that must be faced.[3]

Too many conservative politicians and pundits, on the other hand, have by and large remained impervious to evidence and are still in denial. This stance is completely unconservative. After all, William F. Buckley Jr. himself once claimed, if memory serves me correctly, that the one unforgivable sin in political life is invincible ignorance.

What in the world could possibly be "conservative" about rejecting the conclusions of scientific studies? The *science* of climate change is not liberal or socialist or leftist or conservative or monarchist or anything else on the political spectrum. What distinguishes principled conservatives from liberals and the left is the kind of *policies* they advocate to solve our social, economic, political, and environmental problems, not the acceptance or rejection of evidence that a problem exists.

There is now a consensus within the scientific community that the ways in which we are changing the atmosphere have already begun to increase the global average temperature. This could have such major impacts that it is of great concern and should be dealt with. Rather than persist in irrational denial, American conservatives should take their cue from Margaret Thatcher's speech on the day the first Intergovernmental Panel on Climate Change report was published in 1990: "We have an authoritative early warning system: an agreed assessment by some three hundred of the world's leading scientists of what is happening to the world's climate. . . a report of historic significance . . . what it predicts will affect our daily lives."[4] She, quite properly, did not see any reason for conservatives to reject scientific evidence.

Admittedly, a scientific consensus does not always establish the truth. But the only sensible way of making public policy is to use the weight of the evidence, and on global warming the overwhelming weight of evidence is clearly expressed in the consensus. As Paul and Anne Ehrlich admonish, "Scientific research is not properly carried out by consensus . . . but *science policy should be.* That is, in most cases, society's best bet is to rely on the scientific consensus."[5] Likewise, conservative environmentalist Gordon Durnil contends that the "weight of evidence" criterion is the proper basis for political action.[6]

So let us look briefly at the scientific evidence: at what is known, at projections for future climate, and at possible impacts. In this chapter, of course, I can only give a basic summary. For an authoritative account of the scientific evidence written for nonscientists, I strongly recommend Sir John Houghton's book, *Global Warming: The Complete Briefing.*[7]

Let us begin with what we know, what is indisputable, before we turn to more complex material with significant uncertainties.[8] In our atmosphere there are several trace gases that are "greenhouse gases." If they were not there, our planet would be a very cold place, with average temperatures below freezing.[9] These trace gases allow solar radiation to come in, which warms the earth, but they keep some of the infrared radiation (heat) from the earth from escaping back into space. Hence the analogy with a greenhouse—sunlight warms the interior and the glass panes then keep much of the heat inside. In the jargon, this heat-trapping power produces "radiative forcing" of the climate. As a result of this forcing, our planet is a pretty comfortable place to live. The issue over global warming comes from the fact that we have raised the concentrations of these gases in the atmosphere far above their natural levels, so they trap more than natural amounts of the sun's heat.

Water vapor is the most important greenhouse gas, but its level in the atmosphere is not being changed directly by our actions. Its role in the global warming question comes through feedback mechanisms, to be considered below.

Carbon dioxide is the greenhouse gas that is of greatest concern. Although some other gases are more powerful, molecule per molecule, in trapping heat, CO_2 is present in far greater concentrations. And once in the atmosphere it stays there for a very long time—about a hundred years.[10]

There are two major ways in which we are releasing unnatural amounts of CO_2 into the atmosphere: by burning fossil fuel and by destroying forests, primarily in the tropics. In the jargon, greenhouse gases released by human actions are "anthropogenic" (an ugly word but that is the term scientists use). Burning fossil fuel, which is the main source, released some 5.5 (± 0.5) gigatons of carbon per year in the 1980s. (One Gt = one billion tons.) In 1990, this had risen to 6.1 (± 0.6) Gt. Deforestation, primarily the destruction of tropical rain forests, adds some 1.6 (± 1.0) Gt each year. The ranges given here indicate that the amount of fossil fuel burned can be calculated rather closely, but estimates of the extent to which tropical forests are being destroyed are not nearly as precise.[11]

Once carbon dioxide is emitted into the atmosphere, some of it is absorbed in the ocean,[12] and some of it is taken up by plants through photosynthesis. These are the major "sinks" for this greenhouse gas.

Before the Industrial Revolution, there was a balance between the sources of carbon dioxide and the sinks, and the concentration in the atmosphere had remained stable for thousands of years. The natural sinks for CO_2, however, are not able to take up the increasing amounts of the gas we are emitting and some 3.3 Gt per year remain in the atmosphere.[13] As a result, CO_2 levels in the atmosphere have been increasing over the last couple of centuries and have accelerated rapidly since World War II. More than half of the anthropogenic CO_2 has accumulated since 1950.[14] Preindustrial levels were about 280 ppmv (parts per million by volume) and have risen to about 360 ppmv by the mid-1990s, far higher than any natural level in at least the past 420,000 years. The level is increasing at a rate of about 1.5 ppmv per year and by 1999 had risen to just over 368 ppmv.[15]In other words, for well over a century, our additions of CO_2 have exceeded the planet's ability to absorb them, and we are now basically overwhelming the capacity of the sinks.

Methane (CH_4) is another greenhouse gas, much more powerful than CO_2. Its heat-trapping power is much greater than that of carbon dioxide, but once emitted it does not remain in the atmosphere for nearly as long, only about twelve years, giving it a "global warming potential" of 6.5 times CO_2.[16] Primary anthropogenic sources for methane in the atmosphere are agriculture (rice paddies emit it and it is produced in cows' guts) and leaks from natural gas lines; landfills release the gas and so do coal mines. Methane is eventually removed from the atmosphere and destroyed by chemical processes. Anthropogenic sources have steadily increased the methane in the atmosphere, from a preindustrial level of about 700 ppbv (parts per billion by volume) to over 1,700 ppbv today—some two and a half times higher than any natural level in the past 420,000 years.[17] Additional emissions of methane seem to have slowed down in recent years, probably because of decreased production of fossil fuels in the states of the former Soviet Union. If this trend continues, total concentrations should stabilize at about 1,800 ppbv.[18]

Nitrous oxide (N_2O) is another greenhouse gas that human activities are releasing. Its concentration is now about 16 percent greater than in preindustrial times, a level unprecedented in at least the past 45,000 years.[19] Emissions of N_2O stay in the atmosphere about 120 years and are 170 times more effective than carbon dioxide in trapping heat. The sources for nitrous oxide have not all been identified, although we know that the chemical industry, deforestation, and agriculture play some role. The main sink is in the stratosphere.[20]

Chlorofluorocarbons (CFCs) are extremely powerful greenhouse gases, and they are all produced by industry. They are used in refrigeration, as solvents, for making insulation, as aerosol propellants, and so on. These chemicals are notorious because when released they eventually rise to the stratosphere where they break down and destroy the ozone that protects all life on earth against ultraviolet light. Ozone is itself a greenhouse gas, but its impact varies with altitude and therefore it is very difficult to determine the net effect of the interaction between it and CFCs. The international community, through the Montreal Protocol and its amendments, is phasing CFCs out, so eventually they will no longer be of concern. A lot of these chemicals are still around and all of them will eventually get released into the atmosphere, but their levels in the stratosphere have already stabilized and should slowly diminish. Their replacements, however, are also greenhouse gases and do not destroy ozone. The net impact over the next several decades will be to increase warming potential, both from the CFC replacements and from recovered ozone levels.[21]

Other trace gases have warming potential, but the ones I have described are of the greatest concern. There is no question that they are greenhouse gases. As of 1997, CO_2 has contributed some 70 percent of the added radiative forcing, CH_4 is responsible for about 24 percent, and N_2O for 6 percent.[22] As we release more and more of them into the atmosphere, we increase the level of unnatural forcing on the climate. But, as we will soon see, the global climate is enormously complex, and it is influenced by many factors. Moreover, the added gases and their radiative forcing will produce complicated "feedbacks," some that will add to the warming and some that will reduce it.

One additional fact seems to be well established by now—the surface air of the earth is getting warmer.[23] The 1980s turned out to be the warmest decade since detailed record keeping began a century ago and the 1990s were even hotter. The warmest years on record were in the 1990s. Since the late nineteenth century, global average surface air temperature has increased by nearly 0.6°C.[24] (Scientists use the Celsius or centigrade scale; to translate to Fahrenheit, multiply by 9/5, so 0.6°C is 1.08°F.)

Half a degree or so may seem to be rather small. And when we turn to the projections for the future, we will find that scientists are worried about increases of 2 or 3°C over the next century. Those numbers look insignificant, of course, because we daily experience changes far greater than that. But those numbers are global averages and in that context they are actually very large. In global average terms, there is a difference of "only" 5 or 6°C between the middle of an ice age with glaciers covering the northern United States, and our world today.[25]

To this point, we have been considering facts that are clear and undisputed. We know that there are greenhouse gases; we know that we are increasing their concentrations in the atmosphere; and we know that the earth's surface air temperature is getting warmer.

To go beyond these facts, we need to ask two critical questions: How much have our greenhouse gas emissions added to warming the earth, and how much is natural climatic variation? And what does the future hold if we keep dumping more of these gases into the air? Now things get much more complicated, with many uncertainties, which scientific studies acknowledge at every step.

To answer these questions, scientists obviously cannot conduct experiments by manipulating a variable with a control group for comparison—we do not have a spare earth of, say, 1850 vintage to set aside and observe as the control. Just as economists make predictions using econometric models, climate scientists construct computer models of the earth's climate, incorporating known physical relationships of the atmosphere with its different dynamics at different levels, and the influences of the oceans and the continents. The most complex of these models are called general circulation models and there are several of them, developed by climatologists at different universities and research institutes. Scientists run the models—which requires enormous amounts of supercomputer time—to see what they project will happen. If a model can accurately produce or "predict" today's climate and climate patterns from the past, there is confidence that it can take past and present data, plus specified assumptions about future greenhouse gas emissions, and predict future climate trends.

But the real climate is enormously complex and even our best supercomputers have limited capabilities, so even the best models are still simplifications of the real world. Nevertheless, they offer the best evidence available. They are being refined all the time. Although predictions of future climate at local levels are beyond their capacity, they are now able to produce general predictions in which climatologists have considerable confidence.[26]

We need to take a brief look at some of the important uncertainties and complexities that the models incorporate, and then we can turn to some of the major projections. One must follow the scientific literature to appreciate the enormous amount of effort devoted to sorting out the various causal factors and confronting the uncertainties in order to reduce them with more precise information and calculations. Only a short summary of the major points can be given here.

First, we need to know what will happen to the composition of the atmosphere in the future if we continue with "business as usual." The answer to that question is not a simple extrapolation from the present; it depends on assumptions about such things as future population trends, worldwide economic development, future technologies, governmental policies, and the like over the next century or more. Obviously, there is much here that simply cannot be precisely predicted. So the Intergovernmental Panel on Climate Change has developed a number of scenarios, rather than a simple prediction, which are now widely used. The different possibilities incorporate higher and lower estimates for the variables and therefore result in different projections for greenhouse gas emissions and their levels of concentration in the atmosphere.

Given a scenario for the atmospheric composition over the next century or so, what will happen to the climate? To answer that, we must determine how sensitive the earth's climate is to forcing by greenhouse gases. How much will the climate heat up if we double the level of greenhouse gases? No precise answer can be given at this time; rather, a range of sensitivities is used, which produces a range of projections for future climate. Emphasis is then usually placed on the projection based on the current best estimate of climate sensitivity.

There are also a number of "feedbacks" whose effects must be taken into account. Feedbacks are effects of greenhouse gases that in turn either increase their impact (positive feedbacks) or reduce it (negative feedbacks). A positive feedback increases global warming, and a negative feedback reduces it. There are many of these and only the more important examples can be summarized here.

As noted above, the ocean is a sink for CO_2. But as the ocean warms, it can absorb less and less, so it becomes less and less effective as a sink. And as atmospheric levels of CO_2 increase, the oceanic sink becomes less effective.[27] The ocean eventually becomes a net source, emitting CO_2 and so increasing the level of greenhouse gases and their forcing (a positive feedback).

The reflectivity of the earth (its albedo) is another feedback. During an ice age more of the earth is covered by snow and ice, so it reflects more of the solar radiation back into space (a negative feedback). As greenhouse gases warm the earth, the area covered by glaciers and polar ice will be reduced and so will winter snow cover, leaving darker surfaces that absorb more radiation and add to the warming (a positive feedback).[28]

A warmer world may have more violent storms, such as hurricanes (discussed below). But hurricanes significantly increase the flux of CO_2 from the ocean to the atmosphere.[29]

The arctic tundra is another major concern, for two reasons. It contains an enormous amount of methane and nitrous oxide. As greenhouse gases warm the world, the tundra's permafrost will melt, releasing the gases, which will add to the warming.[30] The tundra has also been a sink for CO_2, but as it gets warmer the rate of soil decomposition will increase and it will be transformed into a source. In fact, release of CO_2 has already begun, both in the tundra and, for a similar reason, in the boreal forests.[31]

A consequence of global warming that will have both positive and negative feedbacks is water vapor. A warmer world will increase evaporation. Water vapor is itself a powerful greenhouse gas, and increasing levels of it in the atmosphere will be a powerful positive feedback, but it will also produce more clouds. In general, high-level clouds trap heat, a positive feedback, but low clouds increase reflectivity and so are a negative feedback. What the overall effect may be is an area of great uncertainty, but it is likely to be a net positive feedback.[32]

For a final example, we need to consider a major aspect that is extremely complicated, with both positive and negative feedbacks—the reaction of plants, especially forests, to elevated levels of carbon dioxide and a warming climate. As a generalization, it is often simply asserted that the higher levels of CO_2 projected for our future will stimulate plant growth, so they will take up more CO_2 from the atmosphere—a negative feedback.

It is, unfortunately, more complicated than that. Some plants respond to higher levels of carbon dioxide but for other plants the response is much less.[33] Most of the CO_2 fertilization studies have been done under carefully controlled conditions in greenhouses, and their results may not translate to the real world, where many other variables also affect plant growth. "CO_2 fertilization is highly contingent upon ideal conditions in a non-ideal world."[34] The few field studies that have been done indicate that some kinds of ecosystems will take up more carbon dioxide, but others will not. Moreover, the sources of enhanced CO_2 are also sources of air pollutants that decrease plant growth.[35]

In many plants, an initial stimulus to growth and greater use of CO_2 in photosynthesis does not last. They acclimate to the enhanced CO_2 and return to their normal rates of growth.[36] And warmer temperatures cause plants' respiration to increase, giving off CO_2—a positive feedback.[37]

Most terrestrial carbon is stored in the soil, not in plants. Warmer temperatures increase the rates of decomposition in the soil, releasing carbon dioxide to the atmosphere, another positive feedback.[38]

A changing climate will destabilize entire ecosystems, which will result in changes in vegetation and soil types. As we will see in more detail below, with global warming the climatic ranges for ecosystems will shift toward the poles. This may be especially harmful for forests, which store most of the terrestrial carbon. A serious loss of forests, which cannot migrate fast enough, would be another very large source of carbon dioxide—another positive feedback. Moreover, "the loss of carbon through enhanced respiration due to disturbance is commonly much more rapid than the accumulation of carbon through regrowth of the plant population of disturbed ecosystems."[39]

Overall, as Smith and Shugart contend, even with the fertilization effect, a doubling of CO_2 could cause the release of an enormous amount of CO_2 in the first 50–100 years, "increasing the atmospheric concentration by up to a third of the present level."[40] That would be a huge addition to global warming. But, unfortunately, it would not stop there. Under business as usual, the earth would not reach an equilibrium, as at the end of Smith and Shugart's study, with new ecosystems happily in their new places. If we continue with business as usual, we are not looking at "transitions to a new stability, but to continuous instability."[41]The totality of feedbacks from plants and the soil in a warmer world is still in question, but the ones that increase global warming seem likely to predominate.[42]

One last major influence on climate must be mentioned: the emissions known as aerosols. These do not fall under the category of feedbacks because they are independent of greenhouse gases, but their effect is to counteract greenhouse warming, a "negative forcing" of the climate. The most important aerosols are sulfates, and although there are natural sources, by now the most important source is sulfur dioxide air pollution from burning coal. These aerosols in the atmosphere affect climate in two ways. They reflect the sun's rays (a direct cooling effect) and they also act as nuclei for water vapor to condense and form clouds (an indirect effect that can either cool or warm, depending on the kinds of clouds).

However, aerosols are short-lived—they stay in the atmosphere for only a few days. And, as might be expected, they are largely regional, above and downwind of heavily industrialized areas such as the northeastern United States and Europe. Moreover, their cooling effect operates only during the day, whereas the greenhouse gases trap heat all the time.

The impact of aerosols has only recently been studied, although it was long suspected. The climate models in the 1980s produced current climates that were somewhat warmer than current observations. The primary source of the error was that they did not include the offsetting effect of aerosols. Much research on this question has been done since the late 1980s and continues today. The general conclusion is that, averaged over the entire earth, aerosols offset half or less of the impact of the greenhouse gases.[43] Today's models that incorporate forcing by aerosols as well as greenhouse gases produce results for current conditions that are close to actual observations.

Sulfur dioxide pollution has significant harmful effects, so we are making major efforts to reduce it. To the extent that we are successful in solving air pollution, we will reduce the emissions of these aerosols and their cooling effect will decrease.

In sum, to model the earth's climate and to predict future conditions, the climate models now incorporate the atmosphere, the land, and the ocean, and they take into account aerosols and all of the important feedbacks. Projections for the future usually use 1990 as the base year and extend to 2100. They are typically based on the IPCC "business as usual" scenarios with their emission projections based on assumptions about population, the global economy, technical developments, and governmental policies.

Before we see what they conclude, we need to consider one last item of importance: climatic projections for the future are almost all based on a doubling of CO_2, that is, increases in all greenhouse gases equivalent to twice the preindustrial level of carbon dioxide. This doubled CO_2 scenario is arbitrarily chosen but it is now the canonical convention. What we must always keep in mind if we are interested in policy is that greenhouse gases will *not* stop increasing at that level unless policies are radically changed. If they keep increasing, global warming will also keep increasing, beyond the arbitrary stopping point of the studies.

As mentioned earlier, in 1992 the IPCC developed a series of scenarios for the future. The one the IPCC believes is most likely, labeled IS92a, uses medium-level projections for the socioeconomic variables and assumes that current governmental policies will continue. What results is that by the year 2030 or so, combined greenhouse gas concentrations will have reached a level twice as high as they were in preindustrial times. Using the most likely estimate of climate sensitivity, much of the warming that has already happened must be attributed to our greenhouse gas emissions. By the year 2100 global mean temperature will be 2.5°C higher than it was in preindustrial times, or 2.0°C higher than today.[44] It is also important to note that the warming will not be evenly distributed around the world: high latitudes will warm more than lower latitudes.

In sum, the general consensus based on climate models is that our actions have already produced an unnatural warming of the earth and if we continue with business as usual, by the end of the twenty-first century—within the lifetime of many of our grandchildren—we will have raised global mean temperature by about half the difference between the middle of an ice age and the warm period between ice ages. That is both a rate of change and a degree of change much larger than any in the last 10,000 years.[45] It is also important to note that because the ocean takes a long time to warm, there is a delay between the emission of greenhouse gases and the total impact that they make on the climate. The doubling of greenhouse gases will arrive before the middle of the century, but the full climatic impact will not be felt until the end of the century, or even beyond. Meanwhile, if we continue with business as usual, the greenhouse gases just keep on increasing. By the year 2100, under the IS92a scenario, CO_2 levels will be about two and a half times their preindustrial level and still going up. The climate, in other words, will not stop warming at 2.5°C; all those additional emissions commit the earth to even greater and more drastic warming. These are the projections we get from climate models: Human emissions of greenhouse gases have already warmed the earth and will accelerate the warming unless we take actions to reduce the emissions.

These projections come from computer models. But there are independent studies of past climate changes that empirically validate the kinds of causal connections and climate sensitivities incorporated in the models. There have been many studies of the climate going back as far as the ice ages by analyzing indicators found in deep sea sediments, continental deposits, and ice cores drilled in Greenland and Antarctica, with direct evidence of greenhouse gases coming from the ice. The longest of these cores was extracted over several years by a team of Russian, French, and American scientists at Vostok in Antarctica. The drilling is now finished and the 3,300 meter core provides evidence for climatic changes over the past 420,000 years, that is, over the last four cycles of ice age and deglaciation. Analysis of air bubbles trapped in the ice shows that during the ice ages the level of greenhouse gases in the atmosphere was very low. As the earth warmed and the glaciers receded, the levels of CO_2, CH_4, and N_20 increased, but often with a delay. The force that triggers ice ages and deglaciations comes from variations in the earth's orbit, not from the greenhouse gases. But those orbital changes by themselves are insufficient to explain the amount of warming as an ice age ends. Once the earth begins to warm, greenhouse gases increase and contribute to the warming, causing about half of the increase in temperature that brings the earth out of an ice age. This gives us empirical evidence of climatic sensitivity to greenhouse gas forcing.[46]

Another kind of empirical study appeared recently in which the authors reconstructed the climatic record from A.D. 1400 to the present from a wide variety of proxy indicators. They found relatively cold periods during the seventeenth and nineteenth centuries, warmer intervals in the sixteenth and eighteenth centuries, and then a distinct and pronounced warming in the twentieth century. "Almost all years before the twentieth century [were] well below the twentieth century mean."[47] The authors tracked changes in solar irradiance, volcanic aerosols, and greenhouse gas concentrations and concluded that each factor "has contributed to the climate variability of the past 400 years, with greenhouse gases emerging as the dominant forcing during the twentieth century."[48] As Gabriele Hegerl points out in a review of this study, "their results support, independently of climate models, the conclusion that anthropogenic influences have dominated the evolution of temperature in the twentieth century."[49] Others have extended this climate record back to about the year A.D. 1000. They find that the recent warming of the earth is unprecedented and, as Thomas Crowley concludes, "natural variability plays only a subsidiary role in the 20th century warming," which can best be explained as "due to anthropogenic increase in GHG [greenhouse gases]."[50]

The account of global warming outlined above is now accepted by the vast majority of climate scientists. The most authoritative reports are those issued by the Intergovernmental Panel on Climate Change, previously mentioned several times. The IPCC was established by the World Meteorological Organization and the U.N. Environment Program, and its Science Assessment Working Group was charged with reporting on the current status of scientific knowledge. Its cochairman, Sir John Houghton, explains its procedures:

> In preparing these reports we realized from the start that if they were to be really authoritative and taken seriously, it would be necessary to involve as . . . many as possible of the world scientific community in their production. A small international organizing team was set up at the Hadley Centre of the United Kingdom Meteorological Office at Bracknell and through meetings, workshops and a great deal of correspondence most of those scientists in the world (both in universities and government-supported laboratories) who are deeply engaged in re-

> search into the science of climate change were involved in the preparation and writing of the reports. For the first report [in 1990], 170 scientists from 25 countries contributed and a further 200 scientists were involved in its peer review. For the second comprehensive report in 1995, over 400 scientists from 26 countries submitted draft text and over 500 reviewers from 40 countries participated in its peer review.[51]

Houghton concludes, "The IPCC reports can therefore be considered as authoritative statements of the contemporary views of the international scientific community."[52]

Each report also includes a summary for policymakers, "the wording of which is approved in detail at a plenary meeting of the Working Group, the object being to reach agreement on the science and on the best way of presenting the science to policymakers with accuracy and clarity. The plenary meeting which agreed unanimously the 1995 SPM was attended by 177 delegates from 96 countries, representatives from 14 non-governmental organizations and 28 lead authors of the scientific chapters."[53]

The 1995 Summary for Policymakers of the Science Assessment Working Group, after summarizing the evidence to date and taking into account the uncertainties, concludes:

> THE BALANCE OF EVIDENCE SUGGESTS
> A DISCERNIBLE HUMAN INFLUENCE ON GLOBAL CLIMATE
>
> . . . Since the 1990 IPCC Report, considerable progress has been made in attempts to distinguish between natural and anthropogenic influences on climate. . . . Most of these studies have detected a significant change and show that the observed warming trend is unlikely to be entirely natural in origin. . . . Our ability to quantify the human influence on global climate is currently limited. . . . Nevertheless, the balance of evidence suggests that there is a discernible human influence on global climate change.
>
> CLIMATE IS EXPECTED TO CONTINUE
> TO CHANGE IN THE FUTURE
>
> . . . For the mid-range IPCC emission scenario, IS92a, assuming the "best estimate" value of climate sensitivity and including the effects of future increases in aerosol, models project an increase in global mean surface air temperature relative to 1990 of about 2°C by 2100. . . . The lowest IPCC emission scenario (IS92c) leads to a projected increase of about 1°C by 2100. The corresponding projection for the highest IPCC scenario (IS92e) . . . gives a warming of about 3.5°C. In all cases the average rate of warming would probably be greater than any seen in the last 10,000 years. . . . Because of the thermal inertia of the oceans, only 50–90% of the even-

> tual equilibrium temperature change would have been realized by 2100 and temperature would continue to increase beyond 2100, even if concentrations of greenhouse gases were stabilized by that time.[54]

That is the best scientific conclusion available. As Klaus Hasselman observes, "It is now generally accepted that climate change through human activity is a reality."[55] The only sensible thing for policymakers—including conservative policymakers—to do is to accept global warming as a reality in the same manner that the scientific community has. Of course, we should be alert to new developments in scientific knowledge and be prepared to review and revise policies as needed. But it must be said once again: there is nothing conservative about denying the best scientific studies available nor is there anything conservative about rejecting the overwhelming weight of the evidence.

IMPACTS OF GLOBAL WARMING

The next questions, of course, concern the impacts of global warming. What might happen if the earth becomes significantly warmer than it was before the Industrial Revolution and the age of fossil fuels? Before looking at some specifics, we need to remember that we are *not* faced with a simple, slow increase in temperature everywhere. There will be great regional variations in the changing climate. The warming will be greater at high latitudes than at low latitudes, and there will be considerable changes in precipitation patterns. Nor are we looking at temperatures increasing to a level "just a little" warmer than today. The 2.5°C warming projection for 2100 is a global *average* and it is a *very large increase*. It would make the earth hotter than it has been in well over 100,000 years.[56] And the rate of warming we are forcing on the entire globe is *far faster* than a normal, natural rate.

Moreover, climatic reaction to forcing is not likely to be simple and linear—merely getting warmer and warmer. Weather extremes are likely to become more frequent: droughts, heat waves, floods. As thresholds are reached, changes can come in the form of shifts of mode or "surprises" that we may be ill prepared to confront. As Wallace Broecker warns, "Coping with this type of change is clearly a far more serious matter than coping with a gradual warming."[57] In other words, some of the dangers may be completely unpredictable because we are disturbing what is now a basically stable system. And if we continue with "business as usual," the disturbance keeps on increasing and increasing and ever increasing. We always have to keep in mind that we are not causing "transitions to a new stability, but to continuous instability."[58]

In fact, our climate has been remarkably—and unusually—stable for the past 10,000 years or so, and it has been well suited for human civilization to develop and flourish. As Scott Lehman writes, "In geological time, such conditions are a rarity, and they can by no means be taken for granted."[59] It would have been far wiser to leave well enough alone. But we have not, and the changes we will likely face, as Wigley and Raper put it, "are certain to present a considerable challenge for humanity."[60] So let us look briefly at some of the likely challenges.

Human Health

Some direct impacts on people will be on their health, and two in particular need to be noted. In a warmer world, extreme events like heat waves will be more common. [61] During days of unusually high temperatures, death rates can double or even triple.[62] Laurence Kalkstein and Karen Smoyer developed a model to predict how hot spells produced by global warming would affect mortality in large urban centers. Based on threshold temperatures beyond which mortality rates rise rapidly, they project that the mortality rate in New York City would increase from 320 heat-related deaths per summer (the highest in the United States) to at least 880. In Los Angeles deaths would increase from eighty-four to 824 and in Chicago from 173 to 622. For the United States as a whole, we can expect 6,600 to 9,800 additional deaths each year. (In the United States, the frequency of days over the mortality threshold is already increasing.) Worldwide, the climate models indicate conditions that could result in hundreds of thousands of additional deaths each year from heat waves. Heat waves will also make some respiratory ailments worse, such as asthma and bronchitis.[63]

Moreover, in a warmer world, pests and germs will thrive. "Parasites and disease will do well on a warming earth. They are, by definition, organisms that colonize and exploit. Those species of parasite that are already common will be able to spread and perhaps colonize new susceptible hosts that may have no prior resistance to them. . . . In general, these effects are likely to be worse in the temperate zone, where parasites from the tropics can colonize new hosts."[64] Insects that carry tropical diseases will expand their ranges. For example, the mosquitoes that transmit malaria and dengue fever would be able to expand their range from Central America into the United States, which is, in fact, already happening. The portion of the world's population affected by malaria could increase from 45 percent to 60 percent by 2050, causing 50 to 80 million more cases of the disease each year. And many other tropical diseases, such as schistosomiasis, sleeping sickness, and yellow fever would spread as well.[65] Simple numbers cannot convey the magnitude of this problem. As Dobson and Carper stress, "The most striking feature of parasitology is the diversity of parasites in the warm tropical regions of the world and the frightening levels of debilitation and misery they cause."[66]

Sea Level

As the earth warms, the sea level will rise. It is already 10–25 centimeters higher than it was a hundred years ago. A common projection is that with a doubling of greenhouse gases, the sea level will be half a meter higher in 2100.[67] A small amount of this rise will be from melting of glaciers and snowfields, but most of it will be due to expansion of sea water as it heats up.

A rising sea will likely have several undesirable effects. Low-lying coastal areas will flood, so coastal defenses (e.g., dikes and levees) will have to be built to protect many developed areas. This impact, unfortunately, will probably be worst in some developing countries such as Bangladesh, which cannot afford expensive constructions for protection. Some island nations will simply be wiped out.[68] Already, two uninhabited islands in the Pacific have been submerged and two others are about to go under, indicating "that a predicted disappearance of such coral atolls is happening faster than expected."[69]

Moreover, with a rising ocean, storm surges will extend farther inland, causing damage to areas now out of reach. Some 45 million people worldwide now suffer flooding from storm surges. With the projected rise in sea level, by 2100 that number will double and the sea will keep on rising. A one foot rise in sea level—less than that projected by 2100—would increase the area of the United States subject to flooding by sea surges by 3,500 square miles.[70]

A higher sea level will cause salt water to intrude into coastal aquifers, affecting the water supply of many coastal cities. A rising sea will also destroy many coastal wetlands and estuaries, which are biologically some of the most productive places in the world. In many cases, the wetland and estuarine ecosystems cannot simply migrate farther inland because human developments—cities and agriculture—block the way. The projected sea rise by 2100 would destroy from 17 to 43 percent of coastal wetlands worldwide.[71] Fisheries may be affected because over two-thirds of the fish caught for human consumption depend on those areas for at least part of their life cycles.[72]

These are some of the likely results of the rise in sea level in the most probable scenario. But we also have indications of a potential consequence of global warming that would be devastating: a collapse of the West Antarctic Ice Sheet, which would raise sea levels *much* higher.

Several Antarctic ice shelves are already breaking up as the climate has warmed. The Wordie Ice Shelf, for example, decreased by 1,300 square kilometers in just over twenty years.[73] The northern Larsen Ice Shelf collapsed in catastrophic fashion in January 1995. In just a few days, 4,200 square kilometers broke away.[74] A study of nine Antarctic ice shelves showed that five northerly ones have retreated dramatically in the past fifty years.[75] In early 1999 scientists published satellite images showing that the Larsen B and Wilkins Ice Shelves lost nearly 1,100 square miles in just the past year.[76] (This is also happening in the Arctic, although in a different way: the Arctic ice has lost 5 percent or so of its area but it has lost 40 percent of its mass, as it melts from below.)[77] The breakup of these ice shelves will not affect sea levels because the ice is already floating and displacing water, but they show how much effect warming is already having at the poles.

The Antarctic ice cap as a whole is not of concern because it would take thousands of years to melt. But there is at least one ice sheet on land, the West Antarctic Ice Sheet, that is a concern. If it were to collapse into the ocean, the sea level would rise by *five or six meters* within a century or so, something that may have happened before.

The question is whether this particular ice sheet is now stable or unstable, and what effect a warming climate would have on its current level of stability. Some scientists believe that it is stable and poses little threat, but others believe it is already unstable or could become unstable if the ice shelves that buttress it were to break up. The latest IPCC report concludes that we simply do not know at this point how much of a threat it poses.[78]

The important point for policy is that there are potential disasters out there. The probabilities of these risks might be low but the consequences would be catastrophic. In some cases the probabilities cannot now be determined. We need to be fully aware of the extent to which, by destabilizing the climate, we are playing Russian roulette with the planet and with the lives of our children and grandchildren.[79]

The Hydrological Cycle (Precipitation)

A warming climate will enhance the hydrological cycle, causing more and greater weather extremes. In fact, it is generally the extreme events that will cause much more damage than the increase in the average temperature. The intensity of storms will likely increase and a greater portion of rainfall will come from intense storms. This is, in fact, already happening and it is producing more floods, which cause considerable damage each year.[80] Even worse flooding is probably in our future. The general prediction is that precipitation will increase at high latitudes and decrease at midlatitudes. As the storm tracks move toward the poles, some areas of the world will suffer greater droughts. By the 2050s, in places in the United States where droughts now occur 5 percent of the time, frequency of severe droughts could increase to 50 percent. Water basin models project that, with global warming, in eighteen major water regions covering the bulk of U.S. water resources, water supply would decline by one-third. The impact on the western United States could be especially severe. For example, in California's Central Valley, annual water deliveries would be cut by 7 to 16 percent, in a region where demand for water is expected to increase by 50 percent in the next few years.[81]

Another possibility is that global warming will increase the destruction from tropical storms. There is conflicting evidence from climate models on whether the frequency of hurricanes will increase. But several studies conclude that the intensity of hurricanes could be enhanced once they are formed. More intense tropical storms would not only cause more damage in the areas that are now struck by them, they might be able to travel farther north along our eastern seaboard before losing strength, causing damage in areas currently in little danger from these storms. The IPCC conclusion is that at this point we cannot tell what will happen.[82]

A pattern of increased damage from extreme weather may already have begun. Many of the worst natural disasters have happened in recent years. Four times more natural catastrophes occurred in the 1990s than in the 1960s. Floods from unusually intense or prolonged rainfall have caused enormous damages in recent years. And with drought comes wildfires, which have also caused significant losses. El Niño events that cause many of these kinds of disasters have been increasing in frequency and intensity. El Niño that ended in 1998 caused 23,000 deaths and $33 billion in damages worldwide. Some scientists believe that global warming has contributed to this increase.[83]

The insurance industry, which has taken a severe beating from these disasters, is very concerned. In the 1980s, annual insurance losses due to storms were $2 billion. In the 1990s they were $12 billion.[84] As global warming changes weather patterns, we may face a situation in which premiums skyrocket or insurance companies quit writing coverage for floods and hurricanes or go bankrupt. After Hurricane Andrew, eight insurance companies serving Florida collapsed and many others threatened to quit issuing policies in the state.[85] This would leave private individuals and companies to face increased losses unprotected, or we would have to pass the burden on to taxpayers.

Agriculture

Our food supply depends on the climate, so a good many studies have been done on the possible impacts of global warming on agriculture. The general prediction is that total agricultural production in developed countries will be affected relatively little. These nations can afford to make the adaptations that will be required. But arid and semiarid third world countries could be in even worse trouble than they are now.[86]

Even for the developed countries, however, these rather sanguine predictions should be accepted only with caution because they depend on some very important assumptions. They assume the full effects of CO_2 fertilization, which will almost certainly not happen in real world farming conditions.[87] The results of the Adams et al. studies are so dependent on CO_2 fertilization that if it does not take place, the economic losses increase at least threefold. Rosenzweig and Parry note that their conclusion is strongly dependent on CO_2 fertilization; without it, world crop yields will decrease by 11 to 20 percent while population continues to increase.[88]

Several offsetting impacts can be expected. Many food crops do indeed respond to increased CO_2, at least in greenhouses, by growing faster—but so do weeds. Soil moisture at mid and high latitudes is expected to decrease significantly, especially during the summer. Even though enhanced CO_2 tends to increase the water use efficiency of plants, the decreased soil moisture could offset the fertilization effect.[89] Plant pests and diseases are also expected to thrive in a warmer world.

Those studies also assume significantly increased use of irrigation, with no indication of where the water is to come from. As climate zones shift north, more of the precipitation in mountains in mid and lower latitudes will be in the form of rain, leaving less and less snowpack to provide water for irrigation in the growing season. Already, mountain glaciers are receding worldwide and snow lines are advancing to higher elevations, storing less and less water for the summer.[90] Nor have these studies accounted for the likely increased incidence of drought and other weather extremes in the midlatitudes. They simply assume warmer temperatures.[91]

Even with all the favorable assumptions in their model, which cannot be taken for granted in the real world, and with high levels of adaptation by farmers, Rosenzweig and Parry conclude that the number of people at risk of hunger will increase.[92] In sum, it would be prudent to be more concerned for the world food supply than the typical studies suggest.

Ecosystems

A warming world will have a *major* impact on all natural ecosystems, and here we can expect a lot of trouble. Many plants and animals are very sensitive to climate, both to temperatures and to precipitation patterns,[93] and global warming will alter both. In a warming world, climatic zones will shift toward the poles. Peters gives a conservative estimate that a 3°C temperature increase would produce a general shift of some 300 kilometers in the temperate zone.[94] Davis and Zabinski estimate that a doubling of greenhouse gases will shift geographical ranges for several important types of trees in eastern North America by 500 to 1,000 kilometers to the north.[95] Shifts of habitat zones are already under way.[96]

Unfortunately for many plants and animals, we are forcing a shift that will be not only very large but also *much* faster than happens naturally, and far faster than many of them will be able to migrate or disperse. Even some animals that are quite mobile could be unable to disperse if their paths are blocked by human development—cities, highways, farms. Natural habitat is already so fragmented and surrounded by development that many plants and animals will simply have no chance to find new homes. And in the nature preserves that we have established to protect rare plants and animals, many of these species will find the climate no longer suitable.[97] Thus one almost certain result of global warming will be the extinction of many more species—accelerating the collapse of biodiversity we are already causing (see Chapter 8).

Our forests face disaster. Trees disperse very slowly, so they will be unable to take advantage of many opportunities that may open up as climatic zones shift. Maximum migration rates in eastern North America for trees whose seeds are transported by animals is ten to fifteen kilometers per century. For trees whose seeds are dispersed by wind, maximum migration rates are twenty to thirty kilometers per century, assuming that no barriers are in the way and suitable sites for new trees are readily available.[98] These rates are far too slow for the climate shifts we are causing.

Moreover, within many species of trees, subpopulations are closely adapted to their specific locations, so they may be unsuited to their new environment even if it is still within the general range for that type of tree. A tree in Maine, for example, may now be in the northern part of the range for its species. If climatic zones shift 500 kilometers north, that tree may still be in the middle or southern part of the species' range, but it may be ill adapted to its new environment, and its seeds will be also.[99]

Our forests are already stressed by pollution, poor management, and the like, to which will now be added problems from their new climate. Stressed forests are highly susceptible to attacks by pests and diseases and also to fires. We can expect considerable destruction of our forests by attack from many quarters—all of which will add CO_2 to the atmosphere, another positive feedback. For example, in the southeastern United States, eighteen tree species may become locally extinct and their forest lands replaced by scrub or savanna. The boreal forests of Canada and Alaska, where climate changes will be greater, are at even greater risk.[100] Climatic changes are already causing trees to die in some regions of Canada, where a 3°C warming is expected to decrease the boreal forests by over one-third. Warming has already contributed to an increase in forest fires in Canada.[101] Also at risk are the great forests of the Pacific Northwest, which are very sensitive to climate change, primarily because of the altered disturbance regimes it will produce (increasing the frequency of fires, wind storms, pest outbreaks, and diseases). In northern California and southwestern Oregon, as the climate warms, the ponderosa pine and Douglas fir forests may change into grasslands, and the east side forests in Oregon may become juniper and sage steppe.[102] In the southwestern United States, a doubling of CO_2 could produce a 60 percent increase in the number of fires caused by lightning and an increase in annual areas burned of over 140 percent. The Great Lakes region could lose between one-quarter and one-half of its forest biomass in the next century from dieback, and the western United States could lose 40 percent. [103]

Woodwell and his colleagues predict that this will be widespread. "A warming of tenths of a °C per decade can be expected to outrun the capacity of forest trees to respond in a very few decades. The effect is a transition from forest to shrublands, grasslands, or to more severe impoverishment."[104] Even where the new conditions are suitable, developing the new types of forests requires "decades to centuries."[105] As a result, the EPA "foresees a substantial loss of healthy forest area . . . and a net reduction in U.S. forest productivity for several centuries."[106]

Ocean ecosystems are also in danger. Many fish are very sensitive to temperature. The world's fisheries are already in serious trouble from overfishing (see

Chapter 9), and global warming will add to the stress on fish populations. Coral is extremely sensitive to temperature and to increases in carbon dioxide. Coral reefs are the most productive areas of the sea: some 65 percent of marine fish depend on them for at least part of their life cycles. But coral reefs are already dying worldwide. Rising sea levels, warmer temperatures, and increasing CO_2 will all add to the stress.[107]

Environmental Refugees

Several of these consequences of global warming will, as we have seen, impact many third world countries severely. We can anticipate, as a result, a considerable increase in the numbers of environmental refugees as people in poor countries flee from droughts, lost coastal lands, flooding, and the like. Houghton cites one study which predicts that 150 million people could be displaced by 2050.[108] Any consideration of global warming has to take into account not only the financial costs of helping these refugees and the potential effect on developed nations of new waves of "boat people" but, above all, their misery and suffering as refugees.

Surprises

By definition, of course, we cannot predict what surprises might confront us in a warming world. But we know that the climate for the last 10,000 years has been unusually stable. And it has given us conditions very favorable for the development of human civilization. We know that we are disturbing this stability by changing one very important climatic determinant: the composition of the atmosphere. We also know that past climates have been very different, most notably, of course, during the ice ages. Climate usually changes very slowly—it takes thousands of years for the glaciers to advance in an ice age and thousands of years for them to recede. But we also know that within those long-term cycles there are instabilities that can change conditions rapidly. We have already considered one, the risk of collapse of the West Antarctic Ice Sheet, which would raise sea levels drastically over just a century or so. There is at least one other known phenomenon that concerns climate scientists—the circulation system in the North Atlantic, known as the thermohaline circulation.

The thermohaline circulation is a critical part of the worldwide oceanic circulation. Upper ocean currents now flow in the Atlantic from south to north, bringing enormous amounts of heat that gives western Europe its moderate climate. As water evaporates from the surface of the sea, seawater becomes saltier and therefore heavier. By the time the current reaches the North Atlantic, the surface water cools and becomes so much heavier than water beneath it that it sinks and forms what is known as the North Atlantic Deep Water. This water then travels south in the depths, and the worldwide circulation, of which this is a crucial part, perpetuates itself.[109]

A disturbance of this circulation would have drastic regional consequences for the climate, especially in the area around the North Atlantic. We know that in the past, this system has been disturbed and it may have shut down altogether. The latest example, and the most studied, is an event that occurred near the end of the last ice age, known as the Younger Dryas. About 12,000 years ago, the retreat of the glaciers came to an abrupt halt, the North Atlantic climate cooled, and the glaciers in the region advanced once again for some 1,500 years, before finally retreating to their current location. The transitions bracketing the Younger Dryas were very abrupt—it may have taken only five years or so for the cooling to begin and fifty years or less for it to end.[110]

What apparently happened was that the thermohaline circulation had shut down suddenly, so heat from the tropics was no longer carried north to warm the North Atlantic and Europe. As the glaciers melted from North America, the water was channeled down the Mississippi River system. But as the ice retreated north, what is now the St. Lawrence River opened as a channel for the meltwater. This released so much freshwater into the North Atlantic that its salt concentration was reduced and it no longer sank, no longer formed the North Atlantic Deep Water, and the engine for the oceanic circulation system was turned off. The North Atlantic area lost an important source of heat, and ice age conditions returned to the region until the amount of freshwater coming through the St. Lawrence River was reduced and evaporation once again could increase the salinity of the sea water so it would sink. Although the primary climatic effects were on the North Atlantic area and Europe, the Younger Dryas affected a much wider area. Its impact was felt in the interior of North America, in the tropical Atlantic, in the northern Andes Mountains, and even in Africa, where prolonged droughts resulted.[111]

The question now, of course, is whether it could happen again. In the global warming scenario, is there anything that could affect the thermohaline circulation? And the answer from two different angles is, unfortunately, yes.

First, one of the predictions of climate models is that in a warmer world, precipitation increases at high latitudes (and decreases at lower latitudes). Syukuro Manabe and Ronald Stouffer found that with a doubling of CO_2 the thermohaline circulation slows down; if CO_2 reaches four times the preindustrial levels, the thermohaline circulation nearly stops. The climatic changes produce so much extra precipitation at high latitudes that the freshwater runoff "caps" the North Atlantic Ocean. It is even possible that the double CO_2 scenario could stop the circulation. The authors note that the model run did not include freshwater from melting continental ice sheets. If that is also taken into account, the doubled CO_2 condition, which we are rapidly approaching, could induce the disappearance of the thermohaline circulation.[112]

A second threat to the thermohaline circulation comes from the melting of the Arctic ice. If this process speeds up in a warming world, the fresh meltwater could cap the North Atlantic and slow or stop the circulation system.[113] Shutting off the oceanic circulation system would produce a drastic change in climate around the North Atlantic—in Europe, and probably a much wider area as well. It would have severe impacts on marine ecosystems and fisheries. It also would severely weaken the oceanic uptake of CO_2, making the earth even more susceptible to our emissions—another positive feedback.[114]

According to Stefan Rahmstorf, the thermohaline circulation in the North Atlantic is a nonlinear system with several equilibrium states. In geological time, it has changed states frequently, producing substantial changes in the climate in a wide region. Its stability for the past 10,000 years is the exception.[115] It would be only prudent not to disturb it. But, unwittingly, we may be doing just that, and Rahmstorf warns us that this involves "risks that no nation bordering the North Atlantic would willingly take."[116]

The collapse of the West Antarctic Ice Sheet and weakening of the thermohaline circulation are risks that scientists have identified. We have no idea how many completely unforeseeable "surprises" might be lurking in global warming, but Jerry Franklin and his colleagues warn that "it is certain that there will be many surprises as we are blindsided by unknown interactions between shifting environments and biota."[117]

This is merely a brief summary of potential impacts of global warming. The uncertainties are great, but the summary indicates at least in general terms the major risks scientists see in the climate changes likely to occur. The warmer the climate the greater the impacts. The faster the rate of change the greater the risk of potentially devastating surprises. As Christian Azar and Henning Rodhe contend, with risks such as these, "the burden of proof must lie on those who argue that it is safe and acceptable to cause changes in the global climate system that substantially exceed the natural fluctuations during the last millennium."[118]

Here the conservative political philosophy has a crucial role to play. It is the conservative, perhaps above all, who has a future orientation that extends beyond the next election cycle. The conservative insists that society is intergenerational: We have obligations to future generations and we act as stewards for our descendants. We are forcing our descendants to face these risks—and they were never consulted. It is the conservative, perhaps above all, who sees that prudence is the most important political virtue. And at this point, business as usual is surely imprudent in the extreme. Moreover, we need to act as stewards of God's creation; we have no right to trash the place.

At this point we need to turn from science to policy. And one question that faces us at this transition is, How much might the damages from global warming cost? Several economists have attempted to estimate in monetary terms the value of damages expected from global warming. Since these estimates could influence policymaking, we must look briefly at their results. In Chapter 7 I will outline the general kinds of actions that could be taken to stabilize the climate by reducing anthropogenic greenhouse gases. Finally, I will suggest a number of possible policies that conservatives should consider to implement those actions.

DAMAGE ESTIMATES

We have seen some of the probable consequences of global warming if we continue with business as usual: increasing threats to human health from heat waves and from the spread of diseases and pests into new areas; rising seas threatening coastal areas and the water supplies of many of our major cities; more intense storms and more damages from floods and droughts; threats to food supplies that will hit especially hard nations in which hunger is already widespread; the destabilization of ecosystems, not to a new stability but to continuous instability; and millions of environmental refugees. We have also seen that possible catastrophes may be lurking in business as usual: collapse of the West Antarctic Ice Sheet or major alteration in the ocean circulation. And we have heard from scientists who emphasize that climate change will include nonlinearities that are unpredictable; there may be unforeseeable thresholds that could produce drastic changes if crossed.

We must keep the probable consequences and possible risks clearly in mind as we look at the estimates of damages from global warming that have been produced by economists—for, as we will soon see, they have about them an air of complete unreality. To be sure, economists frequently stress that the study of greenhouse damages is just beginning. As William Nordhaus says, "It must be emphasized that impact studies are in their infancy and that studies of low-income regions are virtually non-existent."[119] In their 1995 report, the IPCC Working Group III, investigating economic and social dimensions of climate change, concludes that "the level of sophistication in socioeconomic assessments of climate change impacts is still rather modest."[120] And Richard Tol characterizes the analysis of the uncertainties in these studies as "based on educated guesswork and heroic *ad hoc* assumptions."[121]

If this field is in its infancy, we can only hope that it will grow up soon because at this point it is seriously misleading for policymakers. Let us look at some of the major studies that attempt to put damages from global warming into monetary terms. I will then consider the kinds of damages they simply omit, some incorrect assumptions they make about climate change, and some of the factors in the studies that predetermine the results they obtain.

William Nordhaus, using a model he calls DICE (dynamic integrated climate-energy model), produced one of the earliest estimates of damages from climate change.[122] His damage estimates have been very influential and they often form the basis for estimates by other economists. Our economy, he asserts, is largely independent from the climate and therefore greenhouse gases will cause very little in the way of economic damage. For the variables he can quantify he calculates that a 3°C increase in average temperature would cost only one-quarter of one percent of our gross national product. These are largely costs of adapting to higher seas (building levees and so on) and from loss of lands to the rising seas. Admitting that there are costs he cannot quantify, he raises his medium estimate for damages to the whole world to 1 percent of gross world product, with maximum costs of global warming at 2 percent of world production.

Given these damages, Nordhaus calculates that the "economically optimal" reduction of greenhouse gases is about 10 percent now, rising to 15 percent later in the twenty-first century. He claims this would produce economic benefits to the world of $270 billion,[123] but it is difficult to see where they could come from since, as he himself points out, his "optimal" greenhouse gas reduction would have virtually no effect on global warming.[124] To reduce greenhouse gases any more than that token amount, he claims, would be extremely expensive. Although Nordhaus frequently calls global warming "the granddaddy of public goods problems" because emissions affect climate all over the globe for centuries to come, he nevertheless concludes that it presents only an insignificant economic problem. He wants to relegate it to a low-level concern, since we have so many pressing problems to deal with—except, he usually concedes as an afterthought, that there are uncertainties. (We will return to this concession shortly.)

William Cline published a study of greenhouse gas costs in 1992,[125] estimating that warming of 2.5°C would cost the United States something over $53.6 billion, or 1.1 percent of our GNP. Although this estimate is not far from those of Nordhaus and others, Cline takes global warming much more seriously than do the other economists. He believes the projected damages justify an aggressive policy to reduce greenhouse gas emissions: a cut of over a third, stabilizing total emissions at 4 Gt per year, would be justified on economic grounds. (For the reasons why he takes climate change most seriously, see note 157.)

Samuel Fankhauser and Richard Tol have published independent estimates of greenhouse damages and have also collaborated on several studies. Fankhauser concludes that a 2.5°C warming from a doubling of CO_2 would cause damages in the United States worth $61 billion, or 1.3 percent of GDP. The world would lose $269.6 billion in damages, or 1.4 percent of gross world product.[126]Tol estimates the costs of doubled greenhouse gases and 2.5°C warming at $74 billion for the United States and Canada—1.5 percent of GDP. The world's total damages would be $315.7 billion, or 1.9 percent of total production.[127]

There are other estimates of damages, but these four economists have done most of the work on the subject and their estimates are the most widely used. The differences among them are, in part, due to different valuations. For example, in estimating the costs of increased mortality, Tol values an American life twice as highly as Fankhauser does. There are also some differences in presumed impacts of global warming. Cline and Tol estimate significant losses in agriculture in the United States, but Nordhaus and Fankhauser assume little effect. Nevertheless, they all come to remarkably similar conclusions on the bottom line: between 1 and 1.5 percent of GDP for the United States.

Producing such a small impact on economic production, global warming is made to seem so insignificant that no serious action should be taken to forestall the climate changes we are creating. This is why I said earlier that these studies have about them an air of complete unreality—and that is the charitable interpretation. Climate change will produce major destabilization all over the planet, with considerable risks and uncertainties. Yet these studies indicate that there is little reason to worry or to make any significant changes in our policies or actions.

It is interesting to note that Nordhaus surveyed in depth several economists and several scientists and found that the scientists were much more concerned about global warming than the economists were.[128] So there seems to be an uncharitable interpretation of these studies of damages as well. Some economists, one might infer, simply believe that "as long as we can still make money, to hell with the planet," which could only be described as moral bankruptcy.[129]

Looking at these studies more closely to find out why their estimates of the costs from a destabilization of our planet are so small and insignificant, we find that they omit many of the expected impacts of climate change, they use some very questionable valuations, and, most importantly, they incorporate a factor that predetermines their results.

First, the cost estimates simply omit many impacts because they cannot be quantified. To a great extent, as the saying goes, economists count what they can count and not necessarily what counts. For example, none of the estimates includes loss of "ecosystem services" which, as we will see in Chapter 9, are of enormous economic importance even though they are almost never marketed or priced. Our economy, in fact, is a wholly owned subsidiary of the ecosystem. As the IPCC warns, a rapidly changing climate will destabilize almost all ecosystems, threatening their ability to provide these services.[130] Our well-being, in other words, is not as independent of the climate as these economists assume.

The IPCC Working Group III that looked at studies of projected damages from climate change found a considerable list of other things that they simply omit, such as increased insurance costs, damage from nontropical storms, river floods, extreme hot spells, and other catastrophes, morbidity, and human hardship. They also found quite a few expected damages that have been only partially estimated, such as damage from droughts, damage to urban infrastructure, damage from species loss, air and water pollution, and forced migration. And they also note that for some kinds of damages, a monetary estimate "only partially reflect[s] the potential welfare loss."[131]

The basic studies also omit any factor for uncertainty and risk aversion. In fact, they do not take into account potential catastrophes at all. Samuel Fankhauser explains that the estimates "are concerned solely with what is currently perceived to be the most likely damage scenario."[132]

Only Tol factors in damages from the rate of climate change. And the estimates arbitrarily stop at a doubling of CO_2, but greenhouse gas emissions will not.

Fankhauser and Tol claim that in more recent studies improvements have been made in estimating nonmarket impacts, but the accuracy of these figures still "remains low." Although they are becoming more comprehensive, especially on health impacts, they are far from complete.[133] As Fankhauser, Tol, and Pearce conclude, "available estimates on the costs of climate change are . . . neither accurate nor complete."[134]

Moreover, in placing only monetary values on the effects of climate change, the economists omit from their final conclusions the rights of anyone or anything but present consumers.[135] "Values" counted for species loss and ecosystem destruction (such as estuaries) are based on their current economic productivity or consumers' presumed "willingness to pay" to preserve them. No obligations of stewardship ever enter their calculations. (Tol at least adds an increasing cost for losses of nonmarketed amenities because as people get richer they place higher "values" on environmental quality.)

These studies treat future people the same way. All damages are reduced to present value to present people. That is, the underlying assumption is that we today have full rights to do harm to our descendants just so we can consume more and more today. (The way in which this is done is by discounting, and we will consider it in greater detail shortly.)

Nothing could be farther from conservatism, with its rejection on principle of just this kind of materialistic calculus, with its insistence that society is a partnership among generations far into the past and far into the future, and with its belief that we must exercise prudent and pious stewardship over what we have inherited and over what God has created.

Another reason the authors of these damage estimates get such insignificant results is that they have misinterpreted the projections of climate change. They believe that the general circulation models project that over the next century the world will slowly and gradually get just a little warmer—like, say, going from early April to mid-April. "Most economic studies of the impacts and policies concerning climate change are based on scenarios [of] . . . smooth and gradual warming." And Nordhaus specifically contrasts the scientists' "specter" of catastrophes to "the smooth changes foreseen by the global models."[136]

But we have repeatedly seen that this is *not* what climate scientists project. In the first place, the 2.5°C warming projected for the next century is a global average and it is *very large*. It is fully half the difference between an ice age and an interglacial period—only it is being added onto an interglacial warm age. A 2.5°C rise in temperature would create a climatic regime that the earth has not experienced in the last 100,000 years.

In the second place, we have repeatedly seen scientists warning that the risks of climate change are not from a slow *average* temperature increase. Most of the risks are from the increase in weather extremes that will likely occur in a warming world: more floods, more droughts, more deadly heat waves, higher storm surges, possibly more intense tropical storms, stronger El Niños, and the like. Likewise, the damages to ecosystems in a warmer world will mostly come as events: more forest fires, wind throw from storms, outbreaks of diseases and pests. Waves of environmental refugees will not result from the earth's being a bit warmer but from loss of land, floods and droughts, loss of fresh water supplies, more frequent destruction by storm surges. The economists seem oblivious to all of this, even though the scientists stress it repeatedly. To look only at global average temperature projections is quite misleading about the future we are creating.

Richard Tol acknowledges that the real damages in a warming world are most likely to come from extreme events, yet his damage estimates are still based the assumption of smooth and gradual change. He identifies this as an important area for future research: "the discipline of economics has paid relatively little attention to the macroeconomic impacts of exogenous shocks."[137] Fankhauser and Tol indicate that scholars are now starting to give attention to this,[138] but very little has yet been done.

Moreover, the damage studies assume that an equilibrium climate will be reached at a doubling of carbon dioxide. But that is not what will happen. The scientists repeatedly warn that greenhouse gases will not automatically stop at a doubling and, consequently, we are not generating what will become a new equilibrium climate. What we are creating will be continuous instability that gets worse and worse.[139]

In the third place, we have seen the scientists warn that climate is exceedingly complex, replete with nonlinearities that are poorly understood. They warn of possible thresholds and unforeseeable "surprises" as we force the earth's climate into new regimes. And they have identified some possible catastrophes in the new territory we are forcing the earth to explore.

None of this is incorporated by the economists into their estimates of costs of projected climate changes. William Nordhaus even acknowledges this on several occasions (his concession mentioned above):

> Given the modest estimated impact of climate change along with our other urgent concerns, we might conclude that global warming should be demoted to a second-tier issue. Yet, even for those who downplay the urgency of the most likely scenarios for climate change, there remains a deeper anxiety about future uncertainties and surprises. Scientists raise the specter of shifting currents turning Europe into Alaska, of mid-continental drying transforming grain belts into deserts, of great rivers drying up as snow packs disappear, of severe storms wiping out whole populations of low-lying regions, of surging ice sheets raising ocean levels by 20 to 50 feet, of northward migration of old or new tropical pests

> and diseases decimating the temperate regions, of environmentally induced migration overrunning borders in search of livable land. . . . Once we open the door to consider catastrophic changes, a whole new debate is engaged.[140]

Nordhaus has, of course, mixed together consequences that are quite likely along with some that are possible risks. Nevertheless, that is precisely what the scientists who are studying the possible consequences of greenhouse gases are saying! And all of it is ignored in the estimates of damages from global warming.

In a recent article, Richard Tol acknowledges some of these hitherto ignored impacts, such as increased storms and the spread of tropical diseases. He then calculates how the damage estimates would change if they were to include aversion to these risks. Adding risk aversion alone, he concludes, triples the damages.[141] We do, indeed, have "a whole new debate" and the damage estimates based on a scenario that is just a little warmer than today have very little to contribute to it.

The economists' damage estimates are also sensitive to at least three things: the value they place on human life, whether or not they include an adjustment for equity, and the discount rate used. The first two can make significant differences, but the last factor completely predetermines the results they get.

The estimates of costs of global warming include a category for increased human mortality. The question then becomes how much monetary value to place on a human life. We have seen that Richard Tol uses a figure that is considerably larger than the other economists use, so this category is much larger in his calculation of damages. But the really serious disputes about this category are over the use of a much higher dollar value for a human life in rich countries than in poor countries.[142] When the same value is placed on a human life in all countries, the damage estimates are much larger. In Fankhauser's model, for example, if the same rich country valuation is applied to all human lives, the costs of human mortality increase from $89 billion to $344 billion, which raises the total damages to 2.8 percent of the world's gross product.[143] (And even with this, as Fankhauser admits, "the distress and mental health effects suffered by survivors remain uncounted.")[144]

Many of the direct damages from a warmer climate will fall disproportionately on poorer countries. Rising sea levels will be an especially large problem for Bangladesh and Egypt. The studies of food production in a warmer world indicate that agriculture is likely to be affected much more seriously in less developed countries than in rich countries. Total economic losses are expected to be a considerably larger portion of GDP for developing countries than for rich countries—up to nearly 9 percent for Africa and South and Southeast Asia.[145] In other words, the impacts of global warming will fall disproportionately on those least able to afford them and least able to pay for adaptation and mitigation. Besides, the marginal value of a dollar is larger for a poor person than for a rich one.

Therefore, some analysts have adjusted the damage estimates by including an equity factor, weighting damages in poorer countries more than in rich countries. The result is significantly larger total costs of global warming.

Christian Azar and Thomas Sterner estimate global damages of greenhouse warming, with the assumption that costs in poor countries should be weighted to reflect the welfare losses actually felt. Taking this into account raises the damage estimate by a factor of three.[146] After the same fashion, Richard Tol factors income inequalities between rich and poor countries into his estimates and likewise finds that adjustment makes the total damage estimate fully three times larger.[147]

Although these two factors make significant differences in estimates of the dollar value of damages from global warming, the decision on what discount rate to use completely predetermines the outcome. Even those limited categories of damages the economists have counted produce such low totals only because they have chosen to use positive discount rates. A different choice, as we will see, produces a vastly different bottom line.

Discounting the future works just like calculating compound interest only in reverse.

> The basic principle underlying discounting is simple: A dollar today is worth more than a dollar at some time in the future. This is the same "time value" principle that underlies the concept of interest. Suppose lender L loans borrower B $100 in year one, to be repaid in year two. L will forego current use of the $100 only if B pays her a premium for that foregone use when B repays the loan in year two. That premium is interest. If B and L agree that B will pay $110 in year two for the use of L's $100 in year one, the simple interest rate is ten percent. If we asked L how much $110 in year two is worth to her today, she would presumably answer "$100." L "discounts" the money she will receive in the future by ten percent. This reflects the time value of money principle: X dollars one year from now is worth less than X dollars today.
>
> The term "present value" describes the current value to the recipient of a benefit that will be conferred in the future. In the above example, the present value to L of $110 in year two is $100. The ten percent rate L uses to discount the money she will receive in year two is called the "discount rate." Note that this analysis also applies to *costs* to be incurred in the future.[148]

A small amount of money invested at compound interest becomes a considerable fortune in 150 or 200 years. In similar fashion, reversing the process, "any positive discount rate causes catastrophes in the distant future to be reduced to insignificant factors in the present decision making process."[149] Ralph d'Arge and his colleagues calculated in 1982 that "if world GNP in 100 years were the same as today's GNP, discounting [at the current prime rate] would reduce the value to only approximately $70,000."[150] This offers a good illustration of the extent to which discounting makes even the greatest catastrophes in the future irrelevant today. If we knew for certain that in 100 years there would be a catastrophe which would cause damages equal to the entire 1982 world production, it would be "worth" spending only $70,000 today to prevent it. This is precisely what the economists have done in estimating damages. The major damages from global warming will occur several decades from now and increase in the more distant future. The economists are calculating how much those distant damages are "worth" today.

Discounting has long been controversial even among economists. In earlier times, Frank Ramsey contended that discounting "is ethically indefensible and arises merely from the weakness of the imagination." Roy Harrod believed that it is a "polite expression for rapacity and the conquest of reason by passion."[151]

Discounting is now standard operating procedure in cost-benefit analysis. In general, it makes sense when a business or an individual is deciding whether or not to make a particular investment, with a normal time horizon of ten or fifteen years. But discounting becomes completely inappropriate when we have to deal with environmental problems such as global warming that have time horizons of a century or more and that affect all of society. These problems are far outside the realm for which the economists' analytical tools have been designed.

> The longest "normal" time horizons in economics are those for infrastructure investments such as power stations, where thirty years is a possible horizon. . . . Horizons of this type are not long enough to raise one of the issues central to long-term environmental problems, namely equity between generations. For five to fifteen years in the context of business plans, the efficient use of money is the central issue. For a century or more, when considering the planet's life support systems, equity and "sustainability" naturally come to the center of the stage (although efficient use of capital does not leave the stage). So normal intertemporal economic problems and long-term environmental problems have quite different time scales and involve different issues. It is natural that the methodology for one does not fit the other perfectly.[152]

These economists have used standard cost-benefit analysis with discounting in a realm far outside the one for which it was designed. Granger Morgan and colleagues warn that "in such cases, the uncritical application of conventional tools can violate the assumptions on which they are based [and] produce silly or misleading findings. . . . One can expect that the straightforward application of standard ideas and methods will often fail."[153] And that is precisely what we see here.

In estimating damages from global warming, the choice to use a positive discount rate is a decision from the very start that the future does not matter. "Use of a social discount rate greater than zero leads ultimately to a disenfranchising of future generations."[154] It predetermines the result of the calculations because it will *always* produce a damage estimate that is insignificant. Consider how much difference the discount rate can make in some of the studies we are dealing with here. William Nordhaus uses a 3 percent discount rate for future greenhouse damages and concludes that they are so small that it is "economically optimal" to reduce emissions by only about 10 percent. But if the discount rate is 0 percent, by his own calculations damages are so much greater that "economically optimal" reductions are three times as large.[155] Peter Schultz and James Kasting make some adjustments to the DICE model, updating estimates of peak greenhouse gas concentrations. With their adjustments, using Nordhaus's preferred 3 percent discount rate, "optimal" emission reductions by the year 2045 are still only 13 percent—just a bit more than Nordhaus's original estimates. But, after eliminating all discounting, the damage estimates are so much larger that optimal reductions of greenhouse gases are *79 percent* by 2045 and *97 percent* by 2200![156]

William Cline finds a similar difference using the DICE model. He compares a 3 percent discount rate with 0 percent and finds that eliminating discounting multiplies the damage estimates nine times.[157] Richard Howarth gets similar results with DICE[158] and so does Fankhauser with his own model.[159]

In a recent study, Richard Tol uses his model to calculate marginal damages for the three main greenhouse gases with a time horizon extending to the year 2100 and with five different discount rates. For CO_2, for example, without discounting, the damages are seven times greater than when discounted at 3 percent and some fifteen times greater than when discounted at 5 percent.[160]

Christian Azar and Thomas Sterner use a much longer time horizon and an updated carbon cycle model. They combine equity weighting with a discount rate of zero and conclude that damages from greenhouse gases are fifty to 100 times greater than Nordhaus estimated.[161]

Clearly, the decision to use a positive discount rate predetermines the outcome of the damage estimates. And these different results, of course, have major implications for public policy. The greater the estimate of greenhouse damages, the more important it is to take actions now to prevent them. The decision to use a positive discount rate in calculating these damages is an a priori value judgment that future people simply do not matter at all in our policymaking today; that the only thing of any importance is to maximize consumption by the present generation. It cannot be stressed enough that a conservative must reject this as morally repugnant. As Derek Parfit contends, "The moral importance of future events does *not* decline at *n* percent per year."[162]

But it is also clear from these comparisons that if you approach the problem of estimating damages of global warming on the basis of conservative principles, and do not write off future generations, *even the few and narrow categories these economists use produce results that are compelling for aggressive policies to reduce greenhouse gas emissions today.*

Finally, one more argument must be confronted. Some economists justify discounting costs of climate change and deferring any action to solve it on the grounds that the future generations who will suffer these damages will be richer than we are today—so we should let them pay. This claim must be rejected on many levels. We have no right now to harm others just because they are richer than we are. Nor can I require Bill Gates to make my car payments just because he has more money than I do. We have no more right to harm future people, or to impose burdens on them, just because we think they will have more money.

The assertion that future generations will be richer than we are assumes that economic growth will continue, which may be likely for some time to come. But it also assumes that economic growth is the same thing as an increase in welfare, an assumption that is no longer valid, if it ever was. Any previous connection there may have been between gross domestic product and welfare has long since been broken. Ever increasing percentages of our GDP are going to purely defensive expenditures that do not make us any better off. Ever increasing amounts of money are being spent on "positional goods," status symbols that do not increase overall welfare because they are a zero sum game. Several attempts to measure true economic welfare have concluded that *since the 1970s economic growth has in fact been leaving us less well-off than we were.*[163] So we have no reason to assume, even if future generations have higher economic production per capita than we do, that they will be "richer" in any real sense.

The argument that we should let future generations pay for climate change assumes that all goods are interchangeable. Climate change is, on any normal human time scale, irreversible. So this argument assumes that it is acceptable to deny our relatively benign climate, in which our civilizations and economies and current ecosystems have developed, to future generations because they will have more televisions and stereos and stuff to keep them happy. As Tol points out, in the economic models, "biodiversity and videorecorders, say, are fully substitutable."[164] But it is very questionable whether more gadgets can compensate for environmental degradation.[165] As William Cline asks, "How many video cassette recorders will the future generations really consider an adequate compensation for 10°C or more global warming—especially if there is catastrophic risk?"[166] We must take seriously Robert Lind's admonition: "Perhaps even a tenfold increase in GDP would not be sufficient to compensate for a significant, but not catastrophic, reduction in non-market goods such as environmental quality."[167]

Moreover, since climate change is essentially irreversible, under business as usual greenhouse gas emissions keep on increasing and their concentration in the atmosphere keeps on increasing. Consequently, *now* is the best and most cost-effective time to reverse the trends. It may be the *only* time we can reverse the trends because the longer they continue and the farther they go the more difficult reversal will be, no matter how much money our descendants may have.[168] Thus the escapist notion that we should consume now and let our "richer" grandchildren pay for solving climate change must be rejected.

Overall, what must we conclude about the economists' estimates of greenhouse damages? Although the most recent ones are undoubtedly better than the earliest ones, as they stand they are still inadequate for making policy. A different approach to estimating damages is needed because the costs they have given us so far are calculated on assumptions about climate change that are not supported by the science of climate change. For example, instead of looking only for changes in agriculture that would be necessary if the weather were a couple of degrees warmer (such as increased costs for irrigation), economists might calculate the impacts on agriculture from greater frequency of droughts and strong storms or from increasing strength and frequency of El Niños—where we have actual costs from past experience and projections that these unfortunate experiences will happen more often. But that would still give us just one kind of information to take into account.

On the other hand, as we have clearly seen, *when you act on conservative principles and eliminate the pernicious effects of discounting, even the few and narrow categories the economists have included in their studies produce a compelling case for reducing greenhouse gases.* But those categories are so few and narrow that they still give a very inadequate indication of the seriousness of the problem we are causing. From a conservative point of view, we need to deal with this problem on moral and scientific grounds, not just on economic ones.

Even William Nordhaus in recent days concedes that with global warming conventional cost-benefit analysis produces "results that are ethically unacceptable." Rather than tinker with discount rates, he now advises, "we are better served by looking at the ultimate objective—in this case, global climate change—and setting our policies with this objective in mind." We should "*focus directly on the ultimate objective*" and try "*to achieve concentrations or temperature objectives.*"[169]

After the climate goals and emission standards are set, economists can then usefully use their tools to do cost-*effectiveness* (as opposed to cost-benefit) studies to determine the best ways of meeting the standards or achieving the goals.[170] And that is precisely the approach conservatives should take, as we turn to solutions in the next chapter.

7

Global Warming: Conservative Solutions

ULTIMATELY, ANY SOLUTION TO GLOBAL WARMING must be international and worldwide in scope. But the United States contributes approximately one-fourth of the world's annual greenhouse gas emissions from human activities,[1] even though we have less than 5 percent of the world's population. Moreover, our emissions of CO_2 per capita and per dollar of production are among the largest in the world.[2] Clearly, we have an obligation to take the lead in reducing these emissions. But the other OECD nations emit a large portion of total greenhouse gases and the developing nations, especially China and India, will in the near future be the major contributors if they industrialize on a "business as usual" basis.

Efforts to address this issue at the international level include the Framework Convention on Climate Change, which was adopted at the Rio summit in 1992. There have been several subsequent sessions, leading up to the meeting at Kyoto in 1997 to draft a treaty that would commit the nations of the world to take the first concrete steps toward controlling greenhouse gas emissions. That treaty exempts the developing countries and limits required actions to the developed nations. The U.S. Senate was right to insist that the developing nations must also commit themselves, and soon, to solving this problem. But the Senate was clearly wrong to reject the treaty because the third world is not yet on board. Since when does the United States refuse to do what is right until, say, Kenya or Bangladesh takes the lead?

The international actions to solve global warming, however, are beyond the scope of this book. I will focus on what we could do in the United States—although many of the policies to be suggested here could be used elsewhere or included in treaty commitments.

The goal is to reduce the rate of climate change to more natural levels. In the terms of the Framework Convention:

> The ultimate objective of this Convention and any related legal instruments that the Conference of the Parties may adopt is to achieve . . . sta-

> bilization of greenhouse gas concentrations in the atmosphere at a level that would prevent dangerous anthropogenic interference with the climate system. Such a level should be achieved within a time frame sufficient to allow ecosystems to adapt naturally to climate change, to ensure that food production is not threatened and to enable economic development to proceed in a sustainable manner.[3]

FOUR WAYS TO REDUCE THE RATE OF CLIMATE CHANGE

In general terms, there are four ways of advancing toward this goal: stop destroying the tropical rain forests, expand the sinks for CO_2, generate our power from low carbon or no carbon fuels, and use energy much more efficiently. I will look at each of these and then turn to specific policy options to implement them.

Preserve Rain Forests

Destruction of the tropical rain forests worldwide is a major source of greenhouse gases. Stopping this deforestation would have a significant effect on greenhouse gas levels, but this requires action by the countries involved. We in the United States can have only an indirect influence on their policies, so although I do not mean to diminish the importance of preserving tropical forests, I do not consider it further here.

Expand CO_2 Sinks

A second way of reducing greenhouse gases is to enhance or expand the sinks for these gases. The key here is CO_2 because there is much that we can do. Among the exotic and expensive engineering proposals that have been made, one suggestion is to capture CO_2 from utilities and factories and inject it into depleted oil and gas wells. This is not really a sink but a means of storage because the CO_2 is still there as a gas and it would be critical to ensure that it could not escape back into the atmosphere. It is, however, worth investigation.[4]

Another possibility is to capture CO_2 from utilities and plants and inject it into the deep ocean, where it would be sequestered for at least several centuries. There are experiments going on now, but it is extremely expensive, the process takes a lot of energy, it may have adverse impacts on marine life, and it is not yet certain that the CO_2 will stay in place.[5]

Another suggestion is to fertilize the ocean with iron. There are areas of the Pacific in which low iron levels limit phytoplankton growth. Fertilizing them with iron would stimulate large blooms of plankton, which take up CO_2. Organic debris from these living systems sinks in the ocean and is removed from the carbon cycle for thousands of years or more. This would, in other words, enhance the "biological pump." Experiments have demonstrated that spreading iron in proper areas does indeed produce rapid growth of plankton. But the practicality of doing so on a very large scale is rather questionable. And even if it were practical, it would probably not have a large effect.[6]

We might wonder why such expensive and questionable engineering projects are being considered, when there is a perfectly obvious and cheap means of extracting and storing carbon that also has many side benefits: plant trees. Forests are the major terrestrial sink for CO_2. They store enormous amounts of carbon in the trees, and even more in the organic matter in the soil. And there are millions and millions of acres available for planting. Of course, trees only take up CO_2 while they are growing—a mature tree respires as much CO_2 as it takes up in photosynthesis. But large-scale reforestation offers us an inexpensive and immediate solution for several decades. It can buy us time to develop and introduce more permanent solutions into our economy. Besides, forests have environmental advantages independent of the carbon they store, protecting watersheds, as habitat for wildlife, and as pleasant places for recreation. So it will be worthwhile to look at a few specific calculations by proponents.

Increasing the forested area of the world by just 10 percent would be sufficient to sequester virtually all of the annual increment of carbon in the atmosphere. Roger Sedjo calculates that forestation of 465 million hectares would remove 2.9 billion tons of carbon each year until the forests are mature.[7] Worldwide there is much land available. The greatest potential for reforestation is in the tropics, because so much of the land that is cleared is only marginally economic at best and often is quickly abandoned. But forestation to sequester carbon in many tropical, third world countries is problematic. It is difficult to guarantee that the new forests will not be illegally logged or burned, just as the original forests so often are.

Reforestation projects in the United States and other developed countries would avoid most enforcement problems. And there is plenty that could be done here. Dixon et al. calculate that in high and mid latitude countries there are several hundred million hectares of economically marginal lands that could support forests.[8] As for the costs, "a host of studies have suggested relatively low carbon sequestering costs through tree planting."[9]

In Chapter 5 we saw that the U.S. Forest Service for years has lost hundreds of millions of taxpayer dollars selling trees below its cost. We could spend the money that is now wasted destroying our forests on replanting the many areas of national forests that have been devastated by unsustainable logging. Since the national forests now produce only a minute portion of our annual harvest of wood, they could all be managed in the future for greenhouse mitigation. (Their environmental values would also be protected this way.)

Some have suggested planting trees on land now in the Conservation Reserve Program. The land is already being "rented" by the federal government to take it out of crop production and forestation would enhance the conservation purposes of the program as well.[10] For permanent storage of carbon, however, these lands would have to be placed in a set-aside program for much longer than the current ten-year contracts.

Planting trees would not have to be limited to marginal agricultural lands (although that is where vast acreages could be found). Planting trees in cities has multiple relevant benefits besides taking up CO_2 directly. Urban trees reduce energy requirements for cooling and heating homes and buildings.

> The energy conservation effects of a single tree are estimated to prevent the release of 15 times more fossil fuel carbon, than the amount of carbon that trees can sequester. . . . Three properly placed trees around homes and small buildings can cut air conditioning demand by as much as 50 percent, while large-scale tree plantings in cities can save as much as 44 percent in residential cooling demand on a hot day. . . . Windbreaks can save about 15 percent of the heat energy used. . . . [and] properly placed trees can reduce winter heating energy costs by 4 to 22 percent.[11]

Recycling paper can also help forests store carbon, at least in the short term. It would reduce the demand for wood and allow the forest stock to expand.[12] In the same way, reusing wood products that are now wasted would allow the forests to expand. Some 40 percent of hardwood lumber harvested in the United States (4.6 billion board feet) is used to make shipping pallets. These are normally used just once and then dumped into landfills, although they could easily be reused many times.[13]

While we plant trees to take up carbon, we should also preserve all of the old-growth forests that remain. Occasionally some politician, usually one beholden to the timber industry, will suggest that since young trees grow rapidly and take up carbon dioxide quickly, we should chop down all the old growth and plant seedlings. Timber companies would love that, but in fact it would be counterproductive.

When old growth is harvested, much of the bark and broken and defective wood is burned for fuel or used for paper, which has a very short lifetime. Of the wood removed for lumber 35–45 percent is lost as sawdust or scrap, most of which is also burned as fuel or used to make paper. All of this carbon is quickly sent back into the atmosphere as carbon dioxide. Overall, harvesting old growth adds so much CO_2 to the atmosphere in the short term that it takes at least 250 years for the new forest to take up as much carbon as was stored in the original forest.[14] So for one more reason, storing carbon, we need to preserve all of the old-growth forests.

Clearly, a well-designed program for planting trees and preserving our current forests could significantly reduce the problems from global warming by storing massive amounts of the main greenhouse gas at modest cost, and with numerous side benefits.

At the very least, for several decades new forests could take up a very large portion of our CO_2 emissions, buying us time to become more energy efficient and moving toward energy sources that do not add to global warming.[15] However, an even more effective long-term use for many of these newly planted forests, instead of merely storing carbon in them, is to use them for biomass energy to produce electricity and heat and fuel (discussed below).

Low or No Carbon Energy

A third way to reduce greenhouse gases is to reduce or, better, eliminate carbon in generating heat and electricity and powering our vehicles. There are many things that could be done. One is to switch from high-carbon to lower-carbon fuels in our power plants. Of the main fossil fuels now used, coal has the highest carbon content, oil has less, and natural gas has the least. Retiring older coal-burning plants and replacing them with natural gas power generation significantly lowers the amount of CO_2 released in producing a given amount of electricity or heat. A study of thirty-one states in the East concludes that if utilities switched from fuel oil to natural gas, their CO_2 emissions would be reduced by over 46 million tons per year at an average cost of $4.48 per ton reduced.[16]

Where coal is used, the efficiency of power plants can be improved. Older coal-fired power plants now convert only 32 percent of the energy stored in their fuel into electricity. Newer technologies can increase that to 42 percent or so.[17] This reduces the amount of CO_2 released from the production of any given amount of electricity. But coal is still a high-carbon fuel and as demand for power increases so will greenhouse gas emissions. (Natural gas wins out here too. The efficiency of a modern gas plant is 60 percent.)[18]

Another way to increase efficiency is to locate electricity generating plants in places where their otherwise wasted heat can be used to make steam for heating, a process known as cogeneration, or combined heat and power (CHP). Industrial plants can also do their own cogeneration. Both the electricity and the otherwise lost heat can then be used productively. This is already done in many places in Europe. The efficiency of using the energy potential in the fuel goes way up—total system efficiencies of some cogeneration plants are well above 80 percent.[19] But greenhouse gases are still emitted.

It would be much better to turn to sources of energy that do not emit greenhouse gases at all or do not add to their concentrations in the atmosphere. Advocates from the nuclear power industry point out that fission reactions release no greenhouse gases, so building more reactors could replace a lot of fossil fuel. But of course nuclear power plants produce dangerous wastes of their own. We clearly should not build any more nuclear power plants until we have found a definitive solution to storing the industry's highly radioactive waste, which is extremely dangerous for thousands of years. Besides, because of the high investment costs and huge subsidies involved, expanding nuclear power production would be a most inefficient use of our money. As Amory Lovins often reminds us, nuclear power in America "ate $1 trillion, yet delivers less energy than wood."[20] Several other technologies, however, could be used.

Solar power has enormous potential. Passive solar systems can heat buildings while producing no greenhouse gases at all. Space heat demand could be reduced by 90 percent or more in residential and commercial buildings by using passive solar designs and good insulation.[21] The Rocky Mountain Institute building in Snowmass, Colorado, needs no furnace at all, and it was built with old technology. They grow bananas inside year-round and pay an electric bill of about $5 per month.[22]

Photovoltaic systems could supply a considerable portion of our electricity demand. Photovoltaics only generate electricity during the day, but that is when demand peaks. And utility grids can incorporate intermittent sources as long as they are not more than about 30 percent of the total.[23]

The costs of solar electricity have come down drastically in recent years, even though federal research and development funds have virtually dried up since the Reagan administration.[24] Solar energy plants have had their problems in recent years, especially from losing research and development money, tax breaks, and long-term contracts that were needed to attract investors. But now major energy companies are becoming interested; BP Amoco and Shell are investing seriously in photovoltaics. As pressure to reduce greenhouse gases increases, others will likely get involved as well.[25] There is, in other words, not only a vast potential but also a realistic future for solar energy.

Wind is another renewable source. In the 1970s a lot of research was done to discover the best designs for windmills to generate electricity. Utility regulations, especially in California, encouraged this development and costs came down rapidly.[26] But, as in the case of solar electricity, the tax breaks and long-term commitments needed for further development and economies of scale disappeared. Today, the initiative for developing wind power has shifted to Europe, where Denmark especially has taken the lead.[27] The Danes already generate 8 percent of their electricity from wind.[28] The European Union plans to generate 10 percent of its electricity from wind by the year 2030.[29]

Wind power is now the fastest-growing electricity sector, increasing by over 25 percent per year through the 1990s.[30] And wind power is once again increasing in the United States. Its potential for protecting climate is demonstrated by Tom Gray of the American Wind Energy Association: "In the 12 months ended June 30, [1999] more than 1,000 megawatts of wind equipment were installed in the U.S., enough to serve more than a quarter of a million households and reduce emissions of air pollutants by more than 50 tons a day. It's also enough to cut emissions of carbon dioxide, the leading greenhouse gas, by 6,000 tons daily. To get the same CO_2 reduction benefit by cutting auto use, we would have to take roughly 250,000 sport utility vehicles (or 800,000 fuel-efficient cars) off the roads."[31]

A third renewable is biomass, typically the use of plant material as fuel to produce power. Burning biomass releases CO_2, of course, but the farms or plantations where the trees or other plants are grown would be replanted, and the new growth would take up the CO_2. The use of biomass thus does not add to the total of greenhouse gases and the fossil fuels that it replaces stay in the ground, safely storing their carbon.

As already noted, planting trees can take up and store a lot of carbon, offsetting much of our greenhouse gas emissions. Where the land is not very productive, it is best to plant trees and let the new forests mature and store the carbon. But on productive land, the best greenhouse policy is to harvest the trees for fuel and then replant. This has the added advantage of providing income from a crop for the owners of the land.[32] In the United States, according to Wright and Hughes, the best option would be to use some 100 million acres of cropped bottomlands that are marginal for grain crops but would be suitable for tree plantations.[33]

Electricity from biomass is already a commercial technology. Sweden gets 14 percent of its energy from biomass and plans to increase this substantially.[34] In the United States, the paper and pulp industry gets power from burning the waste generated by its production processes.[35]

Another type of biomass is cellulosic ethanol. Ethanol can be made from waste materials produced by agriculture and forest products industries or from energy crops grown for the purpose of fuel production. Ethanol could displace 10–15 percent of our gasoline immediately, since cars sold in America today can burn gasohol. With modifications, cars can burn fuels with higher proportions of ethanol, and even ethanol straight.[36]

As a final example—and this is by no means meant to be a complete list of all possible renewable energy sources—the future may belong to hydrogen. Hydrogen can be burned as a gas, producing water and no greenhouse gases at all, but storage and transportation would require a completely new infrastructure.

The technology currently receiving the most attention, however, is the fuel cell.[37] A fuel cell works like electrolysis, only in reverse. Instead of passing an electric current through water, producing hydrogen and oxygen, a fuel cell combines hydrogen and oxygen, producing water and electricity.

Fuel cells have long been used for power in the space program. For generating useable amounts of power on earth, fuel cells have been bulky. This would not be a problem for generating electricity for a building, and one scenario for the future is for individual buildings to have their own fuel cell generators. (Fuel cells can also be used in central generating plants.) Size has been a barrier, however, for powering vehicles, but development has been very rapid, with Ford, General Motors, DaimlerChrysler, and others actively involved. There are several buses in Chicago that run on fuel cells, and a fleet of forty-five cars and buses powered by fuel cells will be tested in California in the next four years.[38] Mercedes-Benz has now reduced the size of a fuel cell motor so that it fits into its smallest compact car.[39] Several companies plan to begin selling fuel cell cars by 2004.[40]

Since a fuel cell generates electricity, the vehicle does not need the bulky storage batteries and long recharge time that limited the usefulness of previous electric cars, and its range is comparable to a gasoline-powered car. Storing the hydrogen is a major problem. Rather than keep the gas in a heavy, bulky tank in the vehicle, Mercedes-Benz stores the hydrogen as a liquid. Most projects, however, produce hydrogen on the vehicle by chemically extracting it from something else, such as methanol. Chrysler's system extracts hydrogen from gasoline. These systems do emit some CO_2, but this would still be a major improvement because a fuel cell operates at much higher levels of efficiency than a gasoline-powered internal combustion engine. A gasoline engine uses only 20–22 percent of the energy in its fuel; the Mercedes-Benz fuel cell car uses 36 percent.[41] General Motors now has a prototype in a compact car that operates at between 53 percent and 67 percent efficiency.[42]

This short summary is only meant to indicate the possibilities for reducing CO_2 emissions from generating electricity with fossil fuels and, more importantly, the tremendous potential of renewable energy sources that do not emit greenhouse gases at all (or at least do not add to them). Renewables could supply more than half of the entire world's energy needs by 2050.[43]

The major reason renewable sources of energy have not penetrated the market much beyond token levels is that, even though their costs have come down dramatically, they are still more expensive than energy from fossil fuels. But—and this is the critical point for policy—the only reason fossil fuels are so cheap is that the costs of the externalities they impose on society are not included in their prices. If the prices of fossil fuels included the costs of the damages they cause—as both economic theory and conservative principles say that they should—then renewable energy sources would already be competitive or very close to it.[44]

With all the emphasis on carbon dioxide, methane often gets overlooked. But CH_4 contributes nearly a fourth of the added greenhouse effect. And methane emissions can easily be reduced, often at a profit, because methane is a valuable resource. Since methane is a much more powerful greenhouse gas than carbon dioxide, efforts here can pay big dividends in protecting the climate.

Methane that leaks from landfills can be captured and burned to generate electricity. The technology is available and proven. Waste Management has thirty-five facilities in the United States using captured landfill gas to generate electricity for nearly 100,000 homes.[45] From older landfills, 30–60 percent of emissions can be recovered. New landfills that have liners to protect groundwater allow up to 90 percent reduction in methane emissions.[46]

Coal mining accounts for a significant amount of methane emissions, but they come from relatively few mines. Again, with proven technology, over half of this methane could be profitably recovered.[47]

Agriculture contributes the largest portion of anthropogenic methane. Much of it is produced in cows' guts, but specific feeds have been identified that reduce this and increase productivity. Recovery systems can profitably capture 50–90 percent of the methane produced by waste lagoons on dairy and pig farms. And changes in the cultivation of rice could reduce emissions by 10–30 percent or more.[48]

Energy Efficiency

Another key to reducing greenhouse gases is to increase our energy efficiency. This does *not* mean we have to sacrifice and "freeze in the dark." It means getting our desired level of energy *services* with less input of *energy* and therefore less use of fuels and lower emissions. No one wants electricity for its own sake; we want light and heat and power. If we can get that light and heat and power in ways that use less electricity, then we can burn less fuel at the generating plants. And we are so extremely wasteful in our use of energy that there are endless opportunities to become more efficient. Numerous studies have concluded that the United States could cut its energy use substantially, in ways that *save* money even at current low energy prices.

By far the most outspoken and dynamic advocate of energy efficiency is Amory Lovins and his Rocky Mountain Institute. He contends that we are now wasting $300 billion every year and that "over half of the threat to climate disappears if energy is used in a way that saves money. In general, *it's far cheaper to save fuel than to burn it*."[49] Opportunities abound in every sector of our economy. And even in a period of falling energy prices, the cost of saving energy is falling faster.[50] Consider some of the examples and general conclusions Lovins presents.

> Buildings could be much more efficient. Pacific Gas and Electric Company wanted to retrofit a 1,900-square-meter office building. The designs submitted would save 67–87 percent on electricity, with paybacks of about three years.[51] A new energy-efficient tract house in Davis, California, needs no heating or cooling system, costs $1,800 less than average to construct, and should save about 80 percent of the space and water heating, air conditioning, refrigeration, and lighting energy allowed by the nation's strictest energy code.[52]
>
> Raising energy prices is not the essential key to getting people to save energy, not even with today's low prices. Seattle has the cheapest electricity of any major city in the country. Between 1990 and 1996, residents saved electric loads nearly twelve times as fast as those in Chicago, where electricity costs twice as much. The difference was that Seattle's utility company showed its citizens how to do it.[53]
>
> Choosing energy-efficient office equipment and commercial and household appliances would save over two-thirds of their energy use with equal or better service and equal or lower costs.[54]

> Proper retrofits of office and retail lighting save 70–90 percent on electricity, with better-quality light. The Rocky Mountain Institute has a fluorescent lighting retrofit package for commercial buildings that saves about 90 percent of lighting energy. But the savings do not stop there. Lower maintenance requirements alone cover half of the cost of the retrofit. And cutting lighting energy means the building can also save on air conditioning. This bonus averages another 35 percent of the direct lighting savings.[55]
>
> Most electricity is used by motors. Proper retrofit saves half of the energy and pays back in less than a year and a half. More than a fourth of the entire world's electricity could be saved this way.[56]
>
> Compact fluorescent lightbulbs (CFLs) can save a lot of energy and money. An 18-watt compact fluorescent bulb produces as much light as a 75-watt incandescent bulb and lasts twelve times as long. In commercial buildings, the long life alone saves enough installation labor and replacement bulbs to more than pay for the CFLs. Pollution reductions are significant. One CFL saves 1,600 pounds of carbon dioxide and eighteen pounds of sulfur dioxide. "The bulb more than pays for itself and pays environmental dividends as well. A utility can give away the lamp more cheaply than it can fuel its existing power plants."[57]

Amory Lovins is hardly alone. Other estimates of potential cost-effective energy savings are also very large. A study by the Electric Power Research Institute concludes that savings of 24–44 percent in electricity demand are possible with existing technologies.[58] In our residences space heating efficiency improvements of 40–60 percent could be made; water heating could use 30 percent less electricity (and with heat pumps or solar systems, 70 percent less). Add up the separate possibilities, and our homes could save 27–45 percent of the electricity we now use. Commercial buildings could save 22–49 percent and industries, 24–38 percent.[59]

Florentin Krause and colleagues have published a number of studies on energy efficiency potentials. They conclude that all OECD countries could decrease carbon emissions by at least 50 percent of 1990 levels over the next thirty to fifty years at "negative net economic cost" (i.e., savings—economists insist on butchering the English language whenever possible).[60]

Arthur Rosenfeld and his colleagues at the Lawrence Berkeley National Laboratory contend that retrofitting with currently available technology would save 45 percent of the electricity used in buildings in the United States. This alone would reduce our total CO_2 emissions by 10 percent. Additional fuel efficiency measures plus switching from electricity to gas space heating would reduce CO_2 emissions by another 5 percent. These measures would also produce net savings of $56 billion per year.[61]

A study by the National Academy of Sciences concludes that CO_2 emissions could be reduced by 37 percent through implementing cost-effective energy savings technologies.[62]

There are dozens of other studies as well that have indicated substantial potential for cost-effective energy savings.[63] The IPCC's review concludes that, realistically, within the next two or three decades, CO_2 emissions in developed countries can be reduced by 20 percent at costs ranging from negligible to net savings. On a longer term, reductions over 50 percent can be made at net savings.[64]

Many economists, however, simply refuse to believe these studies that find billions of dollars in wasted energy and wasted money. For example, Thomas Schelling derides "enthusiastic portrayals of currently available technologies . . . that for some reason have not been successfully marketed. . . . All of these ideas are completely orthogonal to the econometric estimates."[65] In similar fashion, William Nordhaus refuses to take these studies into account because to do so "requires an act of faith that is not warranted by economic evidence. . . . Much work needs to be done to demonstrate that the negative-cost approaches have some relevance for economic policy."[66]

Their incredulity is a perfect example of the often told story of an economist and his young daughter walking along the street. The girl sees a $20 bill lying on the ground and asks her father if she can pick it up. He says no because it is obviously fake or someone would already have picked it up. The economists' assumption is that since people are profit maximizers and since markets are efficient, individuals and companies have already adopted all cost-effective energy efficiency measures.

Robert Repetto has the perfect response to this belief. As he so correctly puts it, this is "an assumption that absolutely dumbfounds anybody who has actually worked inside a corporation for more than a week." He adds, "if it were true that companies typically operate at maximum efficiency . . . [rather than just] muddling along . . . it would be hard to understand exactly what the hordes of management consultants swarming around them are being paid to do."[67]

Amory Lovins relates the example of Dow's Louisiana Division, which implemented nearly 1,000 energy saving and waste reducing projects over a decade or so. An audit confirmed returns on these investments averaging over 200 percent per year. "Market theory would suggest that such enormously profitable savings should have been automatically achieved much earlier by a sophisticated firm in a cost conscious industry: there should not have been thousand-dollar bills lying all over the factory floor. But Dow . . . found that not only was the floor figuratively carpeted in them, but as soon as [they] picked one up, [they] found two more underneath." Lovins comments, "This is now a common experience for advanced practitioners in almost any electricity using facility. Theoretical economists should get used to it."[68]

This "energy efficiency paradox" has been the subject of a good deal of debate in the literature. Engineers' "bottom-up" studies, as they are called, look at current energy use and compare it to the best available technology. They invariably find billions and billions of dollars of potential savings that would be easy to get.

As Amory Lovins often puts it, energy efficiency is better than a free lunch; it's a lunch you're paid to eat.[69] So saving the climate can be done at a profit.

Economists' "top-down" macroeconomic studies, on the other hand, typically conclude that saving the climate will be costly. For example, Alan Manne and Richard Richels estimate that stabilizing U.S. carbon emissions 20 percent below 1990 levels (which would not stabilize climate) would reduce the present value of consumption by $3.6 trillion dollars over the next century.[70] William Nordhaus estimates that reducing greenhouse gas emissions by a similar amount would reduce consumption by $10.9 trillion over the next hundred years, which is 1.5 percent of the total discounted value of consumption.[71]

These macroeconomic studies reach their conclusions simply because of the assumptions built into their models.

> All of the commonly used top-down models are constructed in such a way as to include the *assumption* that reductions in greenhouse-gas emissions can only be purchased at the expense of a reduction in the output of other goods and services. In all the top-down models, the various sectors and agents in the economy are presumed to be operating in a perfectly efficient manner, so that if an additional constraint is placed on their activities (such as being required to reduce emissions of greenhouse gases), the amount of ordinary goods and services that can be produced must fall.[72]

This is, as the Lovins say, an "assumption masquerading as a fact."[73] As Stephen DeCanio claims, "The body of bottom-up studies remain unrefuted. It is the top-down studies that depend for their conclusions on the validity of the 'inevitable tradeoff' assumption; the bottom-up studies are consistent with the reality that firms and other producers ordinarily fall short of complete maximization."[74]

Clearly, there is an "energy efficiency gap." Our economy is *not* operating as efficiently as possible, and if it could become more energy efficient, we could protect the climate while saving money. So the next question is, Why are cost-effective energy efficiency measures not adopted? A fairly standard list of "market barriers" and "market failures" has been identified in the literature.[75]

Consumers and managers frequently lack information that is important for adopting efficiency technologies. Consumers, for that matter, know little about where they use electricity.[76] And their utility bills do not help because they just give lump sums of which only the dollar amount has any meaning to most people. A frequently used analogy compares our utility bills to a supermarket receipt that does not itemize the items purchased but only includes a grand total bill, and nothing is priced on the shelves. Smart shopping would be simply impossible. And, especially for companies, the information on energy-saving possibilities is often technical and out of the managers' fields of expertise. Moreover, it is often difficult and time-consuming to get the necessary information because of the diverse and fragmented nature of the sources. The theory of marketplace efficiency, of course, assumes perfect knowledge on the part of all buyers and sellers.

Another barrier is the split incentives that are found frequently in energy use. An apartment complex, for example, has little incentive to install efficient but more expensive windows because the tenants pay the electricity bills and they are the ones who would reap the savings. And the tenants have little incentive to replace leaky single-pane windows with efficient ones because they will probably move out before the windows could pay for themselves. Business firms face the same kinds of split incentives. And the building sector of our economy, which accounts for over a third of total energy use, is replete with them. The result is wasted energy and money and totally unnecessary pollution and greenhouse gas emissions.

Another barrier is lack of capital. This is especially significant for low-income people who could save money in the long run by weatherizing their homes but cannot meet the up-front costs. Lack of capital is also a major barrier for small companies.[77]

Several factors keep the price of energy artificially low, which reduces incentives to invest in energy efficiency. Governmental subsidies keep energy prices too low. And there are externalities, such as air pollution and greenhouse gas emissions, that are not included in the price of energy.

Finally, there is a considerable literature on people's "discount rates" for energy efficiency investments. Studies of consumers look, for example, at appliance purchases. When choosing between two refrigerators that are the same size and have similar features, one of which is somewhat more expensive but uses less energy, consumers typically buy the cheaper one even though the more efficient one would pay back the difference in price very quickly and save money in the long run. Consumers seem to require a very short payback time before they will spend the extra money on appliances that use less energy. Studies have been done of consumers selecting different kinds of appliances, air conditioners, home insulation, water heaters, and so on. Economists then calculate how high the consumers' "discount rates" must be to justify the seemingly poor choices, that is, what rate of return they require to invest in efficiency. The results are almost always very high, much higher than market interest rates.[78] The same seemingly irrational phenomenon is seen in companies that typically require much faster payback on energy efficiency investments than they do on any of their other investments.[79]Economists often use these high "discount rates" to explain otherwise irrational behavior and claim that it is justified as "economically efficient." If those really *are* their discount rates, then their choices are rational.

For consumers, this presents a fantastic picture that surely bears no relation to reality. It conjures up an image of a consumer in an appliance store contemplating refrigerators and noting the price differences and the energy ratings. The guy then thinks, "Let's see, my discount rate for refrigerators is 86 percent, so will the energy savings pay off at that rate of return?"

Now wait a minute. We are talking about average Americans—consumers who went through one of the industrial world's worst school systems and have some of the weakest mathematical skills in the world. They have probably never heard of "discounting." Economically, the only clear thing they see is the difference in price.

Compounding this absurd scenario even further, some economists have tried to justify the astronomical "discount rates" by claiming that energy efficiency investments are risky. People try to "diversify" risk in their "portfolios" but if they cannot, they will forgo an otherwise good investment unless its payback is extremely high.[80] Or they defer investing in energy efficiency because they calculate an "option value" of waiting to see what happens to energy prices.[81] The result is still "economically efficient."

So now we are to picture our typical American mathematical genius in the appliance store not only calculating a discount rate but mentally figuring how the risk of refrigerator A or B fits into his or her "portfolio" and recalling past fluctuations in the price of coal or natural gas and calculating the "option value" of waiting.

Paul Stern has the perfect description of these theoretical contortions that try to explain away or justify typical consumer behavior as "economically efficient." They resemble nothing so much as "the strategy of the Ptolemaic astronomers who added epicycles to the planetary orbits to make their models conform to the evidence."[82]

Some economists, on the other hand, now recognize these kinds of human shortcomings as inherent barriers to market efficiency. They have given them the rather euphemistic title of "bounded rationality."[83] Henry Ruderman and colleagues conclude that consumers do not understand nor do they utilize "the sophisticated conceptual operations required to project the value of goods and services across time periods."[84] Simply recognizing this reality, instead of trying to save a theory at any cost, opens the way for policies to help.

In the commercial realm, presumably, we largely have to look elsewhere to discover the reasons for the unreasonably fast payback companies typically require for energy efficiency measures. (Of course, "bounded rationality" applies to managers and workers as well as consumers.) In the first place, in the real world, managers and employees do not spend every working minute maximizing profits and minimizing costs. As Nobel laureate Herbert Simon contends, "satisficing" instead of "maximizing" describes how companies actually operate.[85] Moreover, there are many organizational barriers within the typical firm that stand in the way of economic efficiency as well as energy efficiency. Companies are "shaped by internal informational and incentive factors having little to do with the neoclassical optimization paradigm."[86]

For most companies, energy is a relatively small portion of total costs and is likely to be a low priority for management. The energy manager often has low prestige and correspondingly little clout. Management attention is largely focused on the big things that they think are crucial for the firm: market share, product development, legal requirements, and the like. Some classes of investments, such as energy savings projects, get short shrift.[87]

The technological potential clearly exists for companies to reduce their energy use significantly and save money at the same time. The success of some programs, such as the EPA's Green Lights (see below), shows that it is possible to overcome these barriers within firms and make energy efficiency salient and profitable. The opportunities are there for policy to overcome the barriers to energy efficiency.

Proper kinds of policy interventions are justified here, even in the narrowest economic terms. Ronald Sutherland is perhaps the most outspoken and insistent opponent of the advocates for energy efficiency. He contends that energy efficiency per se is not a proper social goal because it could only be achieved at the expense of "economic efficiency." But even he admits that if there are negative externalities, then policy intervention is justified.[88] And, of course, that is precisely what we are dealing with here—pervasive negative externalities from energy use that could largely be solved by efficiency gains. What policies could improve the working of the marketplace by reducing the externalities from our use of energy? We explore this question in the following section.

SOME POLICIES TO CONSIDER

We need to start reducing greenhouse gas emissions and we need to start now. Current U.S. "policy" is to call for a few voluntary measures and little more. The "justification" often given is that there are still many uncertainties in the science of global warming. But the uncertainties in no way justify doing nothing. As Richard Tol contends, "Uncertainty is never a reason for inaction and often a reason for additional, precautionary action"[89]—surely a conservative point of view. Besides, the uncertainties in dealing with climate change are far less than we face in many everyday business decisions. A business "would be happy to have as much uncertainty about interest rates one year out as the IPCC has about the probabilities of man-made climate change over the next century."[90]

Moreover, in our economy, relying solely on voluntary action by companies will not work. For example, in Germany, the auto industry has committed itself to a 25 percent reduction in average fuel consumption of new cars by 2005. In stark contrast, the U.S. auto industry is pushing ever larger, heavier, and more useless gas guzzlers on the public. Conservatives of a traditionalist bent can only wonder (and wish for) what might have been possible here if more of our companies saw themselves as moral agents and integral parts of society, as they apparently do in Germany.[91]

Our current "policy" is, quite simply, irresponsible in the extreme. As John Houghton says, "to 'wait and see' would be an inadequate and irresponsible response to what we know."[92] Likewise, Wallace Broecker admonishes us that "burying our heads in the sand as we have been prone to do is irresponsible."[93]

Since abdicating responsibility for one's actions violates a basic conservative principle, what policies should conservatives advocate? To meet our obligations as stewards of the earth and as trustees for future generations, we should first set "safe minimum standards" for climate protection. The concentration of greenhouse gases in the atmosphere had been very stable for 10,000 years, during which the earth's climate was also remarkably stable. To be prudent, we should set an ultimate goal close to that historic stability, roughly the equivalent of 350 to 400 ppmv of CO_2. This would probably keep climate change within or very close to the range of natural variability.[94]

This ultimate goal requires us to lower current greenhouse gas concentrations somewhat, so it will require a substantial reduction in future emission trends. Just as it took us a long time to raise the greenhouse gases to their current levels, it will take us a long time to lower them to meet this standard. And there is no way to keep greenhouse gases from increasing in the near future because it will take time to get onto the correct path. Just as it takes an aircraft carrier a great distance to turn around, it will take time to turn our economic structure around. Consequently, we also need to set intermediate targets for greenhouse gas concentrations and we need to map a trajectory in which emissions will increase at slower and slower rates over the next few years and then begin a long and increasingly steep decline toward the ultimate goal.[95]

With our goals set, the next step is to design policies to help meet them. And in doing so, we need to keep several things in mind. First, we will need to be adaptable, always keeping up with the latest scientific findings on climate change and with new technological developments for reducing greenhouse gases. "Climate policy making is not a once-for-all event, but an ongoing process. There will be many opportunities for learning and for mid-course corrections."[96] Since we already know a great deal about what needs to be done, the mode to adopt is "act, learn, act."

Second, we need to adopt serious climate protection policies now. We are almost bound to continue with business as usual for some time to come in many sectors of our economy. We will want, for example, to phase in low carbon and renewable energy sources as current capital investments in utilities need to be replaced. It would be extremely expensive to retire huge investments like these before their normal replacement schedules. This is acceptable, and we can still meet our intermediate and long-term goals.[97]

But this does *not* mean that we can wait to set new policies in place. At the earliest possible date we need to assure that new capital investments do not lock in wasted energy and high greenhouse gas emissions for the next thirty years (for electricity generation) or longer (for new buildings).[98] And there are many "no regrets" things we could do right away to improve energy efficiency and reduce greenhouse gases that would *save* money, even without taking climate into account.

Third, since there are several greenhouse gases with many sources for each, no one single policy will be sufficient or even possible. We will need a mix of policies, all designed to reduce greenhouse gases in the end and meet the intermediate and final concentration standards.

I make no pretense of having definitive answers to climate change. What I propose here is a menu of policy options for conservatives to consider. It is only a partial menu, but I hope it will at least suggest the kinds of thinking that conservatives should be doing in designing and advocating policies.

Climate change is a global issue; it affects all nations, and all nations contribute to the problem. Obviously, any ultimate solution will have to take the form of an international agreement to which all significant emitters of greenhouse gases are bound and which penalizes any who refuse to take part. These negotiations are certain to be complex and difficult, primarily over the issues of equity in allocating allowable emissions between industrialized nations and third world countries.

But progress has already been made. We have the Framework Convention on Climate Change and the first step has been taken toward its goals in the Kyoto treaty. The success of the several rounds of negotiations under GATT shows that extremely complex and contentious issues can eventually be resolved. Presumably the final result will be national quotas for emission of the several greenhouse gases, with provisions for worldwide trading of emission permits and credits among governments and private companies, after the model of our sulfur dioxide trading under the Clean Air Act. The Kyoto treaty already endorses the notion of trading. Even the top-down models show that allowing emissions credit trading internationally reduces costs significantly.[99] But the international issues are beyond the scope of this discussion. The United States, as the largest contributor to the problem, has a moral obligation to take the lead in climate protection.

In analyzing policies to advocate, conservatives generally do not want to interfere with the free market any more than necessary. But what we need to realize at the outset is that we do not now have a free market in energy. Some parts of the energy sector are still regulated, many uses of energy are heavily subsidized directly or indirectly by the government, and there are pervasive negative externalities in energy production and use, beyond the emission of greenhouse gases. If energy prices reflected true market rates, energy would be much more expensive than it is today. People would use energy more conservatively, and that would go a long way toward meeting our climate policy goals. What conservatives should want to do first in each sector of the economy is to correct market failures and help the market work properly.

The first step, as the OECD advises, is to look for subsidies that distort the market and result in added greenhouse gas emissions.[100] This would be a good policy to support internationally because energy is heavily subsidized in many countries. The IPCC estimates that global greenhouse gas emissions could be reduced by up to 18 percent, with increases in real incomes, just by eliminating fuel subsidies.[101]Second, we will look for all externalities (e.g., pollution), not just global warming, that should be internalized in prices. Third, we will see what "no regrets" opportunities there are to save money while protecting the climate.

In many, perhaps most, sectors these three actions will be quite sufficient to reduce greenhouse gases to meet the concentration goals. If they are not, and only if they are not, will conservatives turn to governmental interference and in these cases will try to find policies that use market incentives. Direct regulation would only be a last resort and would seldom be necessary.

Before turning to specific sectors of our economy, there are three general policies we should consider. First, the federal government needs to inform the public about global warming and keep the issue salient. There is a lot of misinformation out there. For example, many people apparently believe that since there are pollution control devices on their cars, they are not emitting greenhouse gases. I have even seen this misconception expressed in newspaper articles by reporters who should know better. And there is a lot of disinformation out there. I have seen newspaper columns in which the authors blatantly assert that there is no evidence for global warming at all. The federal government needs to start a serious campaign to inform the public. People need to be made aware of the damage they cause by their waste of energy and they need to be made aware of things they can do to help, such as how to use energy more efficiently.

Second, the federal government should support a major research and development program on renewable energy sources, on energy efficiency technology, and on ways to reduce greenhouse gases other than carbon dioxide. Supporting research to discover new options is a legitimate function of government, and basic research in these areas is grossly underfunded by private industry. The government should not "select winners," at which it is not good at all, but it can support invention of new technologies that can be tested in the marketplace.[102]

Unfortunately, at the present time Congress is so myopic that it seriously underfunds this kind of research and development program. Congressional staffers told *Science* magazine that since gasoline prices are so low, added spending on alternative energy technology is untenable.[103] But it makes no sense to wait just because gasoline is plentiful and cheap. As the head of Shell Hydrogen says, "The stone age did not end because the world ran out of stones, and the oil age will not end because we run out of oil."[104]

Actually, a research program might not require more money. Federal energy research funds now are very poorly allocated—most of it goes to nuclear power (which has no future until a permanent solution for radioactive wastes is found) and coal (which we need to phase out because it has the highest carbon content of all fossil fuels).[105] At the very least, we need to reallocate our current research and development funding to renewables and efficiency improvements.

Information and research are necessary but indirect measures. There is a third general policy that is designed to reduce emissions directly and should be considered by conservatives. This overall approach can take two different forms: a carbon tax or a carbon cap with tradable quotas.

A carbon tax is generally the policy preferred by economists. The tax is designed to raise the price of fossil fuels and thereby reduce the demand for them sufficiently to meet climate policy goals. The top-down studies of costs of greenhouse gas abatement calculate the level of a carbon tax that would either equal the marginal costs of damages from greenhouse gas emissions or produce a specified reduction in emissions.

The main advantage of a carbon tax is that it is a direct means of internalizing externalities and thus recommends itself to conservatives. Once the harms from burning fossil fuels are determined, taxing at that rate makes those who use fossil fuels pay for the damages they are causing. Producers and consumers then have to face the true, total social costs of their production and consumption, but the carbon tax leaves them free to decide what to do. They can choose to become more energy efficient and thereby reduce the tax they have to pay for a given level of energy services. They can cut back on consumption or production, or they can choose to pay the extra taxes. In the economy as a whole, a carbon tax, by raising the price of fossil fuels, will reduce demand for them and therefore reduce CO_2 emissions.

Since fossil fuels are used for so many things, a carbon tax would raise a lot of money. Conservatives do not want to increase the government's role in the economy, so the tax should be made revenue neutral by cuts in income taxes and payroll taxes. "Recycling" the revenue in this way will also greatly reduce the impact of the tax on economic efficiency.[106] At the same time, a carbon tax would reduce air pollution from burning fossil fuels and produce large benefits for people today. (If we use a carbon tax as our primary climate protection policy, a small portion should be set aside for climate protection projects, as discussed below.)

However, a carbon tax has some disadvantages as well. In the first place, as we have seen from our analysis of damage estimates, it is very difficult to put a price tag on the anticipated consequences of global warming. This makes it very difficult, if not impossible, to determine how high the carbon tax should be, if we try to equate the price of carbon fuels with the damages they cause. And there is still no cap on total greenhouse gas emissions, just a higher price for them.

If we take the approach recommended here and aim for atmospheric concentration targets, it will be difficult to determine how high the tax will have to be to reduce fossil fuel use sufficiently to meet the goals. Tax rates will probably have to be adjusted frequently, based on experiences at different rates and this would add to the volatility of energy prices.

Moreover, in some respects a carbon tax is a rather blunt instrument. Raising the price of fossil fuels will stimulate many people and companies to use energy more efficiently. But they are already passing up countless profitable ways to be more energy efficient. It is doubtful that a carbon tax could come close to making the best use of our opportunities here. The key to energy efficiency is not necessarily higher prices. More direct policies might work better (or at least are needed to supplement the tax).[107]

Given the drawbacks, it would probably be better to take an even more direct approach and simply cap CO_2 emissions as far as possible. As with the Clean Air Act and SO_2, a cap could be placed on emissions from all large sources. Carbon quotas would be allocated to these sources—or they could be sold at auction—and the quotas could be bought and sold and leased.[108] The cap would be set initially and reduced over the long term, based on the emission trajectory chosen to meet our climate goals.

A cap-and-trade system would probably have to be limited to large emitters of CO_2, such as utilities and large plants that generate their own power or process heat from fossil fuels. Some large sources from the transportation sector could also be included in the cap-and-trade system, such as airlines and railroads. But transaction costs would be too high to include small sources, such as cars and trucks or gas and fuel oil heat in homes and offices. They would have to be treated separately. But the large sources themselves account for a significant portion of our greenhouse gas emissions—utilities alone emit 40 percent of our CO_2 (discussed below).

To complement the cap, tradable credits could also be issued for creating carbon sinks. This would stimulate development of a new area in our economy. A company could obtain credits by, say, planting a new forest for the purpose of storing carbon. (If the forest burns down, the company would have to purchase an equivalent quota.)

A cap-and-trade and credit system for carbon dioxide emission control would have all of the advantages we saw in Chapter 4 in regard to air and water pollution. It would allow maximum flexibility for individual plants to select the most economical means of reducing CO_2 emissions. Sources able to reduce emissions cheaply could emit less than their quotas and profit by selling the remainder. Sources for which emission reduction would be very expensive could buy extra quotas. Overall costs of meeting our climate goals would be minimized.

For small sources of CO_2, a climate protection surcharge, based on the carbon content of the fuel, could be collected to be used for projects that offset the emissions (discussed below).

We could also approach climate policy in each economic sector separately, in which case there are many policy options conservatives should consider. I will take just two sectors—electricity and transportation—as examples and look at some of the possibilities. (Some of these policies should be considered if we do not adopt either a carbon tax or cap; others should also be considered as valuable supplements to a tax or cap.)

ELECTRICITY

One of the largest sources of CO_2 is electricity generation, which is responsible for 40 percent of our total emissions.[109] Emission reduction policies that are specific to utilities could be adopted, and other policies could be targeted at consumers.

Climate policy for the utilities that generate our electricity, first, should eliminate current subsidies. This would have a major impact in many European and third world countries, where electricity and coal mining are heavily subsidized. But even in the United States, utilities benefit from some subsidies that should be cut. A study for the OECD of energy subsidies in the United States lists several items relating to utilities, such as low-interest loans through the Rural Electrification Administration ($1.2 billion per year), tax exemption on municipal utility bonds ($1.7 billion annually), energy-specific trust funds ($400 to 700 million), other assorted tax breaks ($2.1 billion), subsidies to power marketing agencies ($2.2 billion), and so on.[110] A study for the London Environmental Economics Center found that if U.S. electricity were priced at its marginal costs with no subsidies (but *not* counting externalities), carbon emissions would be reduced by 100 million tons per year. That alone would cut our total emissions by over 7 percent.[111]

Second, the price of electricity should include the costs of all externalities, even before considering climate change. Utilities continue to pollute the air, causing smog, adverse health effects, and acid rain and degrading the quality of life in our country. According to one study the price of electricity from coal would have to be doubled to cover the damages it causes.[112] In fact, some studies have concluded that significantly reducing the use of fossil fuels would pay for itself just from the benefits of reduced pollution—without even counting benefits from protecting the climate. Paul Ekins estimates the environmental and health benefits of reducing burning fossil fuels at between $250 and $400 per ton of carbon saved, which he concludes are even greater than the benefits from climate protection.[113] A study of the Los Angeles area for the California Energy Commission concludes that for every $1 in benefits from reducing CO_2 there are another $3 in benefits from reducing the other pollutants the fuel emits.[114]

Removing the subsidies and including the costs of air pollution in the price of electricity would by themselves reduce demand and therefore reduce CO_2 emissions. As economists always stress, this would help the market work properly. To go beyond that, the traditional command-and-control bureaucratic approach to the environment would probably try to set CO_2 limits on each and every generating plant or limits on major users of electricity. But there are a number of better policies—including cap-and-trade—that would utilize the forces of the market, which conservatives should consider.

In the long term, one key to reducing CO_2 emissions is to switch from fossil fuels to renewable sources of energy. One emerging development that conservatives should enthusiastically support is "green marketing." Some utilities are generating power from renewable sources and marketing it at premium prices. The municipal utility in Austin, Texas, surveyed its customers and found that 70 percent were willing to pay more for its renewable energy projects. Its program, in which customers volunteer to pay $3.50 per month extra to get solar power, is fully subscribed. In Colorado, more than 16,000 people have subscribed to support the state's first wind farm.[115] Green marketing can be done immediately and can be conducted entirely through the marketplace to expand the role of renewables in electricity generation.

Green marketing, however, will not be sufficient to expand our use of renewables substantially. A mechanism for doing so by using market forces is to adopt a renewables portfolio standard. This policy, like emission quotas, sets a percentage of electricity that must be generated from renewable sources, which then can be increased over the years. Credits for generating electricity from renewables are given to certified sources and those credits can be bought and sold and leased. Each utility can decide how to meet the standard. If the standard is 20 percent, a utility could generate 80 percent of its power from fossil fuel and 20 percent from, say, biomass or wind. It could generate 25 percent of its power from renewables and sell the extra 5 percent credits it would earn. Or it could generate more than 80 percent of its power from fossil fuels and purchase credits to make up the difference. A company could generate electricity strictly from renewables, sell the electricity, and also sell the credits it would earn for 80 percent of its power production. As the standard is raised over the years, renewables would play an ever greater role and use of fossil fuels would decrease.[116]

After energy prices skyrocketed in the late 1970s and 1980s, utilities, as publicly regulated monopolies, were required to adopt demand side management (DSM) programs. Since it is cheaper to save electricity than to generate it, much of the emphasis was on helping their customers become more energy efficient. Utilities offered energy audits of homes and offices, after which the auditor would make recommendations on how to get their energy services more efficiently. Some utilities offered grants or low-interest loans to homeowners to upgrade their insulation. One utility in California gave away more than a million compact fluorescent bulbs.[117]

Some of the DSM programs were very successful. In Osage, Iowa, the local utility launched a program to weatherize homes and control power loads at peak periods. The utility saved enough money to prepay all its debt, accumulate a cash surplus, and cut rates by a third. The low rates attracted two factories to the town, and each household saved more than $1,000 per year, which boosted the local economy.[118] Ted Flanigan and June Weintraub profile, as examples for others, forty successful DSM programs that saved significant amounts of energy at "surprisingly low cost."[119]

Many DSM programs, however, did not succeed. Typically, only a tiny fraction of eligible people or companies participated. Not surprisingly, some of the DSM programs in California were among the most successful, since California was one of only a handful of states that let its utilities profit from getting its customers to use less energy. In most states, the utilities were less than enthusiastic about DSM because they were showing customers how to use less electricity, when the utilities could profit only by selling more electricity. That also made them less than credible sources with the people the programs were trying to reach.

In any event, with deregulation of utilities now in process and utilities restructuring, the old type of DSM programs will probably no longer be viable, although utilities may continue DSM in new forms as service programs.[120] Just what the electricity industry will look like in the future is far from clear. One possibility would be the emergence of energy service companies. These would be "full service" companies that would sell not only power but efficiency as well, to give their customers the best possible deal. They could be one important source for the kinds of things DSM programs were supposed to do.

Another possibility is for the government to expand energy efficiency programs, as represented by some voluntary ones the EPA is running: Green Lights (for lighting), Green Star (for office equipment), Energy Star (for commercial buildings), among others. Green Lights is the EPA's flagship voluntary program, which began in 1991 and is open to private companies of all types, governments at all levels, and nonprofits. Participants join the program to save money on lighting in their buildings and to receive public recognition for being environmentally conscientious. A participant upgrades the lighting in at least 90 percent of the area, where profitable. The EPA provides complete technical assistance, information, and training, plus public recognition. Some 1,800 participants with over 5 billion square feet of space have committed to the lighting upgrades. As of 1997, they have prevented about 1.5 million tons of CO_2 emissions and have saved \$158 million in electric bills. According to an International Energy Agency study, full implementation of Green Lights in all facility space would save over \$16 billion on electricity every year.[121]

These kinds of programs could be greatly expanded. As we have seen, one of the greatest barriers to adopting energy efficiency now, even in large companies, is the lack of information and the difficulty of obtaining it. With programs like Green Lights, companies have a readily available source of information and expertise, all in one place. One way to expand this kind of program would be to use the agricultural extension agent system as a model. Extension agents were originally established to transfer the latest research to farmers in their areas to help them expand production. A similar system of energy efficiency agents would be able to inform and encourage and help people and companies reduce their use of energy while maintaining the desired level of energy services.[122]

Yet even for conservatives, there are some places in which governmental regulations have a role. In regard to energy use, incentives are often split so that the ultimate customer, who has to pay the energy bills, has little or no choice.[123] For example, many appliances are sold to the builders of houses and apartments, rather than to the owners or renters who will use them. The builders have every incentive to install the cheapest acceptable models, no matter how inefficient they are in long-term use. Regulations can also overcome some of the problems of "bounded rationality" on the part of consumers.

The federal government has adopted energy efficiency standards for many appliances. The standards pay back any extra costs within three years.[124] As of 1997 they had already saved consumers $15 billion in electric bills. They are expected to save $46 billion between 1990 and 2015 and $132 billion over the long run. As of 1997 the standards were preventing emissions of 53 million tons of CO_2 per year, as well as other pollutants.[125]

However, appliances could be even more efficient than the standards require. Some standards have not even been issued yet and the older ones need to be updated to reflect current technology. "The appliance energy standards program is years behind schedule. New or updated standards should be adopted over the next few years on a wide range of products including clothes washers, dishwashers, central and room air conditioners, ranges and ovens, water heaters, furnaces and boilers, fluorescent lamp ballasts, and transformers. One analysis estimates that stringent yet cost-effective new standards for these products could . . . cut CO_2 emissions by 78 million tons of carbon annually by 2015."[126]

Building standards present a similar situation. Incentives in the building industry are split so many ways that the ultimate owner or renter, who pays the light and heat bills, has little choice but to take what is given.[127] In cases such as these, the federal or state governments should set high performance standards (not technological requirements) to minimize the use of energy and then let the building companies compete for the best ways to meet the standards. Labeling houses with energy efficiency ratings can be effective in getting buyers to choose homes built to exceed standards. They can also get builders to go beyond minimum requirements in order to get more attractive ratings.[128]

Another problem in the construction sector is the way in which architectural firms are paid, which gives them no reward for efficient design.[129] Efficient buildings do not cost more to build because the extra cost of efficient components is offset by savings in downsizing heating and cooling systems. But efficient buildings cost more to design because making sure that everything is properly integrated takes extra time and attention. Currently, designers get no reward for taking the extra time to make the building efficient.

The Rocky Mountain Institute is experimenting with changing architects' contracts to reward them for designing energy efficiency into buildings from the beginning. Their pay would be based on the performance of the building, not just on its initial price. The RMI scheme is to include in the building contract an extra bonus for the architectural firm if the structure beats a prearranged target, and a penalty if it falls short.[130] Governments could adopt this kind of contract for all their new buildings and could encourage the private sector to do so as well.

For a final example of the sorts of approaches conservatives should consider, there are several kinds of "market transformation" policies.[131] In these programs, governments (or organizations) can use their own market power to raise levels of energy efficiency. Each year governments buy huge quantities of computers, printers, photocopiers, lightbulbs, vehicles (these policies are not limited to the electricity sector), and other kinds of equipment. Governments should maximize their own energy efficiency (on a cost-benefit basis) in all procurement contracts. This would not only save taxpayers money now thrown away on wasted energy but would also raise the standards in the marketplace. The federal government is now setting energy efficiency standards on some of the office equipment it purchases.[132] This policy should be expanded to all purchases of equipment, cars and trucks, and so on. State and local governments should do the same.

After the same fashion, governments should set the best possible examples in their own buildings. New governmental buildings should meet the highest standards of energy efficiency. And governments should begin a serious retrofit program to make their existing buildings as energy efficient as possible. As the bottom-up studies have shown, there are many opportunities here to reduce greenhouse gas emissions and save a lot of money that should then be returned to the taxpayers. The Office of Technology Assessment estimates that the federal government could economically save 25 percent of the energy used in its buildings—$900 million per year—without sacrificing comfort or productivity.[133]

Other possibilities include setting high recycled content requirements for governmental purchases where use of recycled material saves energy overall, such as paper. The World Wildlife Fund claims that producing one ton of 100 percent recycled (postconsumer) paper saves 4,100 kilowatt hours of electricity.[134] The Sierra Club Legal Defense Fund some years ago proposed that the California judicial system should require all documents submitted to its courts to be printed on recycled paper. Market transformation ideas like that could be adopted by state courts and the federal system as well.

Other recycling programs also have a role as part of climate policy. For every 10 percent increase in recycling of steel, some 19 million tons of carbon emissions would be saved.[135] In the United States we throw away enough aluminum to rebuild the nation's commercial airplane fleet every three months. Recycling aluminum takes 95 percent less energy than making it from raw materials.[136] Perhaps we should seriously consider a nationwide deposit and refund for aluminum beverage cans, which some states have now. The states with these "bottle bills" account for the majority of recycled aluminum.[137]

These are just a few suggestions of policies that conservatives should consider in the electricity sector of our economy. Besides expanding generation from renewables, the aim of these policies is to take advantage of the energy efficiency gap. To the extent that policy interventions can speed up the diffusion of innovation and close that gap, we can reduce greenhouse gases while saving money. There are almost limitless possibilities here that conservatives should support because they make markets work better. As Amory and Hunter Lovins put it, climate protection is not about command and control; "it's about helping markets to work properly—and then letting them do their job."[138]

TRANSPORTATION

The transportation sector of our economy contributes 32 percent of our total CO_2 emissions.[139] Transportation will be the toughest nut for climate policy to crack, but since it contributes such a large portion of greenhouse gases and emissions are expected to increase substantially in the future, any serious climate policy must make significant changes in this sector. Here I will focus on use of cars and light trucks for personal transportation, which emit close to one-fifth of our CO_2.[140] This is a difficult sector to deal with for three reasons: because we value our mobility very highly, because we have developed a system in which we need personal vehicles to get where we want to go, and because people put fads over efficiency in choosing their vehicles.

I am *not* about to launch a tirade against cars. I am an auto enthusiast and have been for years. I think Formula One racing is the greatest sport in the world.[141] On the other hand, as a conservative, I think that all of us who drive must face up to the problems we cause. We have a moral responsibility to do our share to solve them. So, let us consider some of the things we should do.

Following standard conservative principles, we should first try to remove all subsidies and internalize externalities so that people face the full social costs of their transportation choices. When they know the truth, drivers will face a considerable shock because the private use of cars and trucks is heavily subsidized and produces serious externalities, even without considering the climate. The Office of Technology Assessment calculates that the price of driving fails to include between a third and a half of all real costs.[142]

Clifford Cobb itemizes a number of direct subsidies for driving and damages that drivers escape from paying. A major item is the cost of building and maintaining streets and highways. Although the federal government's portion of road construction comes from a gasoline tax, state and local governments usually use general revenues. Other items that drivers do not pay include their portion of the military costs of protecting oil supply lines, the costs of death and injuries to third parties (pedestrians and bicylists) beyond what insurance pays, and uninsured medical expenses paid from public funds. Environmental damages caused by passenger vehicles include air pollution, water pollution, and noise. Cobb estimates total subsidies and damages, just from personal use of cars and light trucks, at $171 billion per year, or $1.50 per gallon of gasoline.[143]

A study by the World Resources Institute estimates that the subsidies and damages are even higher: over $270 billion per year. To cover those costs would add almost $2.00 to the current price of gasoline.[144]

In other words, following the basic principle of economic efficiency that prices should include all costs and the conservative principle that people should take responsibility for their actions, the price of gasoline that is now about $1.50 per gallon should be at least $3.00—and that does not include climate problems caused by burning gasoline. (Even at that rate, gasoline would still be cheaper than it is in Europe and Japan.) All of us who drive should face up to the fact that we too are "welfare bums." We should get off the dole and pay our own way.

These costs should be paid by motorists, and the logical way is to raise gas taxes to cover them. (Of course, the subsidies now paid out of general revenues should then be returned to the taxpayers.) Responsibility would be placed precisely where it should be. Even without regard to climate policy, this would improve economic efficiency: since driving is now underpriced, people demand too much of it. People would drive fewer miles and buy cars and trucks that get better gas mileage, so this policy alone would reduce CO_2 emissions to some extent.

Cobb calculates that the long-run price elasticity of demand for gasoline is high enough that phasing in, over ten years, an addition of $1.50 or so to the price per gallon would reduce gasoline use by over two-thirds.[145] This rather optimistic estimate would mostly come from switching to higher mileage vehicles, only a small part from driving less. On a longer time horizon, Carmen Difiglio calculates that a price hike of about that much would reduce carbon emissions by 60 to 130 million tons, a very useful contribution to climate policy—some 5–10 percent of current U.S. emissions.[146]

There are many other studies of the effects of full-cost pricing of driving. They do not always define subsidies and externalities in the same way, and their estimates of the several costs often vary. But in general they find that if drivers had to pay the full costs, greenhouse gas emissions in the transportation sector would be reduced by 10–15 percent from a baseline projection.[147]

At this point we can turn to climate policy itself. The bureaucratic command-and-control approach that is usually discussed is to raise the corporate average fuel economy (CAFE) standards. These require each auto manufacturer to meet an average fuel economy standard for all the cars and trucks it produces. (There are two separate standards: currently 27.5 miles per gallon for cars and 20.7 mpg for light trucks.) But this is an indirect, long-term approach and would not necessarily produce any significant decrease in CO_2 emissions. After the gas crises of the 1970s,

CAFE standards were established to reduce gas consumption. The vehicle fleet did indeed become more fuel efficient over the years, but much of that gain was offset as the number of vehicles increased and people drove more miles. In the booming economy of the 1990s people started buying larger and larger trucks, with very poor gas mileage, so now even our fleet average is getting worse.[148]

Conservatives should look at some different policy options. A general carbon tax would directly raise the cost of gasoline, internalizing the climate damage externalities. Higher gas prices will increase sales of fuel-efficient cars and reduce sales of gas guzzlers, but it takes a long time to change the entire fleet to improve its efficiency.

To speed up the process, conservatives should consider a "feebate" system that is aimed directly at the consumer who buys a vehicle. In this scheme, people who buy gas guzzlers have to pay an extra fee, and the worse the gas mileage the higher the fee. People who buy gas sippers get rebates, and the more fuel efficient the vehicle the larger the rebate. The fees cover the costs of the rebates, so that there is no added burden on taxpayers. Carmen Difiglio estimates that in the long run feebates could reduce annual carbon emissions by 30–40 million tons.[149] Feebate schemes could be adopted whether or not there is a general carbon tax.

Feebates and full-cost pricing would lead many people to give up their gas guzzlers and buy fuel-efficient vehicles. But—and this is a key, at least in America, to the public acceptance of the policies—this does not mean drivers have to sacrifice size or performance. The IPCC found that the technology exists to cut energy use for our vehicles by one-third without compromising comfort or performance.[150] Geller and Nadel list a number of advanced technologies, such as lean burn engines, continuously variable transmissions, and lightweight materials, that could increase gas mileage by two-thirds without reducing vehicle size or performance.[151] Alternative fuels and fuel cells offer other possibilities.

Even if we do not adopt full-cost pricing of driving, conservatives should at least advocate assuming responsibility now for our greenhouse gas emissions. An alternative to a carbon tax, aimed specifically at the transportation sector, would be to add a "damage reduction surcharge" at the gas pump. This would not be a "tax," in that it would not go into the general revenue. It would go into a separate fund that could only be spent on greenhouse gas mitigation. Beginning immediately, a 25¢ per gallon surcharge would be imposed. After five years the program would be reviewed to see if the income was sufficient to fund projects that would offset the CO_2 from burning gasoline, in order to meet our emission trajectory goals. If not, the surcharge would be raised. If it generated more revenue than needed, it would be lowered.

On every gas pump there should be a sign explaining the surcharge: that scientists have determined that greenhouse gases damage the climate; that every gallon of gasoline burned adds twenty pounds of greenhouse gas to the atmosphere; and that all of the money collected from the surcharge will be spent to repair the damage done by burning that gasoline. It should also stress that each person should be responsible for damages he or she causes, and this is how we are all taking responsibility for the consequences of our own actions.

The advantages of a surcharge are that people pay it in direct proportion to the CO_2 they emit. It would keep the problem of global warming salient. And it would generate a considerable income for greenhouse gas mitigation projects.

A CLIMATE PROTECTION FUND

A 25¢ surcharge on gasoline would generate $36 billion per year (based on current sales) to be used for greenhouse gas reduction and mitigation.[152] An equivalent surcharge on other fossil fuels used by small emitters outside a quota trading system would also generate income for the fund.

The money should be spent on projects that produce reductions in greenhouse gases. Projects would have to be monitored and held accountable. There is a virtually endless list of possibilities, and I will just suggest a few to indicate the kinds of things conservative policymakers should be considering.

We have seen that there are enormous opportunities to enhance carbon sinks in the United States. Many of the possibilities are prime examples of climate protection projects the fund could support. After decades of unsustainable logging, many areas of our national forests are in serious need of reforestation. Since they now provide only about 4 percent of our nation's harvest of wood, the national forests could be managed for the future as carbon sinks, with the fund paying the reforestation costs. Forest growth can be monitored to assure that it is contributing to our carbon budget.

Forestation projects could also be done on state and private land. For farmers, the fund could support long-term contracts similar to the shorter-term Conservation Reserve Program agreements. The contracted land would then be managed and monitored as a carbon sink. The fund could also support tree planting in cities. As we have seen, urban trees do far more to reduce greenhouse gases than merely take up and store their own carbon.

A climate protection fund could also support energy efficiency projects. There are all sorts of opportunities to save energy by retrofitting public buildings. A city government or school district, for example, could retrofit some of its buildings to improve their insulation, install superwindows, upgrade light fixtures, install efficient cogeneration units, repaint roofs in light colors, and so on.[153] It would have to document energy savings, producing records of energy used before and after the retrofit. The money saved should be returned to the taxpayers through lowered tax rates. Or the fund could be repaid and then the savings returned to the taxpayers.

Cities could save a lot of energy and money by retrofitting streetlights. Most streetlights are not very efficient to begin with, and they scatter light everywhere, including up. Much of the electricity and the light is simply wasted, for a total of $1.5 billion per year.[154] This also causes light pollution, which is another growing environmental problem. San Diego replaced 31,000 old streetlights with efficient ones and is saving $2.7 million per year. The retrofit paid for itself in about three years.[155] A city could propose a streetlight retrofit project, to be financed by the fund.

The fund could also purchase fuel cell buses for cities to replace fossil fuel ones. It could pay to install solar PV panels on the roofs of public buildings. A climate protection fund, in short, could support all kinds of energy efficiency projects and more, at least on public property and for nonprofit organizations. A continuing and highly visible program to save energy and money in the public sector would make numerous options salient and readily available for the private sector as well.

For private companies, the fund could pay for the administrative costs of federally sponsored voluntary programs, such as the EPA's Green Lights, which could be greatly expanded.[156]

Another possibility is to use the fund to provide incentives ("golden carrots") for commercialization of new energy efficient equipment, to help overcome start-up costs and problems. For example, a consortium of utilities offered $14 million in incentives for manufacturers to produce and sell recently developed gas-fired heat pumps. Once sales reach 50,000 units, the manufacturer begins to reimburse the utilities for the incentives they paid.[157] The fund could finance many programs of this type in the private sector.

Another model could be the Super Efficient Refrigerator Program, which was a competition among manufacturers to design a new refrigerator, without CFCs, that would use 30 percent less energy than the federal minimum standards. The winner had to market more than 250,000 units at a price comparable to that of standard products. The winner (Whirlpool) got a $30 million bonus.[158]

Other kinds of projects that ought to be considered for funding include methane recovery from landfills and retiring older coal-fired electricity plants that are publicly owned and replacing them with renewables and modern gas-fired generators. Even rather speculative projects should be considered, such as weatherization of homes for low-income people and free distribution of compact fluorescent lights. (If CFLs replaced just half of our incandescent bulbs, total savings would be $7 billion per year.)[159]

All of these kinds of projects are cost-beneficial, no-regrets actions. We use energy so wastefully that, with the proper policies, we can go a very long way toward meeting our greenhouse gas targets with efficiency measures alone, keeping our current level of energy services and saving a lot of money in the process. A climate protection fund to pay for expanding carbon sinks and energy efficiency projects could be a huge help toward solving the greenhouse problem.[160]

The few ideas suggested here indicate that we have a very wide range of policy options for climate protection. We will be able to protect the earth's climate if we combine many policies to tax or cap carbon emissions, to eliminate subsidies and internalize externalities, to enhance carbon sinks, to replace fossil fuels with renewables, and to increase our energy efficiency. We also need to keep policies adaptable as new evidence and new technologies emerge.

COSTS OF CLIMATE PROTECTION

Finally, we must consider in more detail some claims about the overall costs of climate protection. All sorts of wild assertions appear in the popular press and on the op-ed pages that reducing greenhouse gases will destroy our economy and put countless numbers of people out of work. Those claims are utter nonsense.

In what must be an absolutely unique event in the history of economics, in 1997 an organization called Redefining Progress circulated an "Economists' Statement on Climate Change." It was drafted by five prominent economists (including William Nordhaus), and it has been endorsed by over 2,500 economists in the United States, including eight Nobel laureates. (When was the last time 2,500 economists agreed about anything?) The statement reads as follows:

I. The review conducted by a distinguished international panel of scientists under the auspices of the Intergovernmental Panel on Climate Change has determined that "the balance of evidence suggests a discernible human influence on global climate." As economists, we believe that global climate change carries with it significant environmental, economic, social, and geopolitical risks, and that preventive steps are justified.

II. Economic studies have found that there are many potential policies to reduce greenhouse-gas emissions for which the total benefits outweigh the total costs. For the United States in particular, sound economic analysis shows that there are policy options that would slow climate change without harming American living standards, and these measures may in fact improve U.S. productivity in the longer run.

III. The most efficient approach to slowing climate change is through market-based policies. In order for the world to achieve its climatic objectives at minimum cost, a cooperative approach among nations is required—such as an international emissions trading agreement. The United States and other nations can most efficiently implement their climate policies through market mechanisms, such as carbon taxes or the auction of emissions permits. The revenues generated from such policies can effectively be used to reduce the deficit or to lower existing taxes.[161]

The cost of climate protection is nothing to fear. First, the bottom-up engineering studies prove beyond any doubt that we waste so much energy that we can meet most of our climate goals in ways that save money. Even the economists who believe high "transaction costs" (i.e., market barriers) now prevent energy efficiency from being economically efficient have to admit that the technical potential is there. The energy efficiency gap is real. Consequently, policy interventions can be designed to reduce those market barriers and close the gap, all the while saving us more and more money. And even economists such as Sutherland agree that policy intervention is warranted where there are negative externalities. Since proper policies in this area will help the market work better, they embody conservative principles.

Second, consider the numbers from the top-down studies. For example, Manne and Richels estimate costs of $3.6 trillion to reduce and stabilize emissions, and Nordhaus calculates that emission reduction would cost $10.9 trillion.[162] Keep in mind that these numbers are not out-of-pocket expenses to control emissions; they are estimates of reductions in future economic growth. If those numbers look enormous and scary, that is only because they are totals projected over many years. In their proper context, even those astronomical numbers are pretty insignificant. These are typical figures from top-down studies, which generally say that climate protection will cause about a 2 percent decrease of potential gross world production.[163] How much is that, really? As Thomas Schelling puts it, reducing gross national product by 2 percent in perpetuity to protect the climate "lowers the GNP curve by not much more than the thickness of a line drawn with a number-two pencil . . . [and] will not send us to the poorhouse."[164] William Nordhaus notes that the "mind-numbing" cost figures are actually "modest relative to the total size of the global economy. . . . Overall economic growth projected over the coming years swamps the projected impacts . . . of the policies to offset climate change." His computer graphics are, he says, better than Schelling's pencil, but even so, on a graph of projected GNP with and without climate protection "we can now barely spot the difference!"[165] Michael Grubb and his colleagues reviewed the top-down studies and concluded that even the "upper bound range" of costs for cutting CO_2 emissions in half "is equivalent to reducing average GNP growth rates over 50 years from 3 percent per year to 2.97 percent per year."[166] *That is, even if you believe only the top-down studies, the economic impact of climate protection would hardly be noticed.*

Third, more recent studies have abandoned some critical assumptions of the earlier top-down approach, especially the assumption that technology remains constant. Many of the newer studies incorporate a factor for autonomous energy efficiency improvement (AEEI) that takes into account the rate at which energy efficiency measures are adopted in our economy independently of the price of energy. The faster technological and structural changes reduce the energy intensity of our economy, the lower the cost estimate for reducing greenhouse gases. Just as the damage estimates are very sensitive to the discount rate used, so, it turns out, are cost estimates very sensitive to AEEI assumptions.[167] And many policies, as we have seen, can be adopted to increase the AEEI.[168] Indeed, this is a prime reason for government to support research and development and to speed up adoption of energy efficiency innovations.[169]

Fourth, virtually all of the estimates of costs or savings from reducing greenhouse gases ignore secondary benefits. Many sources of greenhouse gases are also sources of other nasty pollutants—NO_x, SO_2, volatile organic compounds—which produce smog in our cities, cause or exacerbate respiratory ailments, damage crops, produce acid rain, and so on. Reducing the use of fossil fuels not only reduces greenhouse gas emissions for climate protection but also reduces these other emissions and the damages they cause. These secondary benefits outweigh the climate protection benefits in the short run, and they are benefits for people living right now. If we also take these additional benefits into account, the net costs of climate protection policies come way down, even using the older top-down estimates. Savings and benefits go way up under more realistic bottom-up analyses. (It is important to note here that one industry—coal mining—will be severely impacted by policies to protect the climate. It would only be fair and just for our policies to include generous assistance for those workers and their families.)

The bottom line is crystal clear: attempts to scare the American people by claiming vast economic destruction from solving global warming are simply wrong. Just as we saw in Chapter 2, protecting the environment does not hurt the economy. It is even better in the case of global warming. Internalizing externalities in prices makes the economy more efficient. Diffusion of energy efficiency technologies saves money. Developing new technologies for renewables and efficiency expands markets. Climate policy is not about who should bear any undue extra costs. The Lovins even go so far as to ask, "What costs? The interesting question is who should get the *profits*."[170]

CONCLUSION

In Chapters 6–7 we have covered considerable ground—from climate science to economics to the technologies of renewable energy and energy efficiency and finally to policy. The issues of climate change are extremely complex, especially the science of climate change and the economic estimates of damages and abatement costs. (I have attempted no more than a bare outline here.)

Unfortunately, that complexity makes it far too easy to confuse the issue in the public mind. I think, to take just one example, of a newspaper column in which a scientific study was misinterpreted. The writer even asked the authors of the study about it and, he wrote, they told him he was misinterpreting it. But rather than try to understand the study, he spent the rest of his column attacking the motives of the scientists! This sort of thing is all too common in the popular press.

The problem of global warming is, as we have clearly seen, a very real one. And it is precisely the type of problem that the *conservative philosophy* is needed to confront—especially our core principles of responsibility, prudence, and stewardship for future generations. Our current "policy," if you can call doing virtually nothing a policy, is imprudent and irresponsible in the extreme. And it simply writes off the well-being of our grandchildren and great-grandchildren.

As J. W. C. White writes, "If the Earth had an operating manual, the chapter on climate might begin with a caveat that the system has been adjusted at the factory for optimum comfort, so don't touch the dials."[171] Unfortunately, we have been playing with the dials for a long time and have made major changes in one of the important factors that governs climate: the concentrations of greenhouse gases in the atmosphere. Now we must reverse this process and return toward natural conditions of climate variability.

In doing so, we should *not* see climate protection as a threat to our prosperity—it is not.

Neither should conservatives see it as an attack on our freedoms—it is not. Global warming is *not* a liberal conspiracy to increase the power of government.

Global warming *is* a challenge, one that must be met, and in confronting it we should look on it as an opportunity.

Global warming is an opportunity to boost our economy by becoming more efficient.

Global warming is an opportunity to develop new technologies and better sources of energy.

Global warming is an opportunity to advance toward the future by enhancing the diffusion of innovations.

Global warming is an opportunity to maximize what the United States does best: confront problems and invent ways to solve them and win new markets in the process.

So, as Amory and Hunter Lovins ask, "What are we waiting for?"[172]

8

Saving Species: Doing Noah's Job Today

ONE OF THE MOST REMARKABLE—and, as far as we know, unique—things about our lovely planet is the incredible proliferation of forms of life. The rather bland term "biodiversity" refers to the marvelous myriads of plants and animals of all kinds that have adapted to fill every niche on the globe. We do not even know within an order of magnitude how many different ones there are. Estimates run from about 10 million to 100 million.[1] Only some 1.4 million species have been identified, and only a tiny fraction of those have ever been studied in any detail.[2]

In their interrelationships with each other and with the earth and atmosphere, all these varieties of life form ecosystems, and it is ultimately on these ecosystems that our own lives depend, for everything from the air we breathe to the soil from which we get our food to the systems that must finally recycle or absorb the wastes we produce.

In nature, species become extinct and new ones evolve over the course of millions of years. But we humans are now causing the extirpation of species at a rate 100 to 1,000 times greater than occurs normally.[3]

In the history of our planet, there have been five catastrophic events that wiped out a large number of life forms, from which it took millions of years to recover. The latest one happened some 65 million years ago, when the dinosaurs became extinct. Today we humans are causing a sixth mass extinction. Up to half of the world's species may be lost in the next several decades. This has been called "the most irreversible environmental problem of this era."[4] As Norman Myers says, what we are doing "will impoverish the Earth, and hence the livelihoods and lifestyles of people ahead, for at least 5 million years. While there is some uncertainty about what species are good for, we are effectively saying that we are absolutely certain that people for the next 5 million years can do without maybe half of all today's species. That's far and away the biggest decision ever taken by one generation on the unconsulted behalf of future generations."[5]

By far the main reason for the disappearance of so many plants and animals is that we are destroying the habitat they need in order to survive. Approximately half of all species live in the tropics, so the bulk of extinctions comes from destruction of the rain forests. We have only an indirect influence over the policies of these countries. Although I do not mean to minimize the devastating effect on biodiversity in the rain forests, I will focus here on the United States—our problems and our policies.

The diversity of life is under attack here as well. In the United States, more than 500 species have gone extinct since colonial times, half of those since 1980.[6] As of September 30, 1999, the Fish and Wildlife Service listed a total of 1,197 plants and animals as endangered or threatened in the United States.[7] And those are just the ones the agency has gotten around to listing. It has a huge backlog it needs to investigate. The Nature Conservancy has identified 6,500 species in our country at risk of extinction.[8]

Several forces have been identified that are threatening many of our plants and animals. David Wilcove and colleagues list five major causes. Some 85 percent of all species at risk are imperiled by habitat loss or degradation. Invasive alien species that outcompete natives threaten 49 percent. Nearly a quarter (24 percent) are in trouble because of pollution. Overexploitation affects 17 percent, and 3 percent are at risk because of disease. (These categories are not exclusive, and thus they total more than 100.)[9]

Wilcove and colleagues identify several types of activities that have affected habitat so severely that plants and animals are at risk: agriculture, livestock grazing, mining and oil drilling, logging, infrastructure development, road construction, military activities, water development and dams, pollutants, land conversion for commercial development, disruption of fire ecology, and recreation. In the last category, which imperils 27 percent of species at risk, the worst offenders are off-road vehicles, which account for half of that number.[10]

These forces that threaten our native plants and animals are not acting uniformly across the United States. Although there are species in trouble in every state, they tend to be concentrated in a few "hot spots."[11]

Conservatives should be extremely concerned about the mass extinction we are causing. We are violating some basic principles of conservatism and must work to reverse the trends that have such devastating consequences for many forms of life. Wiping out huge numbers of species is obviously impious in the extreme. In Richard Weaver's terms, it is nothing less than a sin. Piety, by contrast, "admits the right to exist . . . of things different from the ego."[12]

The impiety of our actions is also expressed in traditional Judaeo-Christian terms: "The earth is the Lord's and the fulness thereof."[13] The earth and its creatures are not ours to destroy as we choose. God told Noah to save *all* of the different kinds of animals.[14] There is absolutely no reason to believe any less is expected of us today. The moral of the story is again apparent in its conclusion: the rainbow covenant is with *all* creatures, not just with humans.[15]

Exterminating species also violates the conservative principles of prudence and of stewardship for the future. As we will see, there are many reasons to believe that the diversity of life plays important roles in maintaining the earth as a good place for us to live as well. It is highly imprudent to destroy that system piece by piece, with little knowledge of the consequences until it may be far too late. And it clearly violates our obligations to future generations, if we pass on to them a "depauperate" planet, as biologists might put it—a planet depleted of much of its biological wealth.

Conservative principles support preserving biodiversity. In addition, biologists and environmentalists put forward a fairly standard list of reasons for preserving the full range of life on earth. They fall generally into three categories: instrumental or pragmatic, aesthetic, and moral.

We get many useful products from nature, such as medicines and food. Since so few species have been studied, there is every reason to believe that many more useful products will be found in the future. But that potential is being lost as species vanish at an ever increasing rate.

Nature is a vast chemical laboratory. Over millions of years, plants and animals have invented far more chemicals than we could ever do on our own. We get many important medicines from this laboratory, some of which cannot be synthesized. "Nearly 40 percent of medical prescriptions dispensed annually in the United States are based on substances derived from nature or synthesized in imitation of natural substances."[16] "Many of the most significant medicines we depend on today were extracted from plants: digitalis, from the foxglove plant, is used to treat heart disease; morphine, from the opium poppy, is a highly effective pain reliever, especially useful for terminal cancer patients; quinine, from the bark of the cinchona tree, is used to treat malaria; and aspirin, related to a chemical from the bark of the white willow tree, is perhaps one of the most widely used drugs for treating pain and inflammation."[17] The rosy periwinkle, a tropical plant, provides a treatment for childhood leukemia and Hodgkin's disease. Until just a few years ago, the Pacific yew tree was burned as trash as loggers obliterated the old-growth forests of the Pacific Northwest. Now it is a source for Taxol, which is used to treat ovarian and breast cancer.[18] And the examples could easily be multiplied.

There is every reason to believe that we have only scratched the surface in our investigation of nature's pharmacy. But we need to preserve its full range, or we may lose much potential value. Wild versions of food crops often become critical sources for hybridizing to introduce resistance to new diseases and to pests that become immune to pesticides. For example, a leaf blight that wiped out 15 percent of our corn crop was stopped by introduction of genetic material from a wild species of corn in Mexico.[19]

Other useful chemicals, we believe, will be found in nature—if we preserve it. According to chemical ecologist Thomas Eisner, humans have put "the diverse substances that animals, plants, and microorganisms produce for survival and reproductive purposes . . . to countless medicinal, agrochemical, and other uses, and by doing so have benefited enormously. Thousands of natural products have been characterized, yet the majority, comprising a vast, untapped treasury, remain to be discovered."[20] He contends that "prospecting for biogenetic information could well become a major scientific exploratory venture of the 21st century."[21] But extinction depletes that reservoir of potential wealth.[22]

Perhaps the most important instrumental reason to preserve biodiversity is to retain the numerous "ecosystem services" that nature provides us for free.[23] These services include such obvious things as providing food and raw materials that we use. They also include creating and purifying the air we breathe, maintaining the fertility of the soils, pollinating our crops, purifying our water and reducing the damage from floods (in wetlands), cycling and movement of nutrients, and detoxifying and decomposing the wastes we produce.[24]

These services are not marketed, so they do not appear in our national accounts. That makes it all too easy for us to take them for granted or to overlook them entirely, but their economic value is considerable. A group of economists and ecologists led by Robert Costanza estimated the dollar value of just a partial list of these ecosystem services. Their estimates in each case are very conservative; nevertheless, the total comes to $33 trillion per year, well over the total global gross national product (see Chapter 9).[25]

But the continued provision and the quality of these ecosystem services is dependent on retaining a very wide range of species. Biodiversity, we are learning, is important for ecosystem productivity and also for resiliency. David Tilman and his colleagues, for example, conclude that "the preservation of biodiversity is essential for the maintenance of stable productivity in ecosystems."[26] They contend that "the loss of species threatens ecosystem functioning and sustainability."[27] Similarly, Shahid Naeem and Shibin Li conclude that "increasingly depauperate ecosystems . . . may become increasingly less reliable in terms of the goods, services and economic values they provide."[28]By exterminating species at an ever increasing rate, we are attacking these ecosystems on which we depend and reducing their ability to provide us with their many services.[29]

Ecosystems are, to say the least, complex. As Jack Ward Thomas says, "Ecosystems are not only more complex than we think, they're more complex than we *can* think."[30] That means that we cannot know for sure what the consequences will be of removing component parts one by one, until it is too late. Paul and Anne Ehrlich use an interesting analogy. What would you think, they ask, if you walked from an airport terminal toward your plane and found someone on a ladder popping rivets from the wing? He explains that he works for the airline, which has discovered that it can sell these rivets at a good profit. He assures you that the manufacturer has made the plane much stronger than it needs to be, so it can fly without a bunch of rivets. Would you still take that flight? "Ecosystems, like well-made airplanes, tend to have redundant subsystems and other 'design' features that permit them to continue functioning after absorbing a certain amount of abuse. A dozen rivets, or a dozen species, might never be missed. On the other hand, a thirteenth rivet popped from a wing flap, or the extinction of a key species involved in the cycling of nitrogen, could lead to a serious accident."[31]

If a key species is eliminated, it can have a "cascading" effect.[32] Other species that depend on it will die out, which will affect still others, and so on, until the ecosystem "flips" into another state. That is why some clearcut forests never regenerate but degrade to grasslands.[33] If this happens on a sufficiently large scale, the result will be "a biota increasingly enriched in wide-spread, weedy species—rats, ragweed, and cockroaches—relative to the larger numbers of species that are more vulnerable and potentially more useful to humans as food, medicines, and genetic resources."[34]

Since ecosystems are so complex that we cannot predict with any confidence what species may be discarded, the only prudent course of action is for us to make every effort to preserve the full range of life.[35] In Aldo Leopold's homely analogy, "to keep every cog and wheel is the first precaution of intelligent tinkering."[36]

Moreover, when species become endangered because of our actions, they are often showing us that whole ecosystems are in danger. They serve the warning function of the canaries that miners took into coal mines. "Like the canaries, an endangered species can show by its presence that habitat qualities are in jeopardy, with possible or probable human welfare consequences that may follow."[37] They show us where we have done a poor job of managing our resources.

The heated controversy over the northern spotted owl is not just about some little bird. The fact that this owl is threatened with extinction shows us that the old-growth forests themselves are in danger. All of the species that depend on them are at risk, and so are the services those forests provide: water filtration and storage, watershed protection, recreation, salmon fisheries, and the famous quality of life of the Pacific Northwest. In narrowly economic terms, the little old growth that remains in the region is now far more valuable intact than it would be as stumps and paper and lumber (see Chapter 2). If the Forest Service and the timber companies had managed those forests properly, there would never have been a spotted owl controversy.

> The bald eagle warned us about poisoning the food chain with chemical pesticides; the snail darter warned us that boondoggle water projects were claiming far too many free-flowing rivers and valley farms; . . . the Barton springs salamander warned us that over-pumping the Edwards aquifer threatened the agricultural communities of West Texas that also depended on adequate water levels . . . And so on. Every species tells a

> story about resource depletion and degradation, about waste and abuse, about unsustainable uses of the land.[38]

If we use nature with proper respect, restraint, and humility—as conservative principles tell us to do—we will not threaten whole ecosystems with destruction. If we preserve them intact, with their full range of species, they will continue to provide us, for free, with the many services they have to offer us.

Besides the instrumental or practical reasons to preserve species, there are also aesthetic ones. People enjoy watching wildlife. There are, of course, lots of common creatures that people enjoy—chipmunks and deer and geese. But beyond these common animals, it would certainly be a shame—and a widely felt loss—if we could never again see a bald eagle soar over the mountains, or a peregrine falcon free-falling in a dive, or a whale breaching the surface of the sea.

(Of course, these aesthetic reasons to preserve rare species translate into practical ones as well: people pay serious money to travel to see wildlife. Ecotourism is a booming business worldwide. The number of visitors to Yellowstone increased after wolves were reintroduced to the park. In the Platte Valley of central Nebraska, where I grew up, the local economy benefits to the tune of some $40 million per year from people coming to see the thousands of sandhill cranes in the spring migration.[39] Seeing a whooping crane as well would be an unforgettable high point of their trip.)

For most people, aesthetic values generally do not extend beyond the "charismatic megafauna." But any naturalist, professional or amateur, also appreciates the other creatures, even the little things the general public would never see.

Perhaps to be considered as an aesthetic value is the possibility for study of these rare plants and animals. Many people place a high value on scientific knowledge for its own sake. That value is lost forever whenever any kind of plant or animal vanishes from the earth.

Finally, there are moral reasons to preserve biodiversity, and many of them are closely related to the conservative principles discussed above. Life is sacred, so it is immoral to exterminate species. Wiping out an entire species, unlike killing an individual, eliminates the generative processes, so it is a kind of superkilling. "There is not just death, there is no more birth."[40] This does not imply a biocentric philosophy in which all species have equal rights. A theocentric approach reaches much the same conclusion, for most practical purposes. We may have been given "dominion" over the earth, but only as stewards. It does not belong to us, but to God, and we are accountable to its Creator for what we do to it. We have no right to exterminate countless forms of life the Creator has made.[41]

Environmentalists also stress our moral obligation to future generations. We have no right to leave them a planet depleted of its biological wealth. As biologist E. O. Wilson has said, the mass extinction we are causing is "the folly our descendants are least likely to forgive us."[42]

Thus there are many reasons to preserve the diversity of life as we find it on this earth. Unfortunately, markets do not reflect many of these values.[43] They are not included, for example, in the price of critical habitat that a developer wants to convert into a shopping mall. If they were, the mall, in that location, might well not be an economical investment. In fact, the price of that land does not even reflect its full *economic* value to the public. John Loomis and Douglas White review a large number of contingent valuation (willingness to pay) studies of the economic values, not included in market prices, that people put on saving endangered species. They find that "for even the most expensive endangered species preservation effort (e.g., the northern spotted owl) the costs per household fall well below the benefits per household found in the literature."[44] Saving such endangered species is not putting the well being of birds over people. People *value* preserving them. And those economic values are only part of the total value of preserving the diversity of life.[45]

Unfortunately, we have been doing a terrible job of stewardship and we need to change that. David Quammen fears that in just five or six generations—a short time by the long view of conservatism—the earth will be degraded to a "planet of weeds" in which the rich pay huge sums to replace many ecosystem services and the masses of the poor simply suffer.[46] But as Patrick Parenteau says, "One need not subscribe to doomsday theories of total environmental collapse to question whether a world of vanishing species, unraveling ecosystems, and deteriorating environmental conditions is really a world we want to inhabit, or bequeath to our children."[47]

It is not too late to rectify our actions. Our country, with its enormous wealth, must take the lead in this effort. And we did exactly that, twenty seven years ago, by passing the Endangered Species Act. That act has become by far the most controversial of all the environmental laws, so we need to take a look at it and see just exactly what it does.

THE ENDANGERED SPECIES ACT

In considering the Endangered Species Act (ESA), we need to clear up the confusion about the law. First, we need to cut through the fog of mythology that has obstructed the public debate in the last few years. And, second, we need to look at how the ESA is being administered today because major changes have been introduced in recent years.

The ESA takes pride of place (if one can call such a dubious honor by that name) in the hierarchy of demons that infest the minds of a small but noisy segment of the public. All sorts of horror stories have been spread alleging that the ESA has done nasty things to people. As we saw in Chapter 2, those stories turned out to be fabrications or such gross exaggerations that they shed no light at all on the act. "Stories surround the Act like those found at dockside in the days of sailing ships: of sea monsters that forced ships onto reefs and then attacked their crews. Problematically, few sailors could be found who actually witnessed these events and the stories of those who claimed they were witnesses turned out to be something less on closer examination. The facts, then and now, are otherwise."[48]

We also saw that only a tiny fraction of projects ever run into any trouble because of endangered species, and almost all of those proceed with modifications.[49] In only two famous cases did protecting an endangered species cause a dramatic "train wreck," as Bruce Babbitt calls them. The most recent is the northern spotted owl. Despite all the fury over "jobs versus owls," protecting the owl, as we saw, had *no* overall impact on employment in the timber industry in the Pacific Northwest.

The other case, from an earlier day, is the Tellico Dam in Tennessee, in which a little fish, the snail darter, stopped construction of a dam on one of the few remaining freeflowing rivers in the area. Conservatives—and all American taxpayers, for that matter—should have been on the side of the fish. What the media failed to stress, in all the hype over the case, was that the dam was a complete waste of the taxpayers' money and an example of pure pork barrel politics.[50]

The dam was almost complete when a court injunction stopped construction. The case went all the way to the Supreme Court, which ruled in favor of the fish. Congress then amended the ESA to provide for a cabinet-level committee, usually called the "God Squad," with power to overrule the protection of an endangered species on a case-by-case basis. The God Squad "unanimously decided that in terms of public economics the dam had been an economic non-starter. As Chairman Charles Schultze of the Council of Economic Advisors asserted 'the interesting phenomenon is that here is a project that is 95 percent complete and if one takes just the cost of finishing it against the [total] benefits and does it properly, it doesn't pay, which says something about the original design!'"[51]

Pork won in the end. Congress passed a rider exempting this project from the ESA, and the dam was completed. The taxpayers lost in the end. But the fish, we can be thankful, survives. A few populations were found in other streams in the area. These two cases are perfect illustrations of how "the ESA gets blamed for 'train wrecks' that were actually caused by bad economic and political decisions."[52]

One final myth requires a brief look. The charge has sometimes been made that species are selected for protection that are not really in danger of extinction, and that the agencies abuse their powers to declare subspecies and populations threatened or endangered. An empirical investigation by David Wilcove and his colleagues concludes that the charge is false. When they reviewed the plants and animals proposed for listing or added to the endangered list, they found that the median population size at the time of listing was only 1,075 individuals for vertebrate species, 999 individuals for invertebrates, and fewer than 120 for plants. The plants and animals are, indeed, rare. Moreover, 80 percent of those added to the endangered or threatened lists were full species, 18 percent were subspecies, and only 2 percent were population segments of more widespread vertebrates.[53] (For example, bald eagles were listed as endangered in the lower forty-eight states, even though there were still large numbers of them in Alaska.) In fact, on the basis of what we know about population viability, protection under the ESA is not being given "until their total population size and number of populations are critically low."[54]

Hoping that the ground has been cleared somewhat, we will proceed to look at the Endangered Species Act itself. It was originally passed in 1973, replacing previous statutes that were very limited in scope and had few teeth. It has been amended several times since then, and today it is administered in a way that differs from the method in which it was applied in the 1980s and early 1990s. First we will look at the provisions of the act and their contemporary implementation. Then we can consider some of the problems with the act and finally look at some conservative solutions to solve them.[55]

The Endangered Species Act is administered by two agencies. The Fish and Wildlife Service (FWS) has jurisdiction over plants and animals that live on land and in inland waters. The National Marine Fisheries Service (NMFS) has jurisdiction over fish and other animals that spend all or part of their lives in the sea. The procedures the two agencies use are essentially the same. The vast majority of species at risk are under the FWS.

In protecting a plant or an animal, the agencies follow a set procedure. The first step is listing. The agencies review the scientific evidence to determine whether a species is endangered or threatened, which are the two categories they can use. An endangered species is one that is in danger of extinction. A threatened species is one that is likely to become endangered if it is not protected. At this stage, the agencies can only use biological evidence. Whether a species is endangered or threatened is a scientific question; economic or social factors may not be considered.

In the real world, the agencies have an enormous backlog of species proposed for listing and few resources to devote to the process. The Nature Conservancy estimates that 165 unlisted species became extinct while the agencies focused attention on "higher priority" matters.[56] Many plants and animals at risk of extinction end up in a sort of no-man's-land, a holding category called "warranted but precluded." The evidence indicates that they are indeed threatened or endangered, but the agencies' limited resources have to be spent on more critical cases. Nearly 200 species are in this state of limbo.[57]

In the second stage, the agencies are supposed to identify and designate critical habitat that the species requires for survival. At this point, economic factors come into play. For example, in the case of the northern spotted owl, the FWS excluded 3 million acres of private, state, and other nonfederal land from its critical habitat designation for economic and social reasons.[58] Moreover, the act contains some exceptions for this stage that the agencies have been quick to use.

In the real world, many species that get listed never have critical habitat designated for them. The GAO found in 1992 that less than one-fifth of listed species had critical habitat designated.[59] The situation seems to have gotten worse. Of the 178 species listed between 1996 and 1998, not a single one had critical habitat designated.[60]

The third step is for the agencies to develop a recovery plan for a listed plant or animal, since the goal of the ESA is to recover species so they can survive on their own without the act's protection. In practice, the agencies never get around to developing recovery plans for many listed species. As of 1997, 40 percent did not have recovery plans.[61] And the recovery efforts are highly biased: half of all recovery funds are spent on just twelve species.[62]

The main reason for the backlog of species to be reviewed for listing and the lack of recovery plans for a lot of listed species, of course, is limited resources. Congress has never given the agencies anything close to the money they need to carry out these tasks. As Robert Thornton observes, the money Congress gives for endangered species protection is less than the amount spent on pizza in Washington, D.C.[63]

Once a plant or an animal has been listed as threatened or endangered, it is protected by law, whether or not critical habitat has been designated or a recovery plan has been drafted. There are two provisions in the act under which that protection is applied. Section 7 deals with federal projects that may affect a listed species; sections 9 and 10 concern private activities.

Section 7 requires all agencies of the federal government to consult with FWS or NMFS on any of their actions or projects that may impact a listed species, for example, highway construction and development on federal land. It also includes any private project that requires a federal permit. When a project is proposed or a permit is requested, the agency first consults with FWS or NMFS informally. As we saw in Chapter 2, some 90 percent of all proposals are quickly determined to have no potential impact.[64]

If the informal consultation indicates that the proposed project might affect a listed species, a formal consultation is required in which FWS or NMFS studies the project and writes a biological opinion about its impact. Again, as we saw, some 90 percent of formal consultations find that there will be no impact, and the project proceeds.[65]

In those few cases where the study determines that a project will "jeopardize" a listed species, the formal opinion must also include suggestions for modifying the project that will avoid harming the plant or animal. Many of these are quite simple. A fairly typical example is a proposal to build a boat pier on a Florida river in which manatees live. Manatees swim near the surface and suffer terrible wounds from motor boat propellers. The standard modification suggested is to post warning signs and speed limits so the boats can maneuver around the slow-moving animals.[66] Of the few formal consultations to conclude that a project will harm a listed species, 90 percent are modified and proceed. Under section 7, virtually no developments are actually stopped.[67]

Sections 9 and 10 concern listed species on private land. Whether we like it or not, preserving biodiversity cannot be limited to federal land. We will have to deal with the issue of private property. According to a GAO report, nonfederal lands constitute a considerable portion of the total habitat for listed species. Over 90 percent have some or all of their habitat on nonfederal land, over half have at least 81 percent of their habitat on nonfederal land, and over one-third are completely dependent on nonfederal land. Private landowners are the most prevalent type of nonfederal owners, and 78 percent of listed species have some or all of their habitat on private land.[68]

Section 9 says that it is illegal to "take" a listed species, which is defined as "to harass, harm, pursue, hunt, shoot, wound, kill, trap, capture or collect." As the law is interpreted, "harm" includes modifying its habitat in harmful ways.[69] If a listed animal is living on private property, the owners can use that property, but only in ways that do not harm the animal. (Endangered and threatened plants on private property have no protection under the ESA. Well over half of all listed species are plants: 716 out of 1,197.)[70]

There is no way to tell how many projects have been affected by section 9. Since no prior consultation is required, there are no statistics like the ones for the impact of section 7. Those numbers may give us a hint, however, because many section 7 consultations are for developments on private land that require permits. This would indicate that the overall effects of section 9 may be equally small. Walter Reid contends that "the successful record of section 7 consultation provides evidence that, if properly planned, development can take place with little additional pressure on endangered species."[71] Moreover, three-fourths of all listed species do not have any conflict with economic development.[72]

Nevertheless, landowners found section 9 too restrictive, so in 1982 Congress amended the act and added section 10. This new section provides a way for landowners to modify or develop their property even if it will harm a listed animal. The owner can obtain what is called an "incidental take permit." The landowner who wants one must develop a conservation plan that includes mitigation for the harm to be done.

This addition to the ESA languished for a decade virtually unused. In the 1990s use of what are now called habitat conservation plans (HCPs) exploded. Between 1982 and 1993 only fourteen HCPs were completed. By 1999, some 250 HCPs had been approved, covering more than 11 million acres of private and state land, and at least 200 more were being negotiated.[73] HCPs now are drawn up to cover huge territories with many landowners (public and private), and some are even being drafted to cover entire states. They are now the most important application of the act as it impacts private land, so it is crucial to understand how they work.

A property owner can obtain an incidental take permit by preparing a habitat conservation plan. The HCP describes the damage to be done to the species and its habitat. It must explain alternative actions the owner considered and why they were not adopted. The plan must specify actions that the owner will take to minimize and mitigate harm to the listed animal. This often involves preserving some of the land in its natural state or setting aside habitat elsewhere. The plan must indicate how those actions will be funded. In a large HCP this can take the form of mitigation fees paid into a fund to purchase habitat. The permit will be issued if the proposal will not jeopardize the survival and recovery of the listed creatures that live there, even though it may harm them in various ways.

In the first ten years, the few HPCs were mostly for small projects involving only a few acres. Small-scale HCPs were unattractive because of the expenses involved. Biological surveys are expensive, and so are many mitigation actions. On a small acreage mitigation options may be few or nonexistent. HCPs were unattractive to many owners of large tracts of land because they feared that if they went to the trouble of developing one, no sooner would it be approved than some new animal on their land would be listed and they would have to start all over again. This fear is quite realistic in some areas because rare species are concentrated in "hot spots." As habitat is reduced, one after another becomes threatened with extinction. For large landowners, the uncertainty was the primary obstacle to using HCPs.

All that changed in 1994. Secretary of the Interior Bruce Babbitt began advocating HCPs for very large areas, covering many property owners and a multitude of development projects. He also introduced a new policy called No Surprises, which made HCPs instantly attractive to public and private landowners alike.[74]

Today, many HCPs involve only single owners, but some of them have extensive lands. Several timber companies in the Pacific Northwest have adopted HCPs for managing their forests, and so has the state of Washington for its commercial timberland. But many HCPs now cover large territories with hundreds of thousands of acres owned by a complex mix of private individuals, companies, cities, counties, and state and federal governments. With so many entities involved, the costs of developing the plan are spread widely and public lands can be used as the core of conservation areas for the endangered species.

Today's HCPs cover all listed species in the planning region. They also cover all species in the region that are not now threatened but might be sometime in the future. That is, an HCP can now specify what will be done to preserve all species that have limited or narrow habitat requirements.

Here is where the No Surprises policy comes into play. An HCP that protects habitat for all of these creatures is a done deal. The federal government promises that no additional requirements will be made of any property owner who is part of the plan and is carrying it out. If an unlisted species covered by the plan becomes endangered, the property owners will not have any new burdens because of it. If a listed species declines to the point that extra protection is needed, no new burdens will be placed on property owners without their consent. Any additional expenses will be paid by the federal government. This *certainty* is the key that has gotten landowners interested and involved.[75] (Participation in an HCP is voluntary. But a landowner in the planning area who does not participate is individually subject to ESA regulations, without any of the benefits of the plan.)

Seen as a whole, the HCP process today comes as close to ecosystem management as is possible under the ESA. It is not preventive action for creatures that are declining; something has to be listed already for an HCP to be developed. But the wide scale of current plans may well prevent many other species that live there, including plants, from becoming threatened. Nor does an HCP require any positive actions to help listed creatures recover. Nevertheless, many environmentalists see HCPs in concept as the best tool available to preserve rapidly vanishing habitat and the living things that depend on it.[76]

Now we will take a look at two current examples of HCPs. Plum Creek Timber Company is now operating under an extensive HCP. And in southern California, we find a cutting edge plan under a state law. The Plum Creek Timber Company for years was one of the worst, if not *the* worst, timber company in the Pacific Northwest. Idaho governor Cecil Andrus said in 1987 that the company's "only concern was cashing out their equity at everyone else's expense."[77] The company was liquidating its timber at completely unsustainable rates, and it likewise liquidated small timber-dependent communities in Washington and Montana that had long survived on sustained yield forestry. The company was taking a beating with bad publicity from state governors, members of Congress, and even the *Wall Street Journal*.[78] And then it ran into spotted owl trouble.

Oliver Houck describes the company's turnaround: it has adopted an HCP that is on the leading edge of species protection and timber management. The HCP covers 400,000 acres of spotted owl territory in Washington, almost half of which is company land. Most of the rest is in national forests. The plan designates reserves on both public and private land that are set aside for four indicator species (not just the owl), and it protects streams with buffer areas. It designates areas where logging will be unrestricted and other areas where it will be limited on longer rotations. The HCP lasts for fifty years, with a provision extending it for another half century.[79] Whether or not the many plants and animals that need late successional or old-growth forests will find it sufficient will not be known for many years. But this is the kind of ecosystem protection they are getting under today's Endangered Species Act.

The situation in southern California is much more complicated. The coastal sage scrub ecosystem is a prime example of a "hot spot"—an area with a rich diversity of specialized species that is under intense pressure from development. Only 10–20 percent of the ecosystem remains, and it is highly fragmented. Many of the native plants and wildlife are in steep decline. Governor Pete Wilson feared a major "train wreck" if any of them were to be listed as endangered either by the FWS or by the state. (California has an endangered species act very similar to the federal one.)

In 1991, Wilson proposed and the legislature adopted a system for natural community conservation planning (NCCP). This is a positive and proactive approach to protecting entire ecosystems before their inhabitants become endangered. It is an attempt to avoid the rigidity of endangered species regulation.

The first application was in Orange County, covering some 340,000 acres. Riverside County now has a habitat management plan covering 260,000 acres. Los Angeles and San Bernardino Counties are also developing less extensive plans, and a much bigger one is being prepared for San Diego County.

Under NCCP, all landowners and all levels of government in the planning area can volunteer to participate. (If they do not, and something on their property gets listed, then they are directly subject to endangered species regulations, both state and federal.) An independent scientific review panel studied the area and advised the negotiators on what was required for the native species to survive and the ecosystem to function. In Orange County, the parties to the agreement created a 38,000-acre nature preserve, to which the Irvine Company, the nation's largest landowner and developer, donated 21,000 acres. (The company sees the NCCP as a much more promising alternative to the traditional ESA regulations.) A second plan in the county is being negotiated that would preserve 40,000 acres.

The NCCP process in San Diego County is much more complicated. The area is larger (1.3 million acres) and there are 20,000 property owners, whereas in the first Orange County plan a single company (the Irvine Company) was the primary private stakeholder.[80]

One advantage of the NCCP is that it can be used to avoid endangered species sanctions. It involves all of the property owners and all state and local levels of government in the negotiations from the very beginning, instead of having prohibitions come down from the FWS. (But the NCCP process still requires the ESA's presence to motivate the negotiators to get serious. In 1993, the FWS listed the California gnatcatcher as a threatened species. It chose that classification because it gives greater latitude and flexibility to planners than it would if the bird had been listed as endangered. That listing spurred the parties to complete the plan.)[81] The NCCP ecosystem management agreement then becomes incorporated into the normal local and state zoning and planning processes. The federal government has agreed to approve the NCCP plans as acceptable HCPs, complete with No Surprises guarantees.

Douglas Wheeler, California's secretary for resources, says that NCCP

> fundamentally changes the orientation of wildlife protection from single-species, piecemeal conservation in California to a broader, more integrated approach centered on ecosystems and entire habitats. The

> extensive participation of local government in the NCCP program elevates consideration of regional habitat protection priorities in local planning. . . . Existing public land has been integrated into and made a major component of regional habitat protection strategies. For instance, the Bureau of Land Management and the California Department of Parks and Recreation have committed their lands to the purposes of the NCCP program. . . . In addition, the state has pioneered a new policy of "conservation banking" that offers mitigation alternatives to advance regional conservation priorities. . . . Governor Wilson has created a process for reconciling effective long-term habitat protection with the interests of private landowners and locally controlled land use.[82]

Two other new policies must also be considered because they affect how the ESA is applied today. Small landowners have had problems with the ESA because of their limited resources. Consequently, today's regulations offer some help by excluding them from regulations to protect threatened species. Lots as large as five acres can be developed without an incidental take permit. (The exemption does not apply for endangered species.)

Another new regulation, called Safe Harbors, has been adopted to encourage private owners to enhance habitat on their land. The ESA has no requirements for landowners to take any actions to *help* endangered species; it only prohibits harming them. However, many landowners would be willing to make habitat improvements on their land, out of love for nature, appreciation for its increased amenity values, a desire to profit from such things as charging fees for hunting, and the like. But if some endangered bird or mammal then took up residence on their land, they would suddenly become subject to ESA regulations. The ESA, in other words, long discouraged private owners from making any habitat improvements on their land.

Safe Harbors fixes this problem. A landowner can now obtain a Safe Harbors agreement with the FWS that determines the "baseline" condition of habitat quality and presence of listed species on the land. The owner can then improve the land's habitat and is permitted at any time later to make changes that return the land to its baseline condition without violating the ESA, even if many more endangered creatures are living there.

This provision solves the problem illustrated in one of the believable stories often told about the ESA. An Oregon rancher who is also a conservationist created a lake and riparian habitat on his ranch. A wide variety of wildlife was attracted to the lake, including some bald eagles. He was then restricted in what he could do on his land because of the presence of the eagles.[83] This would not happen today because the rancher would negotiate a Safe Harbor agreement before creating the lake. He could then do anything he wanted, including draining the lake and returning the land to its original baseline condition.[84]

PROBLEMS WITH THE ENDANGERED SPECIES ACT

Despite recent improvements, there are significant problems with the Endangered Species Act. Some of these problems contribute to the furious reaction the act arouses in many. But it must be stressed that *any charges* leveled against the ESA which were made *before 1994* need to be completely reexamined, even if they turn out to be true. The new policies have fixed a substantial number of earlier complaints. "Many observers believe that the 'new' ESA that is emerging is better for species, better for business, and simply a better way to govern."[85]

But problems remain, both for biodiversity and for private landowners. The ESA has significant limitations as a mechanism for preserving biodiversity. Three widely recognized problems are structural, and three are matters of implementation. First, the ESA is limited to protecting individual species. Although the act says that its goal is to preserve ecosystems, its legal mechanisms only apply to individual species. The FWS can list the gnatcatcher; it cannot list coastal sage scrub. The use of large-scale HCPs helps to solve this problem, but it can only go so far. That is why California adopted an ecosystem protection law.

Second, the ESA only comes into play when a plant or an animal has already declined to the point of crisis. The ESA operates as an emergency room for species that have come to the brink of extinction, when there are usually few options left for saving them. The time when there would be a lot of flexibility in designing protection plans has long since passed. That is why listing a species can create so much conflict.

FWS negotiates voluntary "candidate conservation agreements" to protect species that are declining but not yet threatened. These are like HCPs but are intended to prevent species from having to be listed. However, there is no explicit provision in the law for these agreements, and they still focus on individual species.[86]

Third, the ESA is not proactive. It gives no authority for the government to take measures to enhance habitat or improve ecosystems to help declining species before they are in crisis. Nor are there any provisions for improving habitat on private land. Safe Harbors help those who, on their own, want to do that. But no one is required, or even encouraged, under the ESA to take any affirmative action.

Fourth, the way in which the agencies have interpreted the act, and the regulations they use to implement it, often give protection against final extinction only. The regulations for HCPs, for example, explicitly state that they do not have to help endangered species, just prevent their extinction.[87] And for governmental projects, the agencies merely have to avoid jeopardizing listed plants and animals.

Fifth, although some HCPs and recovery plans are carefully developed on the basis of sound biological evidence, many are not. A review by Bruce Bingham and Barry Noon of HCPs for the northern spotted owl and other species in the owl's region concludes that "mitigation solutions are often arbitrary, lacking an empirical foundation in the species' life history requirements."[88] Another review of

HCPs in the Pacific Northwest by Daniel Hall finds that a few "have incorporated specific habitat conservation programs" but that many "suffer from fundamental problems" and do not merit the No Surprises guarantee.[89] For HCPs in the Southeast to protect the red cockaded woodpecker, FWS established habitat standards that are less than a third of the size contained in the bird's recovery plan. And the recovery plan had already been found deficient![90] Defenders of Wildlife conducted a detailed analysis of two dozen representative HCPs and found numerous problems. Some plans are excellent but many are extremely weak.[91]

Likewise, Timothy Tear and his colleagues, after examining recovery plans, conclude that even if they were carried out completely, they would not improve the level of endangerment for nearly three-fourths of the animals involved.[92] Results of another study of theirs are similar. "Recovery goals have often been set that risk extinction rather than ensure survival." Many of the recovery plans even have recovery goals that are lower than the existing population size! They conclude that a fourth to a third of the threatened and endangered species "are being managed for extinction."[93]

Finally, Congress has never provided more than token funding for endangered species. The act says it is important to protect our biological heritage, but at appropriation time it never happens. As noted earlier, residents of our nation's capital spend more on pizza each year than Congress spends on endangered species. Thus we see that the ESA is significantly limited as a tool for preserving biodiversity. Conservatives should support more effective means to protect all of the wide range of life in creation.

There are also real problems in the act for private landowners who find a rare creature living on their property. The ESA is all stick and no carrot and conservatives should work to correct this problem. If a landowner wants to build on his or her property and the project runs into a conflict with an endangered species, the entire burden of protecting it falls on the owner. He or she must (1)redesign the project to avoid harming the creature and its habitat, (2) give up the opportunity to use the land as he or she wishes, or (3) obtain an incidental take permit and pay for the mitigation costs. Widescale HCPs help quite a bit because they spread the costs of biological surveys and preservation areas among many owners, public and private. Large HCPs usually use public land for preserve core areas, and governmental grants can be obtained to cover some of the costs of biological surveys and planning expenses. Sometimes private developers recover the opportunity costs of preserving part of their land because the reserved open space increases the value of the lots on which they build. "In southern California, developers frequently dedicate canyons that are hard to develop and that provide added value as open space to future purchasers. This has been mutually beneficial because these areas provide the most needed habitat."[94]

Nevertheless, even with a very large HCP, the burden of protecting wildlife falls on a relatively few landowners, while the entire public benefits from preserving native plants and animals and their ecosystems. And the problem these landowners are paying to solve was not created by themselves, but by earlier developers who benefited by reducing natural areas so much that rare creatures are now facing crisis.[95]

As a consequence, the ESA produces perverse incentives for landowners. Endangered species are liabilities for them, not assets. Landowners now have incentives to destroy habitat, for fear that some rare animal might take up residence. And if it does, the reaction may be to "shoot, shovel and shut up." Anecdotal evidence suggests that this may be of some significance. A Texas company hired migrant workers to destroy hundreds of acres of golden-cheeked warbler habitat just ten days before the bird was listed.[96] Larry McKinney, of the Texas Parks and Wildlife Department, says, "While I have no hard evidence to prove it, I am convinced that more habitat for the black-capped vireo, and especially the golden-cheeked warbler, has been lost in those areas of Texas since the listing of these birds than would have been lost without the ESA at all."[97] And there were reports in California of "midnight bulldozing" just before the deadline for listing the gnatcatcher.[98]

The extent to which this occurs is, of course, unknown. Patrick Parenteau believes it is small.[99] John Turner, director of FWS in the Bush administration, writes that during his time at the agency "we forged thousands of voluntary partnerships with farmers, ranchers, developers and timber companies to protect listed species, wetlands and other habitat."[100] And that was before No Surprises and Safe Harbors.

Clearly, for many property owners, the ESA turns endangered animals into liabilities. Since so many endangered species depend on private land, the goal of reform is to reduce the liabilities and, if possible, turn them into assets.

Because a burden often falls on private owners who are unlucky enough to have endangered animals on their property, the charge is sometimes made that the restrictions placed on them constitute a "taking" which, if indeed it is, the Fifth Amendment to the Constitution says should be compensated by the government. Of course, any land-use regulation or zoning restriction and many other normal governmental actions will affect the value of property, increasing some and decreasing others. Very few are considered "takings." As Zygmunt Plater points out, "Private property traditionally must absorb some public value burdens so long as they don't go 'too far.'"[101] Justice Holmes observed that "government hardly could go on if to some extent values incident to property could not be diminished without paying for every such change in the general law."[102]

The courts have never established hard-and-fast rules for what will constitute a taking protected by the Constitution. Rather, they consider each case separately, following some general guidelines. It is clear that land-use regulations under the ESA would almost never be so severe as to invoke constitutional protection.[103] In fact, at least as of 1998, there has never been a single taking compensation awarded to anyone because of the ESA.[104] Protecting habitat does not mean people cannot use the land; it means our uses are constrained by the needs of the creatures.

We need to put to rest once and for all the objection that preserving endangered species on private land is a taking. However, that does not mean that we should not reform the ESA as it impacts private property. As Patrick Parenteau notes, "Government restrictions need not rise to the level of a constitutional violation in order to work a hardship on landowners. It may well be that, as a policy matter, it would be fairer, and more politic, to spread the costs of habitat conservation more broadly."[105] That is the real basis for reform of the act.

One more major fault with the current approach to preserving biodiversity is that federal policies are often working at cross-purposes. Numerous federal subsidies perversely encourage behavior that destroys habitat and endangers plants and animals.[106] The ESA then steps in to save them but has no mechanism to block the subsidies that are doing so much harm. The taxpayers suffer a double burden—we are first paying for the subsidies and then paying to counteract their consequences.

Several studies have looked at different subsidies and their impacts on biodiversity. A good example is by Elizabeth Losos and colleagues, who focus on subsidized extractive industries and recreation on federal land as well as subsidized water projects. At the time of the study, 62 percent of listed species

> were known to be threatened at least in part from grazing, hard-rock mining, logging, water development and/or recreation. When potential threats were included, the number rose to 68 percent. . . . Water development affected the greatest number of species, harming 29–33 percent of all listed species. The second most widespread cause was recreation, affecting 23–26 percent. . . . The most destructive recreational practices were off-road vehicle use . . . Grazing contributed to the endangerment of 19–22 percent, logging affected 14–17 percent, and hard-rock mining affected 4–6 percent.[107]

Another study was done by Jeff Opperman for the Thoreau Institute. He concludes that 40 percent of the threats to listed species are federal programs or are federally subsidized. Fifty-eight percent of listed species are at least partly threatened by federal activities; 12 percent are primarily threatened by federal activities, and another 11 percent are primarily threatened by subsidized activities. Agriculture is the main problem, especially grazing on public land. Water projects are the next most serious threats. Recreation threatens more species than mining and timber combined. Although off-road vehicles are the major problem within the category of recreation, hiking threatens twenty-five species.[108]

Opperman features the Everglades, with fifty-six listed species, as a prime example of the destructive effect of subsidies. The taxpayers first paid for the federal government to construct drainage projects to create the Everglades Agricultural Area. This reduced the Everglades ecosystem by over half. The largest crop in the area is sugarcane. But sugar is only profitable there because the federal government limits imports. The public therefore pays a second time, since the quotas keep the price of sugar in the United States two to three times higher than the world level.

Moreover, the sugar growers have wrought havoc with what remains of the Everglades. The flow of water has been drastically altered, and large quantities of fertilizers from the fields and mercury from burning the cane after harvest drain into the Everglades, degrading large areas that were once "a river of grass" so that only cattails will grow there. The ecosystem is in trouble, and its wildlife has been decimated.[109] In Chapter 9 we will take a look at the current program to restore the natural flow of water and its quality. Of course, we taxpayers are paying for that too. In sum, many different kinds of subsidies have destroyed wildlife habitat that would otherwise have been left in its natural state. But the ESA has no power to block these subsidies when they put living things at risk of extinction.

With all these flaws, it is amazing that the ESA works as well as it does. And before we turn to some conservative solutions, we need to realize that *the ESA does indeed work*. The charge is sometimes made that the act has simply failed. Its goal is to recover species, but very few have ever recovered and qualified for delisting. Many listed species continue to decline. Both assertions are true, but neither proves that the act does not work. As David Wilcove says, it is "absurd to expect a [twenty-seven] year old law to undo the impact of three centuries of industrial and agricultural development on our imperiled flora and fauna."[110]

We need to remember that the ESA is like an emergency room that only comes into play when a plant or an animal is on the very brink of extinction. We can hope that now, using as much of an ecosystem approach as can be squeezed out of the act, the FWS and NMFS will be able to help these creatures survive and keep others away from the brink. Nevertheless, the agencies have had their successes.

There are many examples of species that have come back, even if most are not yet thriving. Whooping crane numbers have increased over the years. Condors once again fly over the desert. The sea otter population has increased and alligators are thriving. Bald eagles, peregrine falcons, and brown pelicans have done well. Charges have been made to the effect that the eagles and falcons were not saved by the ESA but by a ban on DDT, which made egg shells so fragile that few hatched. Banning DDT removed the cause of their decline, of course, but the ESA protected the necessary habitat for them to recover and supported captive breeding and release programs. The act should get some of the credit. Of the seventy-eight species originally listed as endangered in 1967 under the predecessor of the ESA, forty-four have recovered, are well on the road to recovery, or are stable because of the act.[111]

Jeffrey Rachlinski conducted a more sophisticated statistical study of the species listed under the ESA. His findings indicate that, although the act has flaws, it has clearly worked to preserve dangerously declining species. Many listed species are declining, but most are declining when they are listed, so just looking at the numbers at one point in time tells little. Rachlinski's study compares the status of listed species over time. Comparing species listed for less than one year with those listed for more than a year, he finds that the longer species are listed, the better is their status. Of those listed for more than a year, a smaller percentage is declining, and larger percentages are stable and improving. Of those that were declining when listed, fewer were declining at the end point of his study, and the percentages that were stable and improving at the end are much larger than they were when the species were listed.[112] Plotting the numbers in decline against the number of years they have been listed shows a distinct trend toward improvement the longer they have been under the protection of the act.[113] Rachlinski concludes, "The data clearly demonstrate that endangered and threatened species are better off with the Act than they would be without it. . . . Listing appears to have turned the fortunes of about half of the species it protects. . . . Most remain in decline for the first year after listing, but as time passes, species populations stabilize and even improve. Each year of protection under the Act improves the prospects for listed species."[114]

He also finds that "recovery plans clearly benefitted species. . . . In fact, recovery plans appeared to be the primary mechanism that set species on the road to recovery. Virtually all improving species had recovery plans. Listing and designating critical habitat stabilized species populations, and recovery plans facilitated improvement."[115] "In sum," he concludes, "the Act saves species."[116]

The ESA also saves ecosystems:

> But for the ESA, the old growth forests of the Pacific Northwest would be all but gone by now; but for the ESA, there would be no agreement to restore water quality in the Sacramento-San Joaquin Delta; but for the ESA, there would be no NCCP in Southern California seeking to conserve the last of the coastal sage scrub community; but for the ESA, there would be no Balcones Canyonlands HCP trying to save some of the natural areas around Austin, Texas; . . . but for the ESA, the few patches of long-leaf pine forest habitat of the red cockaded woodpecker would be even smaller and more isolated.[117]

The ESA has been much more successful in protecting species on federal land than on private land. "For listed plants and animals found entirely on federal land, approximately 18 percent are judged to be improving; and the ratio of declining species to improving species is approximately 1.5 to 1. In contrast, for species found entirely on private property (excluding property owned by non-profit conservation groups), only 3 percent are improving, and the ratio of declining species to improving species is 9 to 1."[118]

Clearly, the ESA has flaws and needs to be reformed. But, equally clearly, it has not been an abject failure, to be dumped overboard entirely. As Hank Fischer says, "Society is just taking its first wobbly steps toward devising a system that will prevent extinction of the varied life forms on earth. We are still investigating new techniques and exploring innovative approaches for making endangered species recovery more successful and more acceptable to all citizens."[119] So let us see what conservative principles can contribute to that investigation of new techniques and innovative approaches.

PROTECTING BIODIVERSITY: CONSERVATIVE SOLUTIONS

The most obvious private action a conservative should take is to support land trusts that protect endangered species. Land trusts deal with willing sellers and donors to acquire land or conservation easements in order to prevent development. There are hundreds of land trusts across the country, and their missions are not all the same. Some preserve open space, regardless of its habitat values; others acquire development rights on farmland in order to keep it in agricultural production and prevent it from being turned into suburbs or malls. Any of those may help endangered species. But the largest land trust of all is The Nature Conservancy, which focuses entirely on preserving habitat for rare plants and animals. It has preserved over 12 million acres in the United States, an area larger than the original national wilderness system. It does not lobby in Congress. It does not file lawsuits. It simply negotiates with owners of critical habitat who are willing to sell their land or development rights in order to preserve it. The Nature Conservancy is now one of the largest environmental groups in the country, with over a million members.[120]

Private action is very important, and the more of it there is the less need there will be for regulation. But for saving biodiversity public policy is also required. And before we examine specific policies that conservatives should consider, we need to be clear at the outset that, whether we like it or not, the federal government will have to retain a substantial amount of regulatory power. The reason is simple: many plants and animals have very specialized habitat requirements. To preserve them, their specific habitats need to be preserved, wherever they happen to be found. We humans are flexible, but many species are not.[121] We do not *have* to put those proposed condominiums precisely *there*, where some animal has its niche; we can put them elsewhere. For that matter, we probably do not need any more suburbs or malls or golf courses at all, sprawling out into the countryside. (In the current election campaign, politicians of both parties are making an issue of stopping urban sprawl and it is a powerful local issue in many places.)[122] In order to be able to preserve specialized habitat, the government will have to be able to use its regulatory power in many cases. One goal of reform is to make the impact of this regulatory power more fair.

So what kinds of policies should conservatives support in order to do Noah's job today? There is no single solution to the problem of preserving biodiversity because the needs of plants and animals are so very different and the areas where they are found are equally diverse, from mostly unpopulated desert to the fringes of booming metropolises. What I will suggest here is a menu of options that attempt to embody conservative principles. In protecting specific species or specific ecosystems, we should choose the appropriate ones—or devise others that likewise apply the principles of conservatism. In light of conservative piety toward nature and our duties as stewards of the earth, we conservatives are obligated to support *effective* policies to preserve God's creatures.

First, in creating habitat preserves for rare plants and animals, we should use public land as the core of the nature reserves. To the extent possible, we should avoid imposing on private owners. This is now being done. In the large HCPs, government land forms the first part of the preserves.[123] Unfortunately, since so many rare species rely on private land, we cannot stop there.

Second, conservatives should support the elimination of the many subsidies that are driving plants and animals to extinction.[124] Moving toward a free market in these cases would help preserve biodiversity. If we quit selling timber from national forests below cost, if we charge fair market value for grazing on public land and for the use of water from governmental water projects, we could go a long way toward helping biodiversity. Most national forests would be left intact, prairie land would not be overgrazed, and water would be left in the streams for the native fish.

At the very least, people and companies who now receive subsidies should lose them if their actions threaten rare species.[125] If a rancher cannot or will not keep his cows from trashing bull trout streams on public land, that rancher should forfeit all subsidies (cheap grazing allotments, emergency feed programs, animal damage control, etc.). If the sugar companies in Florida will not clean up their pollutants that are threatening the Everglades and causing its wildlife to decline, the import quotas should go. There is a precedent for such a policy. The old farm program contained a "swampbuster" provision, under which farmers who drained wetlands would lose their crop price supports.[126]

Third, the approach to protecting biodiversity needs to be expanded to the ecosystem level, beyond what can now be done under the ESA. If the FWS could negotiate conservation plans to protect the functioning of whole ecosystems before they are fragmented or destroyed, we could prevent plants and animals from becoming threatened and endangered. And we could do so in ways that minimize the impacts on private property. In other words, the federal government should add a policy similar to California's natural community conservation planning to the ESA. (Protection for individual species is still a prerequisite for stakeholders to get serious about an ecosystem plan, as California discovered in the coastal sage scrub planning process.)[127]

Focusing on ecosystems will also likely have a political advantage as well. "The public may be more receptive to some of the necessary decisions or sacrifices that may need to be made if these changes are to protect their regional ecosystem, rather than for the protection of an endangered, unknown or possibly uncharismatic salamander, fish, or rat."[128]

These are general policies conservatives should support. To deal with the specific problems of preservation on private land, conservatives should consider a whole range of policy options. David Wilcove and colleagues contend that "protecting rare animals and plants on private land is the greatest challenge for the Endangered Species Act."[129] I believe employing conservative principles can help us meet that challenge. Our goal is to be fair to private property owners and, wherever possible, devise incentives to turn endangered species into assets rather than liabilities.

In the last few years a lot of thought has gone into devising possible incentives that could be incorporated into our current policies.[130] Many interesting suggestions have been made that seem to apply conservative principles. Only a few of them may pertain to the situation of any single species or ecosystem, but if they were all available, they would give a wide range of options to choose from.

In a few cases, the government has arranged mutually acceptable land exchanges, trading federal property with little or no habitat value for critical habitat on private land. Some years ago the Interior Department gave a company 100 acres in downtown Phoenix in exchange for 100,000 acres in Florida's Big Cypress.[131] More recently, the government swapped some of its land in Nevada for 5,000 acres of private wetlands in the Everglades.[132] Land exchanges are fairly frequent, but usually for management reasons. They could possibly be increased, with the focus on preserving rare species.

One frequently offered suggestion is estate tax reform. Ranchers, for example, tend to be land rich and money poor. When the owner of a ranch dies, the family often has to divide the property and sell it for ranchettes or housing subdivisions in order to pay the taxes. The tax law results in fragmentation and development of habitat that the owners might prefer to keep intact. The incentives should be the opposite, so one proposal to consider is to exempt land from the tax if it is critical habitat. A variation of this policy would be to defer the estate taxes as long as the property is kept intact.[133]

Another possibility is to allow landowners to take tax credits for improving and managing critical habitat.[134] This would fit nicely with Safe Harbors agreements as an added incentive to help threatened and endangered species.

In or near expanding urban areas, where there is considerable economic pressure to build on habitat that needs to be preserved, transferable development rights (TDRs) could be incorporated into HCPs.[135] TDRs are now used by some local governments to control and direct density of development and preserve open space. Some areas are designated for low-density housing or open space, and property owners in those areas are restricted in their ability to build on their land. Instead, they are issued TDRs. When owners in designated "receiving" areas want to develop their land beyond the normal density allowed, they can do so by purchasing TDRs from owners in the restricted areas. The land that is restricted from development thus retains economic value for its owners.

TDRs could be adapted to habitat conservation in a regional plan. Consider two possibilities. In one case, say, the plants and animals to be protected simply need open space. (The particular needs of the endangered species will determine how the TDR system can be set up.) A biological survey determines that they can survive there if 70 percent of the remaining habitat is preserved, but it need not be in large blocks. In that case, each property owner could build on 30 percent of his or her land. Someone who wants to build on more of the land could do so by buying a TDR from another owner who agrees to build on less.

In a second case, the species to be protected need a large block of prime habitat preserved, but development could take place outside of it. Landowners whose property is in the area to be preserved could be issued TDRs, which owners outside of the preserve would have to purchase in order to build on their land.

TDRs can be modified in many ways. For example, someone who enhances habitat on land or donates a conservation easement on developable land to a land trust could be given extra TDRs to sell. The point of it all is to spread the cost of protecting habitat throughout the planning area, in a way that retains economic value even for private land that is to be preserved in its natural state.

The most noteworthy application of TDRs to preserve nature is in the Pinelands of New Jersey. The pine barrens of southern New Jersey were still remarkably preserved in the late 1970s. In 1978, the Pinelands National Reserve was created as a joint project by federal, state, and local governments, as well as private landowners, to protect it from encroaching development. A master plan divided the region into several zones: a relatively pristine core and a larger surrounding area with graded intensities of land uses. A development credit system was created to compensate landowners in the preservation core and in the other more restricted zones for rights lost to the zoning restrictions. These credits can be sold to developers in designated growth areas, which allows them to build additional housing units over the normal density limits (up to a maximum ceiling). The Pinelands National Reserve has been called the most successful regional land-use planning effort in the country. Manuel Lujan, secretary of the interior during the Bush administration, called it a model for protecting other places in the United States.[136] The model could be applied to HCPs to protect endangered species habitat.

Another policy to consider is "mitigation banking." The general concept here is for someone—an entrepreneur, a land trust, or a governmental agency—to produce new habitat for rare plants and animals. This could be done by acquiring degraded land and restoring it. It could involve restoring a wetland that had been drained or creating a new one. The owner then gets credits (similar to TDRs) for the new habitat, which developers purchase for mitigation of projects in the area that reduce similar habitat.

This concept has been used for wetlands under the Clean Water Act. The results have been mixed, but not because of any technical infeasibility of creating new habitat. The problems have been caused by poorly executed and administered programs, and by the lack of an established bank for developers to draw from.[137] A new bank has now been established in southern California for gnatcatcher habitat.

Encouraging new markets for the production and "sale" of habitat should appeal to conservatives. And the concept has much more potential for helping endangered species than has been realized so far.

A BIODIVERSITY TRUST FUND

In order to provide positive incentives to private property owners, we should have a source of revenue set apart for the purpose and removed from the political appropriation process. Randal O'Toole has long advocated establishing a "biodiversity trust fund," and the concept deserves careful consideration by conservatives.[138] Several sources of revenue could be used. The federal government loses hundreds of millions of dollars every year on subsidized uses of its lands (see Chapter 5). All users of federal land should pay fair market prices—and that includes hikers and campers, not just loggers and ranchers. O'Toole advocates devoting a portion of the revenue generated by usage fees to the biodiversity trust fund. That would be an excellent source of income for the purposes outlined here.

A federal real estate transfer tax, dedicated to the biodiversity trust fund, would be another source of income. The federal government had such a tax until 1968, and thus it would be nothing new.[139] A very modest tax would raise a significant amount of money for biodiversity. The reason for using a real estate transfer tax is to spread the costs of preserving species, over the long run, among those who have benefited from creating the habitat problem in the first place. The protection burdens now fall on developers who did not cause the threats; they were just unlucky enough to arrive on the scene after previous development had destroyed so much habitat that its native species are now rare.

A third possible source of income would be impact or mitigation fees paid by developers, to be spent in their planning areas. Mitigation fees are widely imposed on developers by local governments to cover the increased public costs imposed by new subdivisions, such as for sewer systems, parks, and schools. Typical HCPs today require developers to pay fees to be used to acquire and manage conservation areas. The developers, after all, are profiting from destroying habitat, so it is appropriate that they contribute something to reducing the harm they do. These fees, in other words, are internalizing an externality. And with the other sources of income for the trust fund spreading the costs more widely, mitigation fees in the future could probably be significantly lower than they are today.[140]

The dedicated revenues in the trust fund could pay for a very wide range of policies that would make endangered species assets for private property owners. FWS and NMFS could purchase conservation easements and buy land outright in order to preserve functioning ecosystems that are still intact. They could rent habitat from private owners through an addition to the Conservation Reserve Program. Currently, the Agriculture Department pays farmers to take marginal and erodable land out of production and put it into ten-year reserve contracts. They are to plant the reserved land with vegetation that will prevent erosion and provide wildlife habitat. The federal government could pay farmers extra to create specific kinds of habitat for endangered species on their CRP land, with longer term contracts.[141]

The agencies could target their land acquisition and protection measures in the relatively few "hot spots" where endangered species are concentrated.[142] As Dobson and colleagues conclude, "If conservation efforts and funds can be expanded in a few key areas, it should be possible to conserve endangered species with great efficiency."[143] A trust fund would give the agencies the means to do so.

FWS and NMFS could also use the money to buy out dams and remove them, where restoring free-flowing rivers would help declining fish or shellfish populations. They could offer rewards to landowners for helping endangered species on their land. For example, they could pay for costs of habitat restoration (e.g., to restore a wetland that had been drained) and for management expenses (such as controlled burns) and add a reward on top of it. They could pay bounties to landowners who participate in captive breeding and release programs, to build up populations of endangered species in the wild. They could develop policies similar to the Defenders of Wildlife program that pays bounties to ranchers who allow wolves to den and raise pups on their land.[144] The fund could pay for technical assistance and low-cost loans to help people profit from ecotourists who want to view rare animals on their land. For example, in Missouri "private landowners who have greater prairie chicken booming grounds on their land are promoting the watchability of this natural event. A bed and breakfast centered on the prairie chicken is rapidly becoming a successful seasonal venture."[145] The great interest in ecotourism should create a lot of opportunities here.

The agencies could buy TDRs and retire them, to preserve even more habitat in HCP regions. They could use the fund to pay for the surprises that will inevitably occur under No Surprises agreements. Additional expenses may be needed to protect plants and animals in case unforeseen circumstances develop, and under those agreements they must be paid by the government. And the fund could be used to improve habitat on public land as well (e.g., for reforestation on our national forests).

There are a myriad of ways in which a biodiversity trust fund could be used to protect endangered species and turn them into assets for private property owners. To the extent possible, we should try to get private property owners to become willing partners in the preservation and recovery of rare plants and animals.

However, the money should *not* be used to compensate landowners for ESA restrictions or HCP planning regulations that fall short of a constitutionally protected taking. That, as Robert Meltz contends, would be a very dangerous precedent to set, since ESA restrictions are not different in principle from other common limitations on the use of property, such as local zoning codes.[146]

Rather, the trust fund money should be available in programs that allow landowners to profit from *helping* endangered species. That is, if property owners choose to do nothing beyond obeying the land use restrictions, they can use their land within the limits of those restrictions but they will receive nothing from the trust fund. However, they would have many opportunities to take active steps to help the endangered species that would also be profitable. The incentives should be offered for actions that *benefit* biodiversity but not for merely doing nothing to put it at greater risk.

The kinds of incentives suggested here, as well as others that might be developed, will work well in conjunction with the federal regulatory powers, which must be retained.[147] However, with proper incentives, the regulatory power can be used with a much lighter hand, and in many cases may not have to be used at all, whereas today regulation is the only option available.

CONCLUSION

In obliterating many forms of life, we are causing devastation that Russell Kirk himself lamented, even before the Endangered Species Act was passed. "Rare, strange, and beautiful animals are shrinking toward extinction in much of the world." He then made a characteristic reference to literature, not biology, to drive home the significance of this. He referred to a story by C. S. Lewis in which a diabolical power seeks to overthrow God's universe. One of its first objectives is "the extirpation of the great diversity and charm of animal life on this planet." Kirk then noted that "preservation of the multitudinous animal species has been enjoined by religion since the dawn of human consciousness." Unfortunately, "we Americans have done our despicable share in decimating the animal kingdom."[148]

The situation since Kirk wrote has become much worse. As Michael Soulé, the "father of conservation biology," warns, the biosphere has never before "been under such savage attack."[149]

The ESA and its regulations form the legal structure we have created to try to reverse this trend. In the last quarter of a century several serious flaws in the act (ignore the imaginary ones) have come to light. The act is "a pioneering law that has revealed its limitations."[150]

It is long past time for conservatives to take Russell Kirk's comments seriously. We should take the lead in reforming our policies to make them more effective in preserving our biological heritage, and at the same time apply conservative principles to make them fairer for all stakeholders.

It is up to us to do Noah's job today, and to do it right. Piety demands it. Prudence requires it. And we owe nothing less to future generations.

9

Sustainability

MUCH HAS BEEN SAID IN RECENT YEARS about "sustainable development" and "sustainability." As defined by the Brundtland Commission, sustainable development is "development that meets the needs of the present without compromising the ability of future generations to meet their own needs."[1] The concept of sustainability is an eminently conservative idea. We have a duty to design our economy in such a way that we can produce our goods and services without impairing the ability of the planet to provide for future generations. Renewable resources, such as fisheries and forests and the soil, should be used no faster than they can be replenished. Nonrenewable resources, such as metals, petroleum, and, in some places, groundwater, should be used carefully with maximum provision for recycling whenever possible, and as we use them we should make provision for replacing them with substitutes.

At the present time, however, our economy is far from sustainable. As Russell Kirk observes, "In America especially, we live beyond our means by consuming the portion of posterity, insatiably devouring minerals and forests and the very soil, lowering the water-table, to gratify the appetites of the present tenants of the country."[2]

We are wasting enormous amounts of nonrenewable resources for merely trivial purposes. Most of our cars still get very poor gas mileage (and this is actually getting worse). In many places we are depleting groundwater that will take centuries to recharge just to water lawns and golf courses in the desert.

Even more inexcusable is our rapid depletion of renewable resources. As Amory Lovins's Rocky Mountain Institute warns, the limits we face first "may prove to be not the Earth's nonrenewable resources but its *renewable* ones. For . . . the Earth's supply of fish, forests, topsoil, and biodiversity is shriveling."[3]

We are not being good stewards of the earth. Conservatives need to take very seriously the words of economist Herman Daly: "There is something fundamentally wrong in treating the earth as if it were a business in liquidation."[4]

Sustainability has numerous implications for a conservative approach to environmental and resource policies. Russell Kirk gives us the basic guidelines: "Turning away from the furious depletion of natural resources, we ought to employ our techniques of efficiency in the interest of posterity, voluntarily conserving our land and our minerals and our forests and our water and our old towns and our countryside for the future partners in our contract of eternal society."[5]

For conservative policy, it is important to note that moving toward a free market—removing subsidies and internalizing externalities—would take us a long way toward sustainability in many areas. If drivers had to pay up front the full costs of driving, including paying for the damages caused by pollution and greenhouse gases, we would drive a lot less and we would buy more fuel-efficient vehicles. Emissions of pollutants and greenhouse gases would be much lower, much closer to the atmosphere's ability to absorb and process them safely. Likewise, if in our monthly utility bills we had to pay for the damages of pollution and greenhouse gases, we would use electricity much more efficiently, and renewables would already be cost-effective alternatives to coal. Emissions of pollutants and greenhouse gases would be much lower. If timber companies had to pay for all of the costs of logging on national forests, including all costs of reforestation and restoration, the harvesting of timber on those forests would have been much closer to sustainable levels.

The free market does not automatically assure sustainability, as we will soon see. Nevertheless, many of the policies to move us toward free markets would also move us toward sustainability—doubly desirable for conservatives.

There are many issues here that are important and even compelling—numerous ways in which our economy is not sustainable. But here I want to consider three examples of the problems we are creating for the future and the kinds of actions conservatives could advocate to correct them. In the collapse of fisheries all over the world we see a classic example of a "tragedy of the commons." Underlying our economy are numerous "ecosystem services" that are vital to our well-being but are not marketed or priced, so our economic system mindlessly degrades and destroys nature's ability to provide them. Finally, a task for this new century—already begun at the close of the last—is to restore, wherever possible, the damage we have caused to nature.

FISHERIES: A TRAGEDY OF THE COMMONS

A situation in which renewable resources are quickly depleted is known as a tragedy of the commons, as described in a famous essay by Garrett Hardin.[6] When there are no private property rights in resources, when they are owned in common, the natural incentives result in the overuse and rapid elimination of the resources. Consider a pasture that is owned in common by the villagers who graze their cattle on it. If a villager adds one more cow to her herd, she gets all of the benefits of raising that extra cow. The effects of overgrazing, however, are shared among all of the other villagers. If she refrains from adding the extra cow, she loses all of its benefits and someone else will likely add a cow and get the benefits.

So each person has the incentive to add one more cow, and another, and another. And pretty soon the pasture has been destroyed by massive overgrazing. This is precisely what happened on the open range in the American West during the late nineteenth century, and much of our rangeland still has not recovered.

A tragedy, of course, is not what normally happens with a commons. If they can, the villagers will develop a community system of managing the pasture. Families will have traditional rights to graze specific numbers of cows and no more, and the resource will be protected for generations. The system will be enforced by social sanctions and defended against outsiders by violence if necessary.[7]

But traditional community management can break down for many reasons, leaving a commons open for tragedy. Our capitalistic system is hostile to traditional governance in general.[8] Community arrangements for fisheries, for example, have been thrown out by the courts as violations of antitrust laws.[9] In that case, a conservative approach to protect the resources would be to establish private property rights in them and a market in which those rights could be bought and sold. In the case of the village pasture, its grazing capacity would be determined. Each family would then get a quota—the right to graze a specific number of cows totaling no more than the pasture can support. The quotas could be bought and sold, so that individuals could increase their herds by buying grazing rights; others could quit raising cattle or reduce their herds and sell their rights. By establishing property rights, the pasture is protected and used in a sustainable manner. John Gray contends that "the central ecological function of market institutions is in the avoidance of the tragedy of the commons."[10]

The prime example of a modern tragedy of the commons is the collapse of fisheries all over the world. "No one 'owns' a fish or a whale until he has captured it himself. On the other hand, the long-term costs of overexploitation, in terms of reduced stocks and losses of productivity, are spread among all exploiters. Acting rationally from his own viewpoint, therefore, each individual exploiter is motivated to deplete the resource. From the viewpoint of society at large, however, such depletion is almost always undesirable."[11]

That is precisely what is happening to fisheries all over the world, including those in U.S. waters. Nine of the world's seventeen major fishing grounds are in "precipitous decline" and four are "fished out" commercially. In the northwest Atlantic, total catch has fallen by one-third in the past twenty years. Canada's Grand Banks, once one of the world's most productive fisheries, has been closed indefinitely. According to the Food and Agriculture Organization, some 70 percent of global fish stocks are "depleted" or "almost depleted."[12] "In a single generation, overfishing and poor management have devastated fish stocks . . . around the world . . . putting an estimated 100,000 fishers out of work and threatening the food supply of millions in developing countries."[13]

American fisheries are likewise in trouble. "Up and down the coasts of North America, once-productive fishing grounds are yielding less every year."[14] Some 40 percent of 200 fish stocks off the U.S. coast are being overharvested.[15] For generations, the Georges Banks was one of the most productive of all our fisheries. Overfishing for just a decade or so destroyed it, and it was officially closed in 1994.[16] In 1999, even more fishing areas in the region had to be closed.[17]

Open access to fisheries has produced these tragedies of the commons. Most attempts at management have failed, in large part because they have not confronted the nature of open access. Several policies have been tried. In some cases governments have attempted to limit the total catch. They have also tried to limit the number of boats by licensing them and to limit the number of days open for fishing. The results are political pressure to keep the catch limits too high and mad races when the season opens as each fisher tries to catch as many fish as possible in the first few days. Moreover, the size of the boats keeps increasing, since each fisher wants to be able to get as much of the catch as possible. Overcapitalization of the industry is a problem worldwide—the fishing fleet is now more than twice as large as needed for what the ocean can sustainably produce.[18]

With all this economic inefficiency, ever increasing pressure on the stocks of fish, and declining catches in many places, the fishing industry has become dependent on massive subsidies from governments. Worldwide, annual subsidies amount to around $20 billion.[19] Fishing is now one of the most heavily subsidized industries in the world.[20]

In the United States, Congress subsidized the expansion of the U.S. fleet after it extended our territorial waters to 200 miles and excluded foreign vessels.[21] Now, "federally funded support for hunting previously untargeted forms of marine biodiversity in New England . . . threatens the populations of a growing list of marine species."[22]

The fisheries within the territorial waters of the United States are governed by a system of regional councils. This system is very unusual because there is no limitation on conflict of interest and although they are under the Department of Commerce, the secretary normally must go along with their decisions. Some of these councils, as the *Economist* puts it, have been "hijacked" by the fishing industry and consist largely of industry representatives.[23] They get advice from scientists on the conditions of fish stocks and the levels of harvest that would be sustainable, but the political pressures on the councils are almost always to keep the catch levels far too high.

The results have been decidedly mixed. Some councils have followed sound scientific advice and others have not. The major disasters in our fisheries have been off the northeast coast, since the New England Council and fishers in the region "routinely ignored" scientific advice.[24] That council has been "'incredibly irresponsible and stupid' for allowing persistent overfishing . . . despite a decade of warnings from fisheries biologists." The North Pacific Council, by contrast, has largely followed its scientists' advice and its fish populations "are still in good shape."[25] As Carl Safina observes, "The bottom line is that in fisheries where people have paid attention to the scientific recommendations, there are still fish around. In fisheries where the scientists have routinely been ignored or the most optimistic gloss has been put on the data, we have declines."[26]

Although some fish stocks will never recover, there is encouraging evidence that many can be restored if fishing is reduced or stopped for several years.[27] When the Barents Sea cod stocks began to collapse, Norway and Russia cut their catches dramatically and instituted a rebuilding program. The cod population rebounded and the fishery is back.[28] The same happened with striped bass off the eastern coast of the United States.[29] If stocks recovered, the fishing industry would, of course, benefit in the long run. "If depleted species were allowed to rebuild to their long-term potential, then sustainable use would add about $8 billion to the U.S. gross domestic product—and provide some 300,000 jobs."[30]

A conservative solution to this tragedy of the commons is to eliminate open access to fisheries and establish property rights in them. These rights would take the form of individual tradable quotas (ITQs) that would give their owners rights to set portions of the annual catch from a specified fishery. Since fish populations naturally fluctuate considerably, the ITQs could not be permanent rights to a specific amount, but rather to a set percentage of the allowable catch. The allowable catch should be determined annually by a scientific board for each fishery. The quotas would presumably be distributed initially based on each fisher's historic portion of the catch. Once distributed, they would be fully marketable. Someone who wants to expand his or her operation could purchase a quota from another fisher who wants to scale back or leave the industry. Someone who wants to enter the industry would have to buy a quota from someone already in that fishery. As Colin Clark contends, "Allocated quotas appear to constitute the only feasible method for dividing up property rights to the fugitive resources of the ocean."[31]

ITQs are currently being used around the world, most prominently in New Zealand. Great Britain is allocating its national quota from the European Union by issuing ITQs. And in the United States, ITQs are being used to manage a few species in a few places.

Once property rights are established, in the form of quotas, there is no more reason for overcapitalization in the industry. No longer can an individual maximize his or her catch by getting the biggest and fastest boat. Once the excess capitalization is gone, the fisheries will yield greater economic returns. In two U.S. fisheries that use ITQs the number of vessels has decreased by more than 50 percent. In British Columbia, the economic return in a fishery that adopted ITQs has gone up 65 percent.[32]

A quota system can also reduce political pressure to set catch limits too high. Quota holders have a guaranteed portion of the catch, so they have a continuing interest in maintaining the stock of fish. "Percentage quotas give them strong incentives to support good management, research and enhancement."[33]

The quota system can also protect the economic interests of fishers even when the allowable catch has to be reduced. The *Economist* reports that the European Union may have to cut national quotas because of declining fish populations. In Britain, which allocates its national catch as ITQs, "this could be quite a boon for fishermen, for although their income might fall because they are able to catch less fish, the value of the quota could well rise."[34]

There is even some evidence that an ITQ system can reduce poaching. In European countries where catch limits are not marketable "there is always the temptation to land 'black' fish—which exceed quota limits. But in Britain, rather than catching fish illegally, a fisherman can go and buy a quota from an owner who thinks he will not catch his quota."[35]

ITQs are not perfect, however, and they can present some problems of their own that policymakers need to consider when designing a fish management policy. Political pressure can still be a problem. When New Zealand first adopted its quota system, the large fishing companies pressured the government into setting the catch limits far too high. Counterpressure from environmentalists and smaller fishers was required to get the limits reduced to reasonable levels.[36] And ITQs can become monopolized unless there is a limit on the number one company and its subsidiaries can own.[37]

To keep a fishery sustainable, an ITQ program needs to include several elements. The history of regional councils in the United States shows that the total catch to be allowed each year needs to be set by an independent board of fisheries biologists. The scientists cannot merely be advisers, as they are now. And the mandate for the boards should be to set limits on the conservative side in order to guarantee the long-term sustainability of each fishery.[38] They should also have authority to restrict gear and types of fishing to protect the fishery. For example, they may need to set limits that let juveniles escape to mature and reproduce.[39] They may need to limit trawling because it causes massive destruction to the sea floor and its ecosystems.[40]

If a council or the industry wants to challenge a scientific decision on the total catch to be allowed or gear that can be used, the burden of proof should be on them to demonstrate that a larger catch will not affect the sustainability of that stock of fish.[41]

And, of course, once all the scientific and political decisions have been made, enforcement is the final key. Property rights in the quotas are worth little if enforcement of the system is weak. New Zealand, for example, had serious problems with a black market. Up to 80 percent of all fish sold in the country were coming in through the black market, thus evading the quota limits.[42]

Another policy conservatives should consider, to supplement an ITQ system, to help rebuild depleted stocks, and to help maintain viable fish populations, is to establish a system of marine preserves. These are designated areas within which no fishing is allowed. Where they have been tried, fish stocks rebounded inside the preserves and fishing improved outside of them. A review of the available evidence found that "eighty-six percent of the studies that tested fishery yields found that catches within three kilometers of the marine protected areas were 46 to 50 percent higher than before no-take zones were created."[43]

Finally, until policies are in place to keep fisheries sustainable, there is a pure free market approach that conservatives should surely support—to buy only fish that is certified to come from a sustainable fishery. The Marine Stewardship Council certifies fisheries that are well managed, maintain healthy stocks, and preserve the surrounding ecosystem.[44] Already, several American companies have signed up to buy only fish that have the council's seal of approval.[45] We can support sustainable fisheries with our own buying power.

NATURE'S SERVICES

Our economic well-being depends on many "ecosystem services" that nature provides for free. Our economy, in fact, is wholly contained within the ecosystem, so "a healthy economy can only exist in symbiosis with a healthy ecology."[46] This has too often been ignored. When the human enterprise and the human population were much smaller, this did not much matter. But that is no longer the case and today we must realize that the economy and the ecosystem "are so interdependent that isolating them for academic purposes has led to distortions and poor management."[47] This is now becoming more widely recognized, and not just by biologists and ecologists. A remarkably interdisciplinary group, including noted economists Kenneth Arrow and Partha Dasgupta, concluded that "the environmental resource base upon which all economic activity ultimately depends includes ecological systems that produce a wide variety of services. This resource base is finite. Furthermore, imprudent use of the environmental resource base may irreversibly reduce the capacity for generating material production in the future."[48]

The services we get for free from nature are extensive and diverse. Consider, for example, Dennis King's list of the economic benefits provided by wetlands in the Chesapeake Bay area. They provide aesthetic benefits, as well as fish and waterfowl habitat, and they protect property from storm surges and waves. They trap sediments, which increases the water quality of the bay and its biological productivity as a fishery. Reducing sediment reduces the need for dredging for navigation and maintains higher productivity of hydropower facilities. It also means fewer erratic shifts in stream flow and more consistent patterns of storm water runoff that protects property and coastal ecosystems. And that is just a partial list of services. "These cascading ecological economic impacts generate benefits to consumers, businesses, taxpayers, landowners, and recreationists."[49] And they are all free.

Overall, nature's services benefit us economically in many ways, providing many kinds of raw materials, purifying air and water, mitigating floods and droughts, generating and preserving soils, detoxifying and decomposing wastes, pollinating crops, cycling nutrients, controlling the vast majority of agricultural pests, protecting coastal shores, shielding from harmful ultraviolet rays, partially stabilizing the climate, moderating weather extremes and their impacts, providing aesthetic beauty and opportunities for recreation, and maintaining biodiversity.[50]

These services are not marketed, so they do not appear in our national accounts or on the books of the businesses that benefit from them. So how much are all of these free services worth? A conservative may well believe that it is too bad—even shameful for a society—that such a question has to be asked. But in our materialistic age, far too often the one and only thing that rules everything else is the "bottom line." Consequently, if nature's services never appear in the calculations, they will usually be ignored—and if ignored, they may be sacrificed without even being considered.

A remarkable attempt to put a price tag on some of nature's services was made recently by a group of economists and ecologists led by Robert Costanza. They considered seventeen categories, which is only a partial list. They included only renewable services and even omitted many of these because they could find no published estimates of value. They also omitted several types of ecosystems for which no valuation studies have been done. Using very conservative estimates of economic value, they calculated that these services are worth $33 trillion per year—well over the total gross domestic product of the entire world.[51]

Even though nature's services are not marketed and do not have explicit price tags, they are nevertheless very important. "The point that must be stressed is that the economic value of ecosystems is connected to their physical, chemical, and biological role in the overall system, *whether the public fully recognizes that role or not*. Standard economics has too often operated on the assumption that the only appropriate measures of value are the current public's subjective preferences. This yields appropriate values only if the current public is *fully informed* (among a host of other provisios)."[52]

Nature's services do not show up in the national accounts. And since the gross domestic product is the idol that our society worships and serves, too often we adopt short-sighted policies that sacrifice ecosystems and the services they could provide in perpetuity. "Economies unwittingly provide incentives to misuse and destroy nature by underappreciating and undervaluing its services. Nature in turn is increasingly less able to supply the services that the earth's expanding population and economy demand."[53]

In this respect, our GDP figures are not only misleading but even perverse. For example, we do not subtract anything for soil erosion from the figures for agricultural production, although it reduces the long-term productivity of the land. The destruction is completely ignored in the GDP.

Moreover, we often see one profitable use for something in the natural world and completely ignore all the benefits that are destroyed by that use because nothing is subtracted from our accounts for that destruction. For example, commercial fishermen have reduced the oyster population in Chesapeake Bay to 1 percent of its historical level. Originally, the oysters filtered all the water in the bay every three or four days, removing excess nutrients and keeping the water quality extremely high and the biological productivity of the bay unmatched. Now it takes 300 to 400 days to filter the same amount of water.

> An ecological economic analysis of oyster values might focus on the resulting build up of nutrients in the bay that lead to frequent algae blooms and localized fish kills, or the multi-million dollar federal and state programs attempting to clean nutrients from the bay, or the cost of restrictions on local agriculture and commercial and residential development to control nutrient flows to the bay. This broader focus on oyster value might also include an evaluation of how excess nutrients in the bay have reduced light penetration to submerged aquatic vegetation resulting in a decline in critical habitat for finfish. The resulting decline and recent closure of important rockfish and bluefish fisheries in the bay, and the loss of associated jobs, incomes, sales, and tax revenues in those fisheries might also be included. Ecological economic analysis might show that the real economic value of oysters—their highest and best economic use—is in their natural role as "the kidneys" of the Chesapeake Bay and not as a temporary source of direct income or recreational enjoyment for fishermen.[54]

People endanger nature's services in other ways as well. In Chapter 6 we saw how perverse discounting the future can be. When confronted with long-term environmental issues, it simply writes off future generations as irrelevant. It does the same with questions of sustainability, making it "profitable" to exploit to extinction resources that would otherwise be naturally renewable. So can current interest rates, which are the other side of that coin.

Paul Ehrlich once expressed surprise to a Japanese journalist that the Japanese whaling industry would exterminate the very source of its wealth. The journalist replied, "You are thinking of the whaling industry as an organization that is interested in maintaining whales; actually it is better viewed as a huge quantity of capital attempting to earn the highest possible return. If it can exterminate whales in ten years and make a 15 percent profit, but it could only make 10 percent with a sustainable harvest, then it will exterminate them in ten years. After that, the money will be moved to exterminating some other resource."[55]

In a case like this, it does not make any difference whether the resource is a commons or is privately owned. Colin Clark demonstrates how discount rates can make it "profitable" for private owners to exploit to extinction resources that could otherwise be sustainable indefinitely. This perverse system makes it "rational" to harvest and exterminate a resource that replenishes itself slowly and then invest the income in something with a higher rate of return, rather than harvest that same resource at a rate that could continue forever.[56]

Clark reaches this conclusion on the basis of mathematical theory, but it unfortunately translates directly to the real world. The World Business Council for Sustainable Development established a task force to investigate the effects of financial markets on the environment. It concluded, "It almost always makes more *financial* sense to destroy a sustainable natural resource by overuse or overharvesting, and to put the money in the bank, rather than to use the resource sustainably. This is true of forests, whales, fish, and usually of topsoil. It is true because, with prevailing interest rates, one can enjoy higher annual returns by harvesting and selling all the trees or fish, and banking the money, than by harvesting sustainably. Interest earned will almost always exceed annual profits from maximum *sustainable* yields of slow-growing creatures like rain-forest trees. Human laws and institutions must be made to reflect this dire reality."[57] In other words, a tragedy of the commons is not the only situation that produces over exploitation of renewable resources; so can private ownership if the owners use typically high discount rates or if current interest rates are more than minimal.

As a result of our short-sightedness—institutionalized in our economy in several ways—our destruction of ecosystems that provide us with valuable services is increasing. "Many of the human activities that modify or destroy natural ecosystems may cause deterioration of ecological services whose value, in the long term, dwarfs the short-term economic benefits society gains from those activities."[58] And often the destruction is, unfortunately, irreversible. As Gretchen Daily warns, "The pace of ecosystem destruction, and the typical irreversibility thereof on a time scale of interest to humanity, warrants substantial caution."[59]

Since society is intergenerational, we have an obligation to future generations to redesign our economy so it will be sustainable. In doing this, ecological concerns will have to play a crucial role, lest we leave our heirs "a ruin instead of an habitation." One possibility that should be considered is to apply safe minimum standards wherever nature's services can be threatened by our actions.[60]

With safe minimum standards, we would first set limits to modifying ecosystems within which they could still provide us with their services. The frequent recommendation that we should not permit any net loss of wetlands is an example. The boards of fisheries biologists that should set allowable catch limits in an ITQ system would be determining safe minimum standards for the renewability of each fishery. TMDLs are safe minimum standards based on the assimilative capacity of a river within which it can safely absorb and process various effluents. Within those ecological limits, the market economy can operate and sustainability will be assured.

RESTORATION

One of the major environmental tasks of this century will be to undo the damage we have done to nature in the past. We have already made a start in several important areas, but there is much work ahead. Conservatives should support these kinds of efforts for several reasons, not least out of piety toward nature. Many projects will restore ecosystem services and will promote sustainability. Restoration projects may also produce immediate economic benefits because in some cases we will be undoing pork barrel boondoggles of the past. Here I can only give a few examples to illustrate the kinds of projects involved.

The Everglades

The best-known restoration project currently under way is in the Everglades. Over the years, fully half of the swampland has been drained and canals have been built to channel water so that most of it no longer flows south through the "River of Grass." Much of this land was converted to agriculture, primarily sugarcane. Growing sugarcane there is only "profitable" because it is highly subsidized. Growers receive low-cost government loans and the federal government restricts imports to keep prices high. Consumers in the United States pay twice the world price for sugar, which costs us at least $1.4 billion extra per year; other estimates are over $3 billion per year.[61] On top of that, the Everglades have been severely degraded by agricultural pollution, primarily phosphorus from fertilizers.

The impact has been devastating in the remaining swampland, including the national park. Pollution and reduced flows of water have severely degraded the habitat, causing a wholesale collapse of wildlife. Populations of wading birds are down by 90 percent. All other vertebrates, from deer to turtles, are down from 75 percent to 95 percent.[62] The agricultural pollution stimulates algae blooms, which threaten the commercial and recreational fishing industries.[63] And because much of the flow of water has been diverted, the aquifers are not recharging and the cities of south Florida face water shortages as a result.[64]

A long-term restoration project is now under way that involves all levels of government, and the sugar industry is even (very grudgingly) contributing to the costs. Sugar growers have reduced the phosphorus loadings they send into the rivers.[65] Several levels of government are buying land to create artificial marshes around the agricultural area, which will filter pollutants before the water reaches the Everglades. The Kissimmee River is being returned to a more natural state.[66] As I write this, Congress and the administration are putting together a long-term restoration program.

Restoration efforts of this sort deserve conservative support—and this one has wide support all across the political spectrum. But it is unfortunate that we need to do this at all. Taxpayers were fleeced to create a boondoggle in the first place, taxpayers and consumers were then fleeced to subsidize a sugar industry that should never have been there at all, and now we have to pay again to fix the damages that the boondoggle caused. With a free market in sugar, much of the environmental havoc would never have happened.

Restoring River Ecosystems

In the United States there are some 75,000 dams higher than six feet and tens of thousands of smaller ones that impede the free flow of many of our rivers and streams. Most of these dams were built long ago, with no consideration for the environmental damage they cause. Most of them may be beneficial, but there are a good many that, we now can see, do more harm than good. Removing them will restore the natural flow of the rivers and their natural ecosystems.

Much publicity was given to the decision to remove the Edwards Dam on the Kennebec River in Maine. The dam came down in July 1999, and both fish and the local economy will benefit. It produced very little electricity, which the local utility was required to purchase at three times the going rate. It also blocked fish from migrating to the upper river system. Removing the dam has already restored fish runs in several miles of the river, with corresponding economic benefits from sport fishing and tourism.[67]

"Since the Edwards dam was removed, two dozen other dams have been removed up and down the country, with another 18 or so set to go this year."[68] Overall, nearly 500 dams have been eliminated, most of them in recent years. For example, in Medford, Oregon, a dam was taken out as part of an urban renewal project, which will also restore salmon runs, and a large dam in North Carolina was removed to open up seventy-five miles of the Neuse River and 900 miles of tributaries for several species of fish.[69]Two old dams in Olympic National Park have received a lot of publicity, and Congress has now authorized their removal. Once the dams are gone, the river in the park can return to its natural state and salmon runs will return as well.[70]

Many other dams, mostly small ones, will be candidates for removal, but four large ones are currently the subject of heated controversy. Four dams on the lower Snake River in Washington have virtually wiped out salmon runs in the river basin. Salmon can pass the dams on the lower Columbia, and there are viable populations there. But the dams on the Snake cannot be modified to help the fish. Very few make it upstream past the dams to spawn, and very few young fish make it downstream to the sea. Instead of a flowing river to help their passage, they face a series of slack water lakes in which predators abound. Many of them get sucked into the turbines at the dams and perish.

These Snake River dams were pork barrel projects that were opposed by the Eisenhower administration and the fishing industry (among others) when they were originally proposed. They generate very little electricity, provide no flood control, and do little for irrigation. Their purpose is for navigation: to make Lewiston, Idaho, a seaport.

If the dams were breached, the salmon runs could be reestablished and sport fishing and tourism would thrive. The *Idaho Statesman* newspaper calculates that the net benefits to the region would be some $183 million per year.[71] The Army

Corps of Engineers estimates that breaching the dams and reestablishing salmon runs would create 12,000 jobs in the area.[72] The agricultural produce of the region could be transported by rail and truck, just as it was before the dams were built.

This is the kind of restoration debate in which conservatives should be actively involved. We have here an antieconomic pork barrel project that has done immeasurable damage to the ecosystem, but it can be largely restored at considerable economic gain.

Rivers and their ecosystems have been drastically changed in other ways besides damming. Many rivers have been straightened and channelized, and levees have been constructed to contain floodwaters. Many rivers have been isolated from their floodplains, preventing flood pulses that were major factors in creating the habitat of the floodplains. Countless wetlands along rivers have been eliminated. Wetlands are among the most biologically productive of all ecosystems, and they are also natural filters to remove nutrients and catch sediment. Without the wetlands, not only has habitat been lost, but the pollution now accumulates and goes down the rivers, contaminating estuaries and causing the huge dead zone in the Gulf of Mexico.

Moreover, development in floodplains has been highly subsidized. Over forty federal programs and agencies encourage building in floodplains and their wetlands.[73] One consequence of all this is that floods and flood damage have become much worse over the years. Nature's flood controls—wetlands that store excess water and floodplains that let the water spread out—have been eliminated. Levees simply cause flooding downstream and raise the levels of major floods.[74] The unnatural constriction of the Mississippi River significantly raised the flood level in 1993.[75] The federal government has spent billions on flood control, yet annual damages have increased dramatically.[76] More federal disaster money is spent on floods than all other types of disasters combined.[77]

The Mississippi River flood in 1993, which caused $16 billion in damages,[78] proved that we need to control floods by restoring nature's system wherever possible. Nonstructural floodplain management is now seen as a major part of the solution, and this basically means preserving existing wetlands, restoring some that have been drained, and reconnecting rivers with their floodplains. For example, in a study of the Charles River, the Army Corps of Engineers concluded that comprehensive protection of existing wetlands in the river basin would produce more protection from floods than $4 billion worth of concrete dams, dikes, and levees.[79]

In the upper Mississippi basin, Donald Hey and Nancy Philippi note, despite the construction of levees, average annual flood damage has increased 140 percent. Clearly, a new approach is needed. Hey and Philippi calculate that all of the excess water in the 1993 flood could have been contained in 13 million acres of wetlands. That is about half of the wetlands in the basin that have been drained, and it is only 4 percent of the area of the watershed. There is at least that much idle agricultural land in the river basin. Restoring 13 million acres of wetlands, they contend, would prevent damages even from a flood the size of the one in 1993.[80] Restoring smaller acreages, of course, would eliminate damages from smaller floods. Restoring wetlands would also improve water quality[81] and produce many aesthetic and recreational values as well.

A beginning is being made in the Mississippi-Missouri watershed. In Nebraska, for example, in several places old channels of the Missouri River and their wetlands that had been cut off from the river are being restored. Not only will this create habitat for wildlife, but it will also put some "flex" back into the river system and reduce damages from future floods.[82]

The 1993 flood broke many private levees that have never been rebuilt, so in those places the river is now reconnected with its floodplain. The advantages became clear just five years later. In 1998, after heavy rains, a warning on the Missouri River predicted a crest 12.9 feet above flood stage. But when it came, it exceeded flood stage by only seven feet. The rest of the water had spread out over the plain where the levees no longer existed.[83]

The federal government is now buying out private properties that repeatedly suffer flood damages. Although this is expensive, in the long run it will save taxpayer money. For every $1 spent relocating buildings, $2 will be saved on federal disaster relief.[84]

Once again, restoring nature so that it can provide its services really just reverses governmental boondoggles from the past. Conservatives should support this sort of program, since it helps nature, protects people instead of encouraging them to move into harm's way, and saves taxpayers money in the long run.

CONCLUSION

Since society is a contract across generations, sustainability is clearly a goal conservatives should support. Although the goal of sustainability raises many economic[85] and ecological issues, the three instances I have cited here—protecting renewable resources from extermination in tragedies of the commons, preserving nature's ability to provide the services that are so beneficial for us, and restoring natural systems that we have degraded—indicate some kinds of policies conservative principles support.

10

"Free Market Environmentalism": Environmentalism for Conservatives?

IN PREVIOUS CHAPTERS WE SAW that market-based mechanisms can be more effective than bureaucratic command and control in protecting the environment. We looked at such possibilities as tradable emission quotas, eliminating harmful subsidies, deposit and refund schemes, full-cost pricing, "marketizing" use of the public lands, and the like. Although many of these ideas have been around for a long time in the footnotes of resource economics texts, it is only within the last decade or so that they have gained prominence in the public debate.

In a very short time, it seems, the concept of market-based environmentalism has all but completely won the field among policy analysts, if not yet among policymakers and politicians. As one might expect, the Heritage Foundation adopted the concept some time ago. In 1989, in the book *Mandate for Leadership III*, they severely criticized the Reagan administration: "Although the Reagan Administration extolled the virtues of a free market, it has not offered a positive, market-oriented approach to environmental protection as an alternative to the command and control approach that it inherited." The chapter on the Environmental Protection Agency recommended "wider use of market mechanisms that protect the environment while avoiding the unnecessary costs associated with a purely regulatory approach." And the chapter on the Department of the Interior recommended curbing the many federal subsidies for "environmentally harmful activities."[1]

Just four years later, the Progressive Policy Institute, with connections to the Democratic Party, likewise advocated market based environmentalism.

> The environmental movement is poised to enter a second generation. For two decades it has prompted significant improvements in the quality of our air, water, land, and natural resources, primarily through "command-and-control" regulations that essentially have told firms

> which pollution control technologies to use and how much pollution they could emit. Now, in an era of new environmental challenges and heightened sensitivity to regulatory compliance burdens, market forces can offer in many circumstances a more powerful, far-reaching, efficient, and democratic tool than centralized regulations for protecting the environment.[2]

Environmentalists have often been suspicious of markets and economics, partly because so much of the damage to the environment is caused by business and industry, and partly because conservatives have been so adamantly and, let us admit it, irresponsibly negative. But the Environmental Defense Fund years ago began advocating use of market mechanisms as incentives for pollution control. In the 1996 edition of *State of the World*, we find that Lester Brown's Worldwatch Institute has been converted to the cause. They argue that our Western market-based economies have a fundamental flaw in that "the prices they use to guide buying decisions and allocate resources rarely reflect the full costs of environmental damage. . . . To make the market system reflect rather than obscure ecological realities, societies need to enforce a principle that is at once radical and obvious: that people and businesses should pay the full costs of the harm they do others."[3] And they proceed to recommend a whole range of market-based solutions, rather than the old regulatory ones. The Sierra Club and other national organizations also see the potential for using market-based incentives, at least in some areas.

Since these approaches to environmental protection embody our principles, conservatives can rejoice in this and work to shift our environmental policies farther away from command and control toward increased use of market incentives. But, as with all good things, some take these ideas to an extreme. Some libertarians now advocate "free market environmentalism," which would leave all, or nearly all, environmental protection to the workings of the free market.

Sometimes the advocates of free market environmentalism are reasonable and recognize that there are limitations in the marketplace as well as in government. Terry Anderson, for example, after discussing private provision of recreational and environmental amenities, asks, "Will this work in all cases? Certainly not. But there are a growing number of examples to suggest that getting the incentives right can go a long way toward improving the environment through the private sector."[4] Some years ago, John Baden and Richard Stroup, arguing that market failures do not necessarily mean that government should step in, admit that "the opposite is also true: an imperfection in collective management should not automatically cause us to avoid governmental action. The grass is *not* always greener."[5]

But often free market environmentalists come across as "true believers," idolizing the free market as savior for all and everything. So once again, even in the environmental realm, we need to continue the long debate within conservatism about the adequacy of the libertarian position. To what extent can free market environmentalism be embraced by conservatives? From the traditionalist or fusionist point of view, what are the limits of free market environmentalism? Let us look at some of the proposals free market environmentalists have made to see if they have merits beyond ideological purity. Then I will examine a few assumptions behind the libertarian approach to environmental protection.

At the outset, we should grant that these libertarians are truly concerned about the environment and actually want it to be protected. They are not advocating any free-for-all to allow corporations to rape the land or turn rivers into industrial sewers. They believe that individuals and companies should be held responsible for the environmental consequences of their actions. But they believe that the workings of the market provide a better way to protect us from pollution and to provide the environmental amenities we want than governmental bureaucracies.

Many of the concrete proposals advocated by free market environmentalists are very good ideas that all conservatives can embrace. We have met many of them in previous chapters: establishing property rights to solve tragedies of the commons, expanding water markets in the West, charging all users of the public lands, and tradable emission permits to reduce air and water pollution.

Unfortunately, these libertarians do not stop there. With their extreme dislike of government and their belief that only individual preferences matter, they extend their faith in the market far beyond what it can in fact do. Consider their proposals in two areas: preservation of nature and pollution control.

Free market environmentalists praise all kinds of private efforts to preserve wildlife and natural habitat and, of course, these efforts are praiseworthy by any standards. They like the many ranchers who charge hunters and anglers for access to their land and streams. These ranchers often make as much or more from preserving high-quality habitat for deer and pristine streams for trout as they make from their cattle.

The International Paper Company is often held up as an example of what private markets can do. International Paper owns a lot of timberland in the South. It harvests its timber carefully to avoid destroying the forests and their amenity values. In fact, it manages its land to improve wildlife habitat. It then makes money by charging for recreational use of its forests. If it did not have a market for quality recreation, it would not have incentives to improve its forests as habitat. Since in the South there is little public land providing free recreation, IP can profit from carefully husbanding its forests, where a timber company in the West could not.

Free market environmentalists like organizations such as the Nature Conservancy that raise money from donations and enter the market to buy land which is important habitat for wildlife or to buy conservation easements that ensure the preservation of habitat. They also like the efforts these organizations are now making to buy water rights in the West in order to leave the water in the streams to preserve aquatic habitat.

They also approve of the movement in some states to make their state parks pay for themselves. Park managers then come up with all sorts of ingenious ideas to attract a paying public to their parks and their new programs. And, to take just one more example, they praise developers who set aside large tracts of their land to be preserved in its natural state—which, of course, raises the prices on the home sites they develop on the rest of the land.[6]

Terry Anderson and Donald Leal call these people and organizations "enviro-capitalists." They are using the marketplace to preserve nature. These examples they cite are, of course, fine. We could only wish more people were as sensitive and concerned about the natural world. Unfortunately, many are not.

Moreover, the name "enviro-capitalist" is somewhat misleading. Some of the examples they cite actually are capitalists using their entrepreneurial abilities to make money by preserving nature. But many of the instances these libertarians point to are not capitalists; they are nonprofits and some are even governmental agencies.

In fact, many of the major private efforts to preserve natural habitats are not capitalistic at all. The Nature Conservancy has preserved over 12 million acres, and other land trusts have protected much land as well. But note what they are doing. They are not acquiring land in order to use it as a resource for production and creation of wealth. They acquire land to guarantee that it will *never become* a factor of production. They are acquiring land for the purpose of permanently *removing* it from the capitalistic sector of our society.

Nevertheless, surely all could agree that private actions to preserve nature are very good. But the next step in the libertarian argument is completely unacceptable. Since these libertarians believe that individual preferences are the only things that matter, they sometimes take the extreme position that all preservation of nature should be left to the market. Jane Shaw and Richard Stroup consider the kinds of decisions that must be made in preserving species: what species should have priority, which populations should be protected, and what level of protection should be provided. "A true free market solution would leave these decisions in the hands of individuals and voluntary associations."[7] Terry Anderson and Donald Leal think that to save the spotted owl, environmental groups should lease the most critical old-growth forests; to save the salmon, people who care about the fate of the fish should lease water for in-stream flow.[8] In other words, those who prefer to preserve God's creatures should have to bid against those who do not. And if they are outbid, so be it and let the devil take the creatures.

No. We have an obligation of stewardship, an obligation to care for creation, that cannot be abdicated just because not enough people happen to give money to a land trust. We cannot abdicate those obligations just because global corporations can outbid environmental organizations for critical habitat—no more than we could abandon human rights if some rich people "preferred" to own slaves and could outbid the donations of those who "preferred" all people to be free. Nor do we require people to donate money to private organizations to buy out the preferences of some businesses for child labor. Conservatives can never accept the proposition that values and rights and obligations automatically end whenever private preferences reach the limits of their financial resources.[9]

Private efforts to preserve nature are excellent and are always to be encouraged. They are a valuable part of our total social efforts of stewardship. But there is no way that environmental organizations, even the very large national ones, can do enough to fulfill our duties of stewardship. Consequently, we can never accept the assertion that private actions alone are the only actions we should ever take.

Anderson and Leal also mistakenly believe that property rights will automatically conserve natural resources. If property rights are clearly specified and transferable, they believe, "owners must not only consider their own values, they must also consider what others are willing to pay. . . . [Consequently,] market processes can encourage good resource stewardship. It is when rights are unclear and not well enforced that over-exploitation occurs."[10] Since owners will consider future demands and the potential for higher prices in the future if they preserve the resources now, private ownership will result in conservation of natural resources such as forests.

Unfortunately, such faith in property rights and the market is unwarranted in this case. As we saw in the last chapter, in our economic system it is often "profitable" for private owners to exploit to extinction resources that are otherwise naturally renewable.[11] That is why, for example, there is virtually no old-growth forest left on private land in the Pacific Northwest.[12] Private ownership of natural resources will not automatically suffice to fulfill our obligations to future generations. Anderson and Leal tacitly recognize this in their proposal to use tradable quotas as a solution to depletion of fisheries: a governmental agency's scientists, not the free market, must determine the allowable level of exploitation.[13]

Second, consider the free market environmentalists' approach to reducing pollution. They support emission caps with tradable permits, such as the ones in the Clean Air Act, as an improvement over bureaucratic command and control. But they would much prefer leaving pollution control entirely to the private sector. Environmental policy, they contend, should ultimately be based on property rights. "At the heart of free market environmentalism is a system of well-specified property rights to natural resources."[14] These property rights should be clear and enforceable. Then if some factory in your neighborhood pollutes your air or dumps toxic waste in a stream that runs through your property, you would not complain to the EPA; you would sue the company for damages and try to get an injunction to make it stop.

Private actions, through tort cases, are the free market environmentalists' ideal for pollution control. Terry Anderson argues that "a return to more reliance on the private law of torts and contracts is crucial to the definition and enforcement of property rights necessary for free market environmentalism."[15] According to

Roger Meiners and Bruce Yandle, "A market approach requires that federal statutes be phased out, and therefore replaced with common law liability rules."[16] These libertarians frequently look to the past, when the courts were the only recourse available for victims of pollution. And they lament that bureaucratic control of emissions into our air and water has displaced private suits against polluters.[17]

This proposal is completely unacceptable—from a conservative point of view, or from any other view short of libertarian ideology. To begin with, their assertion that tort actions have been displaced by bureaucratic regulations is not true. Victims of pollution can still bring nuisance suits against polluters.[18] Tort actions are sometimes valuable means of attacking polluters. They can be used as supplements to governmental rules, but tort cases by themselves are not sufficient to control pollution. As Michael Blumm contends, "Nuisance showed itself to be a spectacular failure in confronting the environmental problems of the Nineteenth and Twentieth Centuries."[19] That is why regulations were adopted in the first place.[20]

Moreover, since free market environmentalists are so often concerned with "efficiency," it is surprising that they would want tort cases to be the means of controlling pollution. It would be difficult to imagine any system of pollution control more inefficient than our court system.

There are numerous reasons why we should not rely on tort cases as our exclusive or even our primary means of pollution control. Here I can consider only the major ones. It is extremely expensive and inefficient to sue for damages from pollution. Besides the court costs, the plaintiff will need to pay for monitoring pollution on his or her property. To establish damages from the pollution will require expensive scientific expert testimony.[21] After a court trial, there will be expensive appeals. By comparison with regulatory control of pollution, "resort to heavy reliance on common law liability rules would drastically increase transaction costs."[22] And the whole process can last years before the victim gets any satisfaction.

Trial attorneys work on a contingency fee basis, so they would take only those cases with spectacular damages. Ordinary victims will have no recourse whatsoever.[23] And in those major cases, even if the plaintiff wins, much of the award will go to the attorneys.[24] "In the asbestos litigation, for example, plaintiffs received on average only 39 percent of the total paid by defendants and insurers, with the remainder going to legal fees and expenses."[25]

Class action suits will do little to help in the normal cases where numerous people each suffer limited damages. "Existing scholarship on the class action does not support reliance on this aggregation tool as a means to achieve private environmental enforcement."[26]In many cases the plaintiff will confront an insurmountable burden of proving causation. Many pollutants are now regulated because they are carcinogens. But in our modern society there may be hundreds or thousands of sources of the pollutants that caused a plaintiff's cancer. It is impossible to identify which source emitted the particular toxics that caused this specific cancer. And cancers usually develop years after exposure to the chemicals that cause them. There is no way, in these cases, for a plaintiff to identify a defendant or to prove the precise extent to which this particular polluter contributed to his or her disease. The result is that the victim gets nothing out of the court system.[27]

Even if successful, tort cases only offer remedies after the damage has been done that applies only to the plaintiffs in the case, not to victims of pollution in general. Torts are not proactive; they do not prevent pollution, except to the extent to which they may act as a deterrent. That extent is, unfortunately, small. All of the barriers plaintiffs face reduce the deterrence value of damage suits. Polluters can declare bankruptcy or go out of business, so they can "discount their potential liability to a level equal to their total assets, which may be less than the full amount of potential harm." Many organizational decisionmakers are "insulated from the full impact of their risk-creation choices because of the time lag between managerial decisions and eventual lawsuits, the long lag between environmental contamination and the onset of many diseases, and the limited liability and insurance protection of corporations."[28] Peter Menell concludes that "the tort system is extremely limited in its ability to deter environmental risks efficiently."[29]

A plaintiff also faces major problems in court because the judge and jury have little or no expertise in the scientific issues that are involved. Even the Supreme Court recognizes the inability of generalist judges to understand many complex scientific issues. In an important environmental nuisance case, the Court acknowledged the trial court's admission that "the arcane subject matter of some of the expert testimony in this case was sometimes over the heads of all of us to one height or another."[30] The result is often confusion on the part of the court and no compensation for the victim.

Moreover, even where they work, tort cases are no way to set policy for society. The outcome of each case applies only to that plaintiff and that defendant. Rulings and awards are often inconsistent. Different jurisdictions will probably develop different standards. As Peter Huber warns, "Millions of small tort and contract decisions will not magically coalesce into coherent public policy."[31]

Finally, even when specific defendants are identified and specific damages are proven, there is no assurance that courts will enforce property rights for the plaintiffs. When torts were the only recourse for victims, some courts ignored property rights and ruled on a cost-benefit basis in favor of the polluters. Even Meiners and Yandle acknowledge that some courts denied plaintiffs compensation because they decided "the economic benefit of a factory that employed many people outweighed the damage to a few property holders. Some courts held that pollution was just a fact of modern life and necessary for progress to occur."[32] The plaintiffs simply had to suffer.

With all the problems involved in using common law to enforce environmental protection, free market environmentalists would leave us with a victim's nightmare. Consider a real-world example of what we would face. In a small

Massachusetts town, some families sued two large corporations, alleging that they contaminated the town's drinking water with toxic chemicals that caused some of their children to contract leukemia. The litigation spanned nine years, involved hundreds of expert witnesses and scientific studies, and ended in a settlement of $8 million for the plaintiffs. Their legal expenses ate up $4.8 million of that. None of the settlement was used to clean up the town's contaminated wells. The water supply was cleaned up only when an EPA action against the two companies ensured the needed funding of nearly $70 million.[33] In the libertarian utopia, the wells would still be contaminated and each new group of leukemia victims would have to bring new suits. In sum, private enforcement by tort cases cannot provide adequate protection from pollution. As Edward Brunet concludes, "The concept of 'free market' enforcement . . . is, in reality, no enforcement at all."[34]

Free market environmentalists sometimes even acknowledge these problems. For example, at one point Meiners and Yandle recognize many of the limitations discussed above and admit that regulation may be necessary to protect the public against pollution, which causes future harms such as cancers.[35] But even then, some libertarians are such dogmatic ideologues that this admission does not change their position. Just six pages later, Meiners and Yandle propose to abolish the EPA and return to the regime of common law enforcement.[36]

Occasionally, free market environmentalists offer vague suggestions that the common law could evolve to do the job.[37] But it would require some truly monumental changes in both the court system and the law of torts to come anywhere close. Without such changes, even bureaucratic command and control is better than the system free market environmentalists would condemn us to suffer under.[38]

Moreover, free market environmentalists themselves are sometimes inconsistent about property rights and environmental protection. There *is* a consistent libertarian position here, clearly expressed by Tibor Machan: If pollution cannot be confined to the property of the polluter, the activity that produces it simply cannot be permitted.[39] A pure, philosophical libertarian would allow absolutely no pollution at all. But the free market environmentalists are not nearly that consistent. On occasion they say that with well-defined and enforced property rights, polluters and recipients of the emissions should bargain over the level of pollution.[40] But they want to require unwilling victims of pollution to prove damages, not just unauthorized and unwanted invasion of their property.[41] Consequently, their property rights are not going to be very secure and enforceable.

Moreover, free market environmentalists often revert to the economist's concept of "efficiency" and analyze "how much" pollution is the "right" amount.[42] They fear, for example, that government may produce "to much" clean water.[43] This is a social cost-benefit calculation that overrules rights whenever the balance falls on the benefits side. But if property rights are to be the basis of their ideal society, then the "right" amount of pollution is always exactly zero if any potential victim objects to being exposed to it. I know of no self-styled free market environmentalist who draws this conclusion. So much for the clear and enforceable property rights that are the foundation for their position. As Carl Pope of the Sierra Club says, "The Sierra Club would welcome testimony from the Competitive Enterprise Institute in Congress that EPA's clean air standards program, by allowing polluters to expropriate the lungs of unwilling breathers, is burdensome to the general public, and should be replaced by a requirement of truly voluntary exchange among emitters and breathers. But I don't think I'll hold my breath."[44]The bottom line here, clearly, is that property rights, privately enforced, are completely inadequate for protection from the kinds of pollution produced in modern society.

We have seen that the extremists' free market policies are inadequate in two major realms of environmental policy: preservation of nature and pollution control. Another way to determine the limits of free market environmentalism is to consider the assumptions underlying its ideology. The position is conveniently summarized in a very interesting, readable, and important book by Terry Anderson and Donald Leal, *Free Market Environmentalism*. They advocate many of the policies reviewed above, but they also discuss several of the theoretical underpinnings for their proposals. These theoretical bases especially need to be viewed critically by conservatives in determining the limits of the free market. Three of them seem to be most important.

One premise behind Anderson and Leal's approach to policy is that knowledge is diffuse rather than concentrated, so "ecosystems . . . cannot be 'managed' from afar." The authors conclude from this that "individual property owners, who are in a position and have an incentive to obtain time- and place-specific information about their resource endowments, are better suited than centralized bureaucracies to manage resources."[45] But the private companies that are most heavily involved in extractive industries—which is what is at issue here—are centralized bureaucracies, just as the governmental agencies are. This premise may support decentralization of management, but there is no reason to believe that a forest manager for Weyerhauser has any better "time- and place-specific information" than a forest ranger for the Forest Service.

Sometimes Anderson and Leal deal with the "knowledge problem" in terms of pricing. Market prices provide crucial information for producers and consumers. But where markets are poorly developed or nonexistent, it may be "costly" to obtain information. For example, "the lumber market provides information on timber value as a commodity, but information about the value of wildlife habitat and environmental amenities is more costly because those markets are less developed."[46] This point raises many vexed issues on how to "price" things that are not bought and sold. But Anderson and Leal do not pursue the disputes over alternative ways of "pricing," such as contingent valuation. This is a point on which conservatives must be especially wary because there is a great tendency in policymaking (e.g., in cost-benefit analyses) to overlook or ignore any values that are not easily and clearly expressed in dollar signs.

Anderson and Leal themselves illustrate the extent to which making an ideology of the market leads to the exclusion of all values except monetary ones. They consider an instance of a forest in the Great Lakes region that was logged a century ago. This forest has never recovered and the place is now wasteland because the soil is so infertile. "This does not imply, however, that cutting the trees was a bad decision," since investing the profits in bonds would have yielded a huge return by now, probably "worth" more than the habitat and environmental amenities the forest could have provided.[47] In other words, permanent destruction is just fine as long as someone makes money on it. A conservative, however, must surely think that this example illustrates the very different conclusion drawn by Russell Kirk: "The modern spectacle of vanished forests and eroded lands . . . is evidence of what an age without veneration does to itself and its successors."[48]

A second premise is that human beings act out of self-interest. Consequently, "good resource stewardship depends on how well social institutions harness self-interest through individual incentives."[49] This is a very important point that environmentalists too often overlook. The self-interest for polluters is to maximize profits and avoid responsibility for the damage they cause. But turning to government does not guarantee a solution because politicians and governmental bureaucrats also act out of self-interest.

In the natural resources sector, "political resource managers make trade-offs in terms of political currencies measured in terms of special interest support. . . . The rewards for political resource managers depend not on maximizing net resource values but on providing politically active constituents with what they want with little regard for cost."[50] That is how the classic examples of the iron triangle have developed. A special interest group (e.g., ranchers in the West) wants to exploit a public resource (grazing lands) with every possible subsidy. The politicians beholden to them get control of key committees, from which they direct the agencies (in this example, primarily the Bureau of Land Management) to serve the special interests at the expense of the general public. And at appropriation time, the politicians reward the agency with a bigger budget and promotions for its employees.[51]

Anderson and Leal's analysis here is very important for environmental policy. As they conclude, even when there is a pragmatic case for turning to the government to protect the environment, "there is still no guarantee that the results from political allocation will work very well."[52] And as we have seen, there are many cases in which, for these very reasons, governments have caused environmental degradation. This dilemma forces us to choose between imperfections of the market and imperfections of government. But after raising this extremely important point, Anderson and Leal ignore it and proceed with the assumption that markets will be able to solve almost anything.

A third premise, never analyzed by Anderson and Leal but clearly expressed, severely undermines the utility of the libertarian approach to many of the major issues of conservation. This premise is behind their argument for leaving preservation of nature to private efforts, so we will have to approach that issue again from a slightly different angle. Their assumption is that whenever we preserve nature it is for the benefit of *environmentalists.* Consider their analysis of the opposition to drilling for oil in the Arctic National Wildlife Refuge: "Organized groups that favor preserving wildlife habitat in the pristine tundra can gain by stopping drilling in the refuge. To the extent that those who benefit from wildlife preservation do not have to pay the opportunity costs of foregone energy production, they will demand 'too much' wildlife habitat."[53] Consider also their conclusion that the current system of water allocation in the West "may not provide a sufficient supply of instream flows for wildlife habitat and environmental quality because owners of water cannot easily charge recreationists and environmentalists who benefit from free-flowing water."[54]

Almost all of the North Slope of Alaska has been opened for oil exploration and drilling. Its wilderness character is gone, and much of it has been severely degraded as habitat for wildlife. The last remaining fragment of the North Slope, a critical calving area for caribou, is in the wildlife refuge. It is true, of course, that a few people (probably not all of them environmentalists) enjoy the use of that pristine area for hiking and camping. But the purpose of saving it is not for the benefit of environmentalists but for the benefit of the caribou and other wildlife. One measures "how much" to preserve by the needs of the *animals,* not by the number of campers or the voluntary donations of environmentalists.

In the case of in-stream flows, the traditional system of water allocation in the West requires those who hold water rights to divert the water from the streams. The flow in many rivers has been reduced so much that native fish and other creatures are endangered. Again, it is true that some people benefit from free-flowing rivers by using them for fishing or rafting. But the purpose of preserving in-stream flows is not only to benefit environmentalists but also to benefit the fish. How much flow to preserve is measured by the habitat needs of aquatic creatures, not by the number of anglers or rafters.

On the basis of this third premise, Anderson and Leal's proposals for preserving habitat center around charging hunters, campers, anglers, and hikers for their use of the land and streams.[55] And to a certain extent these ideas may be valuable. As we have seen in Chapter 5, if the Forest Service charged for recreational use of the national forests and could keep the fees, it would have an incentive to protect the forests rather than chop them down. And if a private landowner could charge campers and hunters for access to his or her land, its natural amenities and wildlife would be seen as assets and the owner would have incentives to preserve them. That incentive is lacking now because the public land is often available for free. These proposals would be valuable in protecting at least some natural areas that are important for game animals and fish.

But there is no reason to believe that charging hunters (a small percentage of the American population) and anglers would protect enough habitat to fulfill our duties as stewards of the earth. Not all of God's creatures are lucky enough to have the same habitat needs as deer and elk and trout. And if there were large enough crowds of hikers and campers to make a critical wildlife habitat "profitable," it would probably not be suitable for the wildlife any longer. Insofar as charging for recreational use would help protect nature, it would be a fine and useful policy option. But we should not accept on faith that it would be sufficient.[56]

Free market environmentalism, according to Anderson and Leal, "emphasizes the importance of market processes in determining optimal amounts of resource use."[57] But markets do *not* determine "optimal" resource use; markets determine the most *profitable* resource use. Only an ideologue who "knows the price of everything and the value of nothing" would equate the two. If recreational fees do not preserve sufficient habitat, only an ideologue would put a cap on piety and our obligations of stewardship simply because that is all the nature the market wants. And the authors' concern that the political process might protect "too much" wildlife habitat is simply absurd. The political and economic pressures are overwhelmingly weighted against preservation in virtually every instance. Preservationists today are merely fighting for the few remaining fragments of nature (e.g., in one of the most contentious cases, a mere 8 percent of the old-growth forests of the Pacific Northwest).[58]

Free market environmentalists' utopian extremism illustrates a most important point for conservatives that has been well developed by John Gray. In the 1980s, he contends, free market libertarians performed a very valuable service in demonstrating the inherent imperfectability of government. But they made a fundamental error in not attributing that same imperfectability to the market. Their political thought "suppresses recognition of the institutions of the market as being as fallible, as frail and as obdurately imperfectible as any other human institution."[59]

In choosing environmental policies, Gray insists that the conservative must face the dilemma libertarians try to avoid through their devotion to the market. The conservative recognizes that we should never attribute "to the market a perfection we have learned not to ascribe to government. . . . For a conservative, political life is a perpetual choice among necessary evils. . . . so he will make a choice between the imperfections of markets and those of governments, in the hope that the resultant mixture will best promote freedom and community."[60]

Gray believes that the conservative, who realizes this dilemma, will therefore be in the best position to make environmental policy, and market-based policies will often be the instruments of choice. "It will be wise to pursue, so far as possible, market solutions to the problems of the environment. . . . However, it will be foolish to suppose that market solutions alone will solve the environmental problems that now loom up as the chief threats to the quality of our lives and to those of our children."[61] Gray faces squarely the fact that "the unfettered workings of market institutions may damage the natural and human environments, even if it is true . . . that in most cases they act to protect them."[62] And where market mechanisms will not work, the conservative "will never concede hegemony to the institutions of the market. The market is made for humans, not humans for the market."[63]

In sum, the proper role for conservatives is to use the market where it would work to protect the environment, as opposed to the liberals' penchant for command-and-control regulation, and to restrain the market where it would degrade the environment, as opposed to the libertarians' penchant for sacrificing anything that cannot be turned to a profit. The principles of conservatism are, in fact, the best basis on which to confront the dilemma of market imperfections versus governmental imperfections in every concrete instance, where liberals and libertarians alike are prone to ignore it and carry on their merry way even in the face of obvious failure. By making use of market-based mechanisms, tempered by the basic virtues of piety and prudence, conservatives could have the best policies for a much cleaner environment, a more efficient system with equitable assignment of costs ("the polluter pays"), sustainable use of natural resources, and a much better quality of life for ourselves and for countless generations to come.

Concluding Thoughts

We have examined (and I hope dispelled) several myths about environmentalism, conservatism, and the economic impact of protecting the environment. We have considered several fundamental principles of the conservative political philosophy and have seen how they support protecting the environment, conserving natural resources, and preserving our natural heritage. We then examined some of the most important environmental problems we face and saw that there are policies to deal with them that embody or apply conservative principles. These policies are often quite different from the ones adopted by the liberals and bureaucrats, but they would be effective—often *more* effective—in solving environmental degradation. And we have also seen that the current anti-environmental stance of many politicians and pundits is entirely unacceptable because it violates fundamental conservative principles.

Of course, I have not attempted to present a complete environmental agenda for conservatives. There are many serious problems that I have not considered, and other conservatives may be able to develop new and different policies for the problems I have examined. All I have attempted to do here is illustrate the kinds of policy thinking conservatives should be doing to solve environmental problems. For example, we have seen several federal subsidies that cause environmental damage. Conservative principles on the free market suggest that these subsidies should be removed. But there are many more harmful subsidies than I have mentioned here. For the past several years a coalition of taxpayer and environmental groups has published the annual *Green Scissors* report, listing dozens of direct subsidies and federal projects that benefit special interests. Eliminating them would benefit both the environment and the taxpayers. The latest report lists seventy-seven boondoggles that waste $50 billion of taxpayer money and hurt the environment.[1] Conservative principles regarding the market should be applied in advocating reforms of all of them.

Although I have attempted to indicate the kinds of policy thinking conservatives should be doing on environmental issues, I do not intend for my suggestions to be taken as dogmatic, all-or-nothing proposals. Rather, I want them to indicate the *directions* in which principled conservatives could work to move public policy. For example, I suggest that we could improve the national forests and protect them by eliminating all subsidies and charging all who use them. But many people believe that public lands should be open for the public to use. They are a heritage for all and should be supported by all. Although those who *profit* from using them should not be subsidized, nondestructive recreational use, many believe, should at least in large part be supported by the government. Accepting that claim would still let us move in the suggested direction. Eliminating subsidies for profit-making extractive industries would do much to protect the forests. User fees to cover just *part* of the costs of building and maintaining trails and parking lots would also be an improvement over what we have now. We can improve our public land policies by moving in that direction.

I do, however, insist dogmatically on one point: *conservative principles support environmental protection.*

When we look at the damage we have caused and are causing to our planet—very often completely senseless damage—it is difficult to be optimistic about the future. The major problems we face are the kinds our government and our society are perhaps least able to deal with. If we are to succeed, principled conservatives must contribute a very great deal to our efforts and policies. I see a number of major obstacles that we must overcome.

First, we will meet several of our greatest environmental challenges only if we adopt a future orientation that thinks in terms of generations. But the orientation in our society seldom extends beyond the next election cycle or the next quarterly business report.

Those who work to restore our devastated national forests will not live to see the final fruits of their labors. For restoration of old-growth forests, not even their grandchildren will see the final results. The collapse of biodiversity is happening right now, but the full impact of the accumulating losses may not become critical for years. To see global warming as a serious problem, you have to think in terms of the next century; moreover, you have to be concerned about the kind of world your grandchildren and great-grandchildren will inherit. Even better, to confront those problems you should see them in terms of stewardship of God's creation for all time.

But that is precisely the orientation of principled conservatives, who see society as intergenerational and who see that we have obligations to our heirs far into the future. If anyone in America is able to deal with these kinds of long-term problems, it is surely the principled conservative, and the conservative philosophy is what is most needed.

Second, solving some of our most daunting problems will require a shift toward a nonmaterialistic outlook. But at this point in time, the favorite game in our hyperconsumerist society seems to be "let's see how much of the planet I can waste in one lifetime"—exemplified by the increasing numbers of grossly and uselessly overweight, gas-guzzling trucks and SUVs. The yuppie slogan "whoever has the most toys when he dies wins" has faded away but the attitude prevails.

As Donald Worster writes, "The only deep solution open to us is to begin transcending our fundamental world-view—creating a post-materialist view of ourselves and the natural world. . . . It seems inevitable that such a shift will occur at some point. No world-view has lasted forever."[2] This, of course, is precisely the worldview of the principled conservative. Richard Weaver admonished us that to save the human spirit we need to recreate a nonmaterialist civilization.[3] That is the same challenge we face in order to save our environment.

Third, the key to solving many environmental problems is to get people and companies to take responsibility for the consequences of their actions. But in our society, evading responsibility is practically a way of life and denial seems to be the favorite escape route—denial of evidence and obstruction of any attempt to gather evidence if the results might be inconvenient. The conservative philosophy *insists* that all of us must take responsibility for the consequences of our actions, whether we like it or not—and that is precisely what we need.

Fourth, a final solution to the devastation we are wreaking on the natural world requires that we adopt an attitude of piety and respect toward nature. But as our increasingly urbanized society becomes ever more detached from nature, people have less and less contact with it. Lots of people go out into the natural world for recreation, to be sure. Yet it seems that more and more of them have no respect for nature but merely use it as a playground for noisy, destructive contraptions such as snowmobiles and ATVs.

It is the principled conservative who retains an attitude of piety in our secular society—and that includes piety toward God's creation.

Fifth, a key to solving many of our environmental problems is to move toward the free market, where prices include the full social costs of products and services and where producers and consumers must pay the full costs of their production and consumption. But in our society, companies and individuals—and especially the middle class!—absolutely insist on keeping their subsidies, no matter how destructive they are to the world. And far too many companies do everything they can to maximize their negative externalities and make their profits at everybody else's expense.

The conservative philosophy rejects all of this out of hand. The principled conservative contends that producers and consumers should pay for all of their costs and that subsidies should go, especially subsidies that produce harms.

Sixth, in sum, to solve our environmental problems—especially the long-term ones—we will need, above all, those principles that conservative scholars have long called the permanent things. But in our society today, far too many politicians and pundits who claim to be conservatives are merely using the label as a flag of convenience. Far too often they act not from principle but to serve their campaign contributors, or they merely play on popular prejudices to win ephemeral popularity. This is precisely what Russell Kirk warned us about: such "time servers" and "placemen" cannot govern well.[4] Conservatives should oppose these opportunists at every hand.

We need to return to fundamentals, to principles. If we can succeed in doing that, we will find that the true color of conservative America is Green.

NOTES

INTRODUCTION

1. For a detailed analysis of these attacks, see Robert L. Glicksman and Stephen B. Chapman, "Regulatory Reform and (Breach of) the Contract with America: Improving Environmental Policy or Destroying Environmental Protection?" *Kansas Journal of Law and Public Policy* 5, no. 2 (Winter 1996): 9–46.

2. Quoted by Theodore Roosevelt IV, speech to the Women's National Republican Club, October 8, 1996. Reprinted in part in *Green Elephant* 1, no. 2 (Summer 1997): 1. The *Green Elephant* is a newsletter produced by the Republicans for Environmental Protection, a grassroots organization that takes a very strong pro-environmental stand and is attempting to influence Republican policies from within the party. Theodore Roosevelt IV (now chairman of the League of Conservation Voters) is a member, as was Barry Goldwater. A recent president of the Sierra Club is also a member. For information on REP, see <www.repamerica.org>.

3. Roosevelt, speech, p. 2.

4. For a discussion of two recent polls, see "Eye on Washington," *Green Elephant* 3, no. 2 (Fall 1999): 6.

5. Russell Kirk, *Prospects for Conservatives* (Washington, D.C.: Regnery Gateway, 1989), pp. 260–261.

6. John C. Vinson Jr., "Conservatives and Environmentalists: Allies, Not Enemies," *Chronicles,* June 1996, p. 30.

7. Paul M. Weyrich, "Cultural Conservatism and the Conservative Movement," in *Cultural Conservatism: Theory and Practice,* ed. William S. Lind and William H. Marshner (Washington, D.C.: Free Congress Foundation, 1991), p. 25.

8. Richard M. Weaver, *The Ethics of Rhetoric* (South Bend, Ind.: Regnery/Gateway, 1953), chaps. 3–4.

9. The title of a recent book, *Conservative Environmentalism,* by James R. Dunn and John E. Kinney (Westport, Conn.: Quorum, 1996), suggests that it may do something similar to what is attempted here. Unfortunately, besides belaboring the obvious—that the third world has serious environmental problems and is on the whole a lot worse place to live than the first world—the book is little more than a collection of non sequiturs and abuse of evidence, and perhaps the less said of it the better. But here I must note that nowhere do the authors even attempt to define what they mean by "conservative." In fact, there is no reason given anywhere in the book for including the term in its title.

10. There are other trends within contemporary conservatism, such as populism and neoconservatism, that will not be considered here. They take their principles from the two basic schools, selectively and often mingled with ideas from other political positions as well. We can take the libertarian and traditionalist as the "pure" forms from which the basic principles of conservatism can be derived.

11. John P. East, "The American Conservative Movement of the 1980s: Are Traditional and Libertarian Dimensions Compatible?" *Modern Age* 24 (Winter 1980): 35; George H. Nash, *The Conservative Intellectual Movement in America since 1945*, 2d ed. (Wilmington, Del.: Intercollegiate Studies Institute, 1996), p. 73.

12. East, "American Conservative Movement of the 1980s," p. 34; Nash, *Conservative Intellectual Movement*, p. 73. Those today who call themselves "social conservatives" are pretty close to the traditionalist position.

13. See, for example, the Winter 1980 issue of *Modern Age* (vol. 24), which was devoted to articles debating traditionalism versus libertarianism. A few follow-up articles appear in volume 26 (Winter 1982). Details of this debate can be found, along with sketches of the personalities involved, in Nash's magisterial history, *Conservative Intellectual Movement.*

14. Frank S. Meyer, "Freedom, Tradition, Conservatism," *Modern Age* 4 (1960): 356–357.

15. Edwin J. Feulner Jr., *The March of Freedom: Modern Classics in Conservative Thought* (Dallas: Spence, 1998), p. 142.

16. Nash, *Conservative Intellectual Movement*, p. 164.

17. Feulner, *March of Freedom*, p. 142.

18. Nash, *Conservative Intellectual Movement*, p. 169. For an analysis of Buckley's "libertarianism of this world," with its commitment to the free market as well as to Catholicism and Western civilization, see Dante Germino, "Traditionalism and Libertarianism: Two Views," *Modern Age* 26 (1982): 52–56.

19. See Carl B. Cone, *Burke and the Nature of Politics* (Lexington: University of Kentucky Press, 1957, 1964): 1:326, 2:146–148, 490–491; Conor Cruise O'Brien, *The Great Melody: A Thematic Biography and Commented Anthology of Edmund Burke* (Chicago: University of Chicago Press, 1992), p. 144; Donal Barrington, "Edmund Burke as an Economist," *Economica,* 2d ser., 21 (1954): 252–258; William Clyde Dunn, "Adam Smith and Edmund Burke: Complementary Contemporaries," *Southern Economic Journal* 7 (1941): 330–346.

20. Nolan Clark, "The Environmental Protection Agency," in *Mandate for Leadership III: Policy Strategies for the 1990s,* ed. Charles L. Heatherly and Burton Yale Pines (Washington, D.C.: Heritage Foundation, 1989), pp. 216–218.

21. Russell Kirk, "Conservation Activism Is a Healthy Sign," *Baltimore Sun,* May 4, 1970, A17.

22. Gordon K. Durnil, *The Making of a Conservative Environmentalist* (Bloomington: Indiana University Press, 1995), p. 186.

CHAPTER ONE

1. Paul R. Ehrlich and Anne H. Ehrlich, *Betrayal of Science and Reason* (Washington, D.C.: Island, 1996); Edward Flattau, *Tracking the Charlatans*, 2d ed. (Washington, D.C.: Global Horizons, 1999).

2. There is a small Green Party in the United States, which has a wide agenda. But it certainly does not represent the Sierra Club, Wilderness Society, National Wildlife Federation, The Nature Conservancy, Audubon Society, or any of the other major environmental groups.

3. Frances Cairncross, "Costing the Earth (Survey)," *Economist,* September 2, 1989, p. 5.

4. Russell Kirk, "Common Reader for Everyday Ecologists," *New Orleans Times-Picayune,* September 20, 1971, sec. 1, p. 11, quoting and endorsing a book by Alan Bock.

5. John Gray, *Beyond the New Right* (London: Routledge, 1993), p. 124.

6. Russell Kirk, "Conservation Activism Is a Healthy Sign," *Baltimore Sun,* May 4, 1970, A17.

7. Marshall Ingwerson, "On the Environment, Americans' Words Are Louder Than Deeds," *Christian Science Monitor,* August 2, 1990, p. 2.

8. Gordon K. Durnil, *The Making of a Conservative Environmentalist* (Bloomington: Indiana University, 1995), p. 121. Gordon Durnil is an attorney in Indianapolis who has long been involved with the Republican Party. During the Bush administration he served as U.S. chairman of the International Joint Commission. The IJC is an advisory agency formed by the governments of the United States and Canada to study and make recommendations about border issues, primarily involving water quality and quantity in the Great Lakes region. During his term, the IJC was involved in researching the effects of persistent toxic chemicals that are widespread around the globe but are especially problematical in the Great Lakes. These chlorine-based

chemicals can disrupt the endocrine system of animals (including humans), causing birth defects, behavioral abnormalities, reproductive disorders, and some kinds of cancers. The results of this research led Durnil to become an active environmentalist.

9. For example, the Political Economy Research Center (PERC) and the Foundation for Research in Economics and the Environment (FREE), both in Bozeman, Montana.

10. Joan Hamilton, "Why Conservatives Aren't Conserving," *Sierra,* September–October 1994, p. 33.

11. Ed Marston and Chilton Williamson Jr., "Environmentalism, Culture and Politics," *Chronicles,* June 1996, p. 25.

12. Durnil, *Making of a Conservative Environmentalist,* p. 46.

13. For example, see William F. Buckley Jr., *The Jeweler's Eye* (New York: Putnam, 1968), pp. 15–31.

14. Trebbe Johnson, "The Second Creation Story," *Sierra,* November–December 1998, pp. 55–56.

15. National Religious Partnership for the Environment, <www.nrpe@nrpe.org>.

16. Johnson, "Second Creation Story," p. 55.

17. National Council of Churches of Christ, news release, August 18, 1998.

18. Johnson, "Second Creation Story," p. 52.

19. For example, *Renewing the Face of the Earth: A Resource for Parishes* (Washington, D.C.: United States Catholic Conference, 1994), which includes the Bishops' Statement.

20. Johnson, "Second Creation Story," p. 52.

21. See, for example, John B. Cobb Jr., *Is It Too Late? A Theology of Ecology* (Denton, Tex.: Environmental Ethics Books, 1995) and the extensive bibliography it contains.

22. Chuck D. Barlow, "Why the Christian Right Must Protect the Environment: Theocentricity in the Political Workplace," *Boston College Environmental Affairs Law Review* 23 (1996): 783.

23. Psalm 24:1.

24. Barlow, "Why the Christian Right Must Protect the Environment," p. 800.

25. Barlow, "Why the Christian Right Must Protect the Environment," p. 805, quoting Jonathan Helfand.

26. Wendell Berry, "Christianity and the Survival of Creation," in *Sex, Economy, Freedom and Community* (New York: Pantheon, 1993), p. 98.

27. C. S. Lewis, *God in the Dock* (Grand Rapids, Mich.: Eerdmans, 1970), pp. 42–43.

28. Ernest L. Fortin, "The Bible Made Me Do It: Christianity, Science, and the Environment," *Review of Politics* 57 (1995): 201.

29. Pope John Paul II, *The Ecological Crisis: A Common Responsibility* (message for the World Day of Peace, January 1, 1990).

30. Luke 12:6–7.

31. Barlow, "Why the Christian Right Must Protect the Environment," p. 811; emphasis in the original.

32. Peter N. Spotts, "Theologians, Scientists Meet on 'Green Ethics,'" *Christian Science Monitor,* November 14, 1995, p. 13.

33. Genesis 1:26.

34. R. V. Young Jr., "Christianity and Ecology," *National Review,* December 20, 1974, pp. 1456–1457.

35. Max Oelschlaeger, *Caring for Creation* (New Haven: Yale University, 1994), p. 130. The internal quotation is from Francis Schaeffer.

36. Lloyd H. Steffen, "In Defense of Dominion," *Environmental Ethics* 14 (1992): 64.

37. James A. Nash, "Ecological Integrity and Christian Political Responsibility," *Theology and Public Policy* 1 (1989): 34.

38. Lynn White Jr., "The Historical Roots of Our Ecologic Crisis," *Science* 155 (1967): 1204.

39. White, "Historical Roots of Our Ecologic Crisis," pp. 1205–1206.

40. Carl Pope, "Reaching Beyond Ourselves," *Sierra,* November–December 1998, p. 14.

41. John P. East, *The American Conservative Movement: The Philosophical Founders* (Chicago: Regnery, 1986), pp. 156–159, 167–170.

42. Pope, "Reaching Beyond Ourselves," p. 14.

43. White, "Historical Roots of Our Ecologic Crisis," pp. 1206–1207.

44. Pope John Paul II, "Inter Sanctos," *Acta Apostolicae Sedis* 71 (1979): 1509–1510.

45. Oelschlaeger, *Caring for Creation.*

46. William A. Rusher, "Conservatism's Third and Final Battle," *Intercollegiate Review* 34, no. 1 (Fall 1998): 4.

47. Michael E. Porter and Claas van der Linde, "Toward a New Conception of the Environment-Competitiveness Relationship," *Journal of Economic Perspectives* 9 (1995): 107.

48. Quoted in B. J. Bergman, "Wild at Heart," *Sierra,* January–February 1998, p. 26.

49. Clinton Rossiter, *Conservatism in America*, 2d ed. (Cambridge: Harvard University Press, 1982), p. 204.

50. Russell Kirk, *The Conservative Mind*, 7th ed. (Chicago: Regnery, 1986), p. 229.

51. Peter Viereck, *Conservatism Revisited* (New York: Scribner, 1949), p. x.

52. Stephen J. Tonsor, letter to the editor, *Reporter,* August 11, 1955, p. 8.

53. Richard M. Weaver, *Visions of Order* (Baton Rouge: Louisiana State University Press, 1964), pp. 32–33.

54. Russell Kirk, "The American Conservative Character," *Georgia Review* 8 (Fall 1954): 252–255.

55. Russell Kirk, "Enlivening the Conservative Mind," *Intercollegiate Review* 21 (Spring 1986): 27–28.

56. Milton Friedman, *Capitalism and Freedom* (Chicago: University of Chicago Press, 1982), p. 68.

57. Francis Graham Wilson, *The Case for Conservatism* (Seattle: University of Washington Press, 1951), p. 65.

58. Adam Smith, *An Inquiry into the Nature and Causes of the Wealth of Nations*, ed. Edwin Cannan (New York: Modern Library, 1937), p. 128.

59. "Leaders," *Economist,* April 13, 1996, p. 14.

CHAPTER TWO

1. Quoted in Eban Goodstein, *The Trade-Off Myth: Fact and Fiction About Jobs and the Environment* (Washington, D.C.: Island, 1999), p. 25.

2. Eugene Linden, "The Green Factor," *Time,* October 12, 1992, p. 58.

3. In this context, polls indicating that Americans believe the environment should be protected *even if* it hurts the economy are interesting. For example, a Roper Starch poll in late 1998 found 15 percent of Americans in favor of protecting the environment even at the expense of economic growth; another 57 percent said we should try to have a balance but that the environment was more important; *Christian Science Monitor,* February 1, 1999, p. 12.

4. E. B. Goodstein, *Jobs and the Environment: The Myth of a National Trade-Off* (Washington, D.C.: Economic Policy Institute, 1994), p. 1.

5. Organization for Economic Cooperation and Development (OECD), *Environmental Performance in OECD Countries* (Paris: OECD, 1996), pp. 21–22.

6. OECD, *Integrating Environment and Economy* (Paris: OECD, 1996), p. 42; emphasis in the original.

7. OECD, *Environmental Performance Reviews: United States* (Paris: OECD, 1996), pp. 135, 151.

8. Stephen M. Meyer, *Environmentalism and Economic Prosperity: Testing the Environmental Impact Hypothesis* (Cambridge: Massachusetts Institute of Technology/Project on Environmental Politics and Policy, 1992), p. iv.

9. Meyer, *Environmentalism and Economic Prosperity*, p. 21.

10. Institute for Southern Studies, *Gold and Green* (Durham, N.C.: Institute for Southern Studies, 1994); Bob Hall, "Gold and Green," *Southern Exposure,* Fall 1994, pp. 48–52. The journal article only gives the statistics for southern states; all indicators for all fifty states appear in the full report, pp. 12–19.

11. Institute for Southern Studies, *Gold and Green*, pp. 1, 6; Hall, "Gold and Green," p. 52.

12. Mary H. Cooper, "Jobs vs. Environment," *CQ Researcher* 2, no. 18 (1992): 418.

13. Roger H. Bezdek, "Environment and Economy: What's the Bottom Line?" *Environment* 35, no. 7 (1993): 9.

14. "Muck into Money Again," *Economist*, September 19, 1992, p. 44.

15. Michael E. Porter and Claas van der Linde, "Green and Competitive: Ending the Stalemate," *Harvard Business Review*, September–October 1995, p. 122.

16. Porter and van der Linde, "Green and Competitive," p. 120; emphasis in the original.

17. Porter and van der Linde, "Green and Competitive," p. 122.

18. Porter and van der Linde, "Green and Competitive," pp. 125–126.

19. Porter and van der Linde, "Green and Competitive," p. 126.

20. Porter and van der Linde, "Green and Competitive," p. 126.

21. Porter and van der Linde, "Green and Competitive," p. 126.

22. Porter and van der Linde, "Green and Competitive," pp. 128–129.

23. Michael E. Porter and Claas van der Linde, "Toward a New Conception of the Environment-Competitiveness Relationship," *Journal of Economic Perspectives* 9 (1995): 101–102.

24. Porter and van der Linde, "Toward a New Conception," p. 104.

25. Porter and van der Linde, "Green and Competitive," p. 127.

26. OECD, *Environmental Performance Reviews: United States*, p. 135; emphasis in the original. Many economists are suspicious of the innovation stimulation thesis. They believe that it is highly unlikely that there are significant numbers of cost-saving measures that remain undiscovered, since managers maximize profits and minimize costs. Porter and van der Linde, however, contend that this "Panglossian belief . . . does not describe reality." "Toward a New Conception," p. 99. Porter and van der Linde are, of course, quite correct. This issue will be analyzed in some detail in Chapter 7, when we examine energy efficiency measures to reduce greenhouse gas emissions.

27. Porter and van der Linde, "Toward a New Conception," p. 110.

28. Adam B. Jaffe et al., "Environmental Regulation and the Competitiveness of U.S. Manufacturing: What Does the Evidence Tell Us?" *Journal of Economic Literature* 33 (1995): 149–150.

29. Meyer, *Environmentalism and Economic Prosperity*, pp. 42–43.

30. Robert Repetto, *Jobs, Competitiveness, and Environmental Regulation: What Are the Real Issues?* (Washington, D.C.: World Resources Institute, 1995), p. 8.

31. Jaffe et al., "Environmental Regulation," pp. 147–148.

32. Judith M. Dean, "Trade and the Environment: A Survey of the Literature," in *International Trade and the Environment*, edited by Patrick Low, World Bank Discussion Papers 159 (Washington, D.C.: World Bank, 1992), p. 27; see also Goodstein, *Trade-Off Myth*, pp. 55–66.

33. Bezdek, "Environment and Economy," p. 11.

34. Repetto, *Jobs, Competitiveness, and Environmental Regulation*, p. 11.

35. Repetto, *Jobs, Competitiveness, and Environmental Regulation*, p. 12.

36. Repetto, *Jobs, Competitiveness, and Environmental Regulation*, pp. 15–16, 18.

37. Repetto, *Jobs, Competitiveness, and Environmental Regulation*, p. 22.

38. Goodstein, *Jobs and the Environment*, p. 4.

39. OECD, *Environmental Performance Reviews: United States*, p. 136; emphasis in the original.

40. Goodstein, *Jobs and the Environment*, p. 4.

41. Goodstein, *Trade-Off Myth*, p. 1.

42. Goodstein, *Jobs and the Environment*, pp. 12–14.

43. Goodstein, *Trade-Off Myth*, p. 45.

44. Goodstein, *Jobs and the Environment*, p. 12.

45. Goodstein, *Jobs and the Environment*, p. 11.

46. Bezdek, "Environment and Economy," pp. 26–28.

47. "California Cashes In on Cleaning Up," *Economist*, November 16, 1991, p. 79.

48. OECD, *Economic Performance Reviews: United States*, p. 136.

49. Michael Silverstein, "In Economics, 'Green' Is Gold," *Christian Science Monitor*, February 18, 1992, p. 19.

50. Curtis Moore, "Green Revolution," *Sierra,* January–February 1995, pp. 51–52; "The Money in Europe's Muck," *Economist,* November 20, 1993, p. 81.

51. Moore, "Green Revolution," p. 126.

52. Bezdek, "Environment and Economy," p. 31.

53. General Accounting Office (GAO), *Endangered Species Act: Impact of Species Protection Efforts on the 1993 California Fire,* RCED-94-224 (Washington, D.C., July 1994), p. 2.

54. GAO, *Endangered Species Act: Impact,* p. 11.

55. Fish and Wildlife Service, *Facts about the Endangered Species Act* (Washington, D.C., July 1995).

56. Paul R. Ehrlich and Anne H. Ehrlich, *Betrayal of Science and Reason* (Washington, D.C.: Island, 1996), p. 116. This story is not included in the FWS list.

57. Rocky Barker, *Saving All the Parts: Reconciling Economics and the Endangered Species Act* (Washington, D.C.: Island, 1993), p. 138.

58. Quoted in Barker, *Saving All the Parts,* p. 140.

59. General Accounting Office, *Endangered Species Act: Types and Number of Implementing Actions,* RCED-92-131BR (Washington, D.C.: Government Printing Office, 1992), pp. 30–32.

60. David Hoskins et al., *For Conserving Listed Species, Talk Is Cheaper Than We Think: The Consultation Process Under the Endangered Species Act,* 2d ed. (Washington, D.C.: World Wildlife Fund, 1994); see pp. 11–12 for a comparison of their statistics with those in the GAO study.

61. John C. Sawhill, "Saving Endangered Species Doesn't Endanger Economy," *Wall Street Journal,* February 20, 1992, p. A15.

62. Barker, *Saving All the Parts,* pp. 140–141.

63. Stephen M. Meyer, *Endangered Species Listings and State Economic Performance,* Project on Environmental Politics and Policy Working Paper no. 4 (Cambridge: Massachusetts Institute of Technology, 1995), p. 2.

64. Meyer, *Endangered Species Listings,* pp. 3, 5; emphasis in the original.

65. Meyer, *Endangered Species Listings,* p. 14.

66. Meyer, *Endangered Species Listings,* p. 15.

67. Meyer, *Endangered Species Listings,* p. 16. See also Stephen M. Meyer, "The Economic Impact of the Endangered Species Act on the Housing and Real Estate Markets," *New York University Environmental Law Journal* 6 (1998): 454–460.

68. Peter H. Morrison et al., *Ancient Forests in the Pacific Northwest* (Washington, D.C.: Wilderness Society, 1991).

69. See William G. Robbins, "Timber Town," *Pacific Northwest Quarterly* 75 (1984): 146–155; William G. Robbins, "The Social Context of Forestry: The Pacific Northwest in the Twentieth Century," *Western Historical Quarterly* 16 (1985): 413–427; William G. Robbins, "Lumber Production and Community Stability," *Journal of Forest History* 31 (1987): 187–196.

70. Michael Hibbard and James Elias, "The Failure of Sustained-Yield Forestry and the Decline of the Flannel-Shirt Frontier," in *Forgotten Places: Uneven Development in Rural America,* ed. Thomas A. Lyson and William W. Falk (Lawrence: University Press of Kansas, 1993), p. 210.

71. William R. Freudenberg et al., "Forty Years of Spotted Owls? A Longitudinal Analysis of Logging Industry Job Losses," *Sociological Perspectives* 41 (1998): 18.

72. Freudenberg et al., "Forty Years of Spotted Owls?" p. 18.

73. Michael Hibbard, "Issues and Options for the Other Oregon," *Community Development Journal* 24 (1989): 147.

74. Hibbard and Elias, "Failure of Sustained-Yield Forestry," pp. 195–196.

75. Robbins, "Timber Town," pp. 152–153; Robbins, "Social Context of Forestry," pp. 418–420; Robbins, "Lumber Production and Community Stability," pp. 192–193.

76. Hibbard and Elias, "Failure of Sustained-Yield Forestry," p. 209.

77. *Northern Spotted Owl v. Hodel,* 716 F. Supp. 479 (W.D.Wash. 1988).

78. *Northern Spotted Owl v. Lujan,* 758 F. Supp. 621 (W.D.Wash. 1991).

79. *Seattle Audubon Society v. Evans,* 771 F. Supp. 1081 (W.D.Wash. 1991). The injunction was upheld by the court of appeals; *Seattle Audubon Society v. Evans* 952 F. 2d 297 (9th Cir. 1991).

80. For a short but detailed history of the spotted owl case, see Steven L. Yaffee, "Lessons About Leadership from the History of the Spotted Owl Controversy," *Natural Resources Journal* 35 (1995): 381–412. The Option 9 plan is still controversial because it allows some old growth to be cut and because it has not yet given the industry the amount of timber promised to them; see Chris Carrell, "A Patchwork Peace Unravels," *High Country News,* November 23, 1998, pp. 1, 8–12.

81. David Seideman, "Out of the Woods," *Aubudon,* July–August 1996, p. 71.

82. Thomas Michael Power, *Lost Landscapes and Failed Economies* (Washington, D.C.: Island, 1996), p. 273.

83. T. H. Watkins, "What's Wrong with the Endangered Species Act? Not Much—and Here's Why," *Audubon,* January–February 1996, p. 38.

84. Power, *Lost Landscapes and Failed Economies,* p. 165.

85. Seideman, "Out of the Woods," p. 71.

86. *Seattle Audubon Society v. Evans,* 771 F. Supp. 1081 (W.D.Wash. 1991) at 1095.

87. Freudenberg et al., "Forty Years of Spotted Owls?" p. 1.

88. Freudenberg et al., "Forty Years of Spotted Owls?" p. 15. This study goes far beyond the employment effects of the spotted owl case. The authors analyze employment trends in logging and milling nationally and in the Pacific Northwest from 1947 to 1994, to see if three key environmental "turning points" had any impact. Besides the spotted owl case, they look for impacts of the Wilderness Act of 1964 and the combination of the 1969 National Environmental Policy Act (NEPA) and Earth Day, 1970. The only change in employment trends associated with any of these environmental protection events was with the Wilderness Act and that change was positive, not negative—precisely the opposite of what critics of environmentalism would expect.

89. Freudenberg et al., "Forty Years of Spotted Owls?" p. 22. They suggest that timber jobs have long been in jeopardy not because of environmental protection regulations but for precisely the opposite reason: the forests have *not* been protected from unsustainable logging—"mining" the trees—over the past century or so; p. 18.

90. Jonathan Rubin et al., "A Benefit-Cost Analysis of the Northern Spotted Owl," *Journal of Forestry,* December 1991, p. 28. Contingent valuation is a standard economic methodology that measures some, but far from all, of the economic value of a resource that is not traded on markets. It determines a monetary price for nonuse values, usually by using surveys of people's willingness to pay to preserve that resource.

91. Daniel A. Hagen et al., "Benefits of Preserving Old-Growth Forests and the Spotted Owl," *Contemporary Policy Issues* 10 (April 1992): 13.

92. Hagen et al., "Benefits of Preserving Old-Growth Forests," pp. 20–21.

93. Hagen et al., "Benefits of Preserving Old-Growth Forests," p. 15.

94. Hagen et al., "Benefits of Preserving Old-Growth Forests," p. 14.

95. Thomas Michael Power, ed., *Economic Well-Being and Environmental Protection in the Pacific Northwest: A Consensus Report by Pacific Northwest Economists* (Missoula: Economics Department, University of Montana, 1995), pp. ii-iii; emphasis in the original.

96. Power, *Economic Well-Being,* pp. 9–11; emphasis in the original.

97. Power, *Economic Well-Being,* pp. 14, 17; emphasis in the original.

98. Quoted in Peter Hong and Dori Jones Yang, "Tree-Huggers vs. Jobs: It's Not That Simple," *Business Week,* October 19, 1992, p. 109.

99. Freudenberg et al., "Forty Years of Spotted Owls?" p. 16.

100. Quoted in Seideman, "Out of the Woods," p. 67.

101. Seideman, "Out of the Woods," p. 67.

102. Power, *Lost Landscapes and Failed Economies,* pp. 141, 158.

CHAPTER THREE

1. William R. Harbour, *The Foundations of Conservative Thought* (Notre Dame, Ind.: University of Notre Dame Press, 1982), p. 15.

2. Christopher Dawson, *Progress and Religion* (London: Sheed & Ward, 1929), p. 9.

3. T. S. Eliot, *Christianity and Culture* (New York: Harcourt, Brace & World, 1949), p. 91.

4. William J. Bennett, *The Index of Leading Cultural Indicators* (New York: Simon & Schuster, 1994), p. 8.

5. Frank S. Meyer, "Richard M. Weaver: An Appreciation," *Modern Age* 14 (1970): 243.

6. Richard M. Weaver, *Ideas Have Consequences* (Chicago: University of Chicago Press, 1948), p. 6.

7. Richard M. Weaver, "Mass Plutocracy," *National Review,* November 5, 1960, p. 273.

8. Weaver, *Ideas Have Consequences*, chap. 6.

9. Richard M. Weaver, *The Southern Tradition at Bay*, ed. George Core and M. E. Bradford (New Rochelle, N.Y.: Arlington House, 1968), p. 392.

10. Weaver, *Southern Tradition at Bay*, p. 392.

11. Weaver, *Ideas Have Consequences*, p. 12.

12. Weaver, *Southern Tradition at Bay*, p. 31.

13. Weaver, *Ideas Have Consequences*, pp. 116–117.

14. Weaver, *Southern Tradition at Bay*, pp. 394–395.

15. Weaver, *Southern Tradition at Bay*, p. 396.

16. Weaver, *Southern Tradition at Bay*, p. 391; emphasis in the original.

17. Russell Kirk, *The Conservative Mind*, 7th ed. (Chicago: Regnery, 1986), p. iii.

18. Kirk, *Conservative Mind*, p. 11.

19. Russell Kirk, "Ideology and Political Economy," *America,* January 5, 1957, p. 389.

20. Russell Kirk, *Redeeming the Time* (Wilmington, Del.: Intercollegiate Studies Institute, 1996), p. 38.

21. Russell Kirk, *Prospects for Conservatives* (Washington, D.C.: Regnery Gateway, 1989), pp. 153–154.

22. Kirk, *Prospects for Conservatives*, p. 172.

23. Quoted in Edwin J. Feulner Jr., ed., *The March of Freedom* (Dallas: Spence, 1998), pp. 297–298.

24. Wilhelm Röpke, "The Economic Necessity of Freedom," *Modern Age* 3 (1959): 234.

25. Wilhelm Röpke, *A Humane Economy* (Chicago: Henry Regnery, 1960), pp. 82–83.

26. Röpke, *Humane Economy*, p. 109.

27. Edmund Burke, *Reflections on the Revolution in France*, ed. L. G. Mitchell (Oxford: Oxford University Press, 1993), p. 76.

28. Harbour, *Foundations of Conservative Thought*, p. 110.

29. Murray N. Rothbard, "Myth and Truth About Libertarianism," *Modern Age* 24 (1980): 10; emphasis in the original.

30. Murray N. Rothbard, *Power and Market* (Menlo Park, Calif.: Institute for Humane Studies, 1970), p. 194.

31. Tibor R. Machan, "Libertarianism and Conservatives," *Modern Age* 24 (1980): 29.

32. Tibor R. Machan, *Individuals and Their Rights* (La Salle, Ill.: Open Court, 1989), pp. xiii–xiv.

33. Tibor R. Machan, *Capitalism and Individualism* (New York: St. Martin's, 1990), p. 77.

34. Machan, *Individuals and Their Rights*, p. 156.

35. Machan, *Capitalism and Individualism*, p. 79.

36. Machan, *Individuals and Their Rights*, p. xiv.

37. Tibor R. Machan, "Pollution and Political Theory," in *Earthbound,* ed. Tom Regan (New York: Random House, 1984), p. 98. Even at the political level, the conservative movement, at least in its early years, stressed this principle. According to Barry Goldwater, "Conservatism is *not* an economic theory, though it has economic implications. . . . It is Conservatism that puts material things in their proper place—that has a structured view of the human being and of human society, in which economics plays only a subsidiary role. . . . Conservatives take account of the *whole* man. . . . The Conservative believes that man is, in part, an economic, an animal creature; but that he is also a spiritual creature with spiritual needs and spiritual desires. What is more, these needs and desires reflect the *superior* side of man's nature, and thus take prece-

dence over his economic wants"; *Conscience of a Conservative* (Shepherdsville, Ky.: Victor, 1960), pp. 10–11; emphasis in the original. Goldwater, it should be noted, was a member of Republicans for Environmental Protection, an organization that takes a very strong pro-environment stand.

38. Machan, "Libertarianism and Conservatives," p. 23.

39. Rothbard, "Myth and Truth About Libertarianism," pp. 9–10; emphasis in the original.

40. Russell Kirk, *Beyond the Dreams of Avarice* (Peru, Ill.: Sherwood Sugden, 1991), p. 166.

41. Kirk, *Prospects for Conservatives*, p. 189.

42. John Gray, *Beyond the New Right* (London: Routledge, 1993), p. 4.

43. Gray, *Beyond the New Right*, p. 80.

44. Gray, *Beyond the New Right*, p. 82.

45. Gray, *Beyond the New Right*, pp. 111–112.

46. Gray, *Beyond the New Right*, p. 60.

47. Gray, *Beyond the New Right*, p. 122.

48. Gray, *Beyond the New Right*, p. 82.

49. Machan, "Pollution and Political Theory," p. 100.

50. Rothbard, "Myth and Truth About Libertarianism," p. 10.

51. "Clean Air, Dirty Fight," *Economist*, March 15, 1997, pp. 29–30; Alexandra Marks, "Nation Holds Its Breath Over Clean-Air Rules," *Christian Science Monitor*, June 12, 1997, p. 1.

52. Gordon K. Durnil, *The Making of a Conservative Environmentalist* (Bloomington: Indiana University Press, 1995), p. 184.

53. Machan, "Pollution and Political Theory," p. 97; emphasis in the original.

54. Friedrich A. Hayek, *The Constitution of Liberty* (Chicago: University of Chicago Press, 1960), p. 71.

55. Kirk, *Redeeming the Time*, p. 33.

56. Harbour, *Foundations of Conservative Thought*, pp. 102–103.

57. Quoted in Feulner, *March of Freedom*, p. 170.

58. Durnil, *Making of a Conservative Environmentalist*, pp. 184–185.

59. Durnil, *Making of a Conservative Environmentalist*, pp. 43, 48.

60. Durnil, *Making of a Conservative Environmentalist*, chap. 7.

61. Carl Pope, "A Good, Clean Fight," *Sierra*, March–April 1997, p. 12. The suit is now before the Supreme Court.

62. Carl Pope, "Unnatural Causes," *Sierra*, January–February 1998, p. 13.

63. John Byrne Barry, "The Club in the Capitols," *Planet*, July–August 1996, p. 4. *The Planet* is a newsletter produced by the Sierra Club.

64. Machan, *Capitalism and Individualism*, p. 79.

65. Kirk, *Conservative Mind*, p. 9.

66. Francis Graham Wilson, *The Case for Conservatism* (Seattle: University of Washington Press, 1951), p. 22.

67. Weaver, *Ideas Have Consequences*, chap. 7.

68. Denis Collins and John Barkdull, "Capitalism, Environmentalism, and Mediating Structures: From Adam Smith to Stakeholder Panels," *Environmental Ethics* 17 (1995): 237, quoting Smith's *Lectures on Jurisprudence*.

69. Mark Sagoff, "Some Problems with Environmental Economics," *Environmental Ethics* 10 (1988): 61. Some libertarians believe that enforcing property rights could be the sole and sufficient approach to pollution control. This contention will be considered in greater detail in Chapter 10, where we will see that it simply would not work in our legal system.

70. John C. Vinson Jr., "Conservatives and Environmentalists Allies Not Enemies," *Chronicles*, June 1996, p. 31.

71. For the whole sorry tale, see David Harris, *The Last Stand* (San Francisco: Sierra Club Books, 1996). Under its new ownership, the company has been so abusive of the forests, committing numerous violations of state regulations, that its operating license was suspended by the California government in November 1998; Stanley Yung, "Loggers Told to Stop Cutting," *High Country News*, December 7, 1998, p. 7.The core of Pacific Lumber's old-growth redwoods is the

famous Headwaters Forest, which environmentalists have prevented the company from cutting. The federal government finally bought it in order to preserve it.

72. Clinton Rossiter, *Conservatism in America*, 2d ed. (Cambridge: Harvard University Press, 1982), pp. 38–39.

73. Chuck D. Barlow, "Why the Christian Right Must Protect the Environment: Theocentricity in the Political Workplace," *Boston College Environmental Affairs Law Review* 23 (1996): 823.

74. Gray, *Beyond the New Right*, p. 62.

75. The OECD has formally adopted this idea, calling it the "polluter pays principle." Unfortunately, member countries, including the United States, have far to go in implementing it.

76. Milton Friedman and Rose Friedman, *Free to Choose* (New York: Harcourt Brace Jovanovich, 1980), p. 32.

77. Friedman and Friedman, *Free to Choose*, pp. 31–32.

78. Milton Friedman, *Capitalism and Freedom* (Chicago: University of Chicago Press, 1982), p. 32.

79. House Republican Conference Issue Brief, "Clinton's War on the West," May 26, 1994; Republican national chairman Haley Barbour, quoted in John Dillin, "For '94 Races, GOP Chief Predicts 'Very' Big Gains," *Christian Science Monitor*, September 9, 1994, p. 2.

80. Erich Pica, ed., *Green Scissors 2000* (Washington, D.C.: Friends of the Earth, 2000).

81. Feulner, *March of Freedom*, p. xix.

82. Frank S. Meyer, "Freedom, Tradition, Conservatism," *Modern Age* 4 (1960): 358.

83. John P. East, *The American Conservative Movement: The Philosophical Founders* (Chicago: Regnery, 1986), p. 200.

84. East, *American Conservative Movement*, p. 146.

85. Richard M. Weaver, *The Southern Essays of Richard M. Weaver*, ed. George M. Curtis III and James J. Thompson Jr. (Indianapolis: Liberty Press, 1987), pp. 220–221.

86. Weaver, *Southern Essays*, p. 197.

87. Richard M. Weaver, *Visions of Order* (Baton Rouge: Louisiana State University Press, 1964), p. 121.

88. Weaver, *Southern Essays*, p. 57.

89. Weaver, *Southern Essays*, p. 24.

90. Richard M. Weaver, *Life Without Prejudice and Other Essays* (Chicago: Henry Regnery, 1965), p. 141.

91. Weaver, *Southern Tradition at Bay*, p. 32.

92. Psalm 24:1.

93. Weaver, *Life Without Prejudice*, p. 143.

94. Weaver, *Life Without Prejudice*, p. 141.

95. Weaver, *Ideas Have Consequences*, p. 172.

96. Weaver, *Southern Tradition at Bay*, p. 32.

97. Weaver, *Visions of Order*, p. 123.

98. Weaver, *Ideas Have Consequences*, p. 171.

99. Weaver, *Visions of Order*, p. 121.

100. Weaver, *Ideas Have Consequences*, p. 171.

101. Weaver, *Life Without Prejudice*, p. 145.

102. Weaver, *Southern Essays*, p. 19.

103. Weaver, *Ideas Have Consequences*, pp. 172–173.

104. Weaver, *Southern Tradition at Bay*, p. 39.

105. Eliot, *Christianity and Culture*, pp. 48–49.

106. Russell Kirk, "Impious Generations Are Often Rebuked," *New Orleans Times-Picayune*, February 26, 1968, sec. 1, p. 9.

107. Kirk, *Conservative Mind*, pp. 44–45.

108. Edward O. Wilson, *The Diversity of Life* (Cambridge: Harvard University Press, 1992), p. 346.

109. Russell Kirk, "Man Undoing Noah's Preservation Program," *New Orleans Times-Picayune*, June 1, 1965, sec. 1, p. 9.

110. Weaver, *Ideas Have Consequences*, pp. 170–171.

111. Wendell Berry, *A Continuous Harmony* (New York: Harcourt Brace Jovanovich, 1972), pp. 174–182.

112. Burke, *Reflections on the Revolution in France*, p. 96.

113. Burke, *Reflections on the Revolution in France*, p. 95.

114. Quoted in Frances Cairncross, "Costing the Earth (Survey)," *Economist*, September 2, 1989, p. 3.

115. Carl Pope, "Clinton's Layaway Plan," *Sierra*, May–June 1994, p. 16.

116. World Commission on Environment and Development, *Our Common Future* (Oxford: Oxford University Press, 1987), p. 43.

117. William S. Lind and William H. Marshner, *Cultural Conservatism: Toward a New National Agenda* (Lanham, Md.: Institute for Cultural Conservatism, 1987), p. 90.

118. Bob Hall, "Gold and Green," *Southern Exposure*, Fall 1994, p. 52.

119. Harbour, *Foundations of Conservative Thought*, p. 63.

120. Burke, *Reflections on the Revolution in France*, p. 62.

121. Russell Kirk, *The Politics of Prudence* (Bryn Mawr, Pa.: Intercollegiate Studies Institute, 1993), p. 170.

122. Russell Kirk, "Prescription, Authority, and Ordered Freedom," in *What Is Conservatism?* ed. Frank S. Meyer (New York: Holt, Rinehart & Winston, 1964), pp. 27–31.

123. Burke, *Reflections on the Revolution in France*, p. 61.

124. Quoted in Steve Stuebner, "Jack Ward Thomas: Hail to the Chief," *High Country News*, December 13, 1993, p. 10. Thomas is quoting ecologist Frank Egler.

125. Weaver, *Life without Prejudice*, pp. 140–141.

126. Edward O. Wilson, "Resolutions for the 80s," *Harvard Magazine*, January–February 1980, p. 21.

127. Gretchen C. Daily et al., "Ecosystem Services: Benefits Supplied to Human Societies by Natural Ecosystems," *Issues in Ecology* 2 (Spring 1997): 1.

128. Gerhart Niemeyer, "Russell Kirk and Ideology," *Intercollegiate Review* 30, no. 1 (Fall 1994): 37.

129. Kirk, *Conservative Mind*, p. iii.

130. Kirk, *Politics of Prudence*, chap. 1.

131. Gray, *Beyond the New Right*, p. 63.

132. Barry Commoner offers an insightful summary of the enormous changes in our economy that began shortly after World War II, many of which involved replacing natural commodities with synthetic ones. These changes, combined with the increase in population and explosion of personal consumption, created serious and increasing environmental problems that became obvious to the general public by the late 1960s; we are still trying to deal with them today; *Making Peace with the Planet* (New York: Pantheon, 1990), chap. 3.

133. Richard M. Weaver, *The Ethics of Rhetoric* (South Bend, Ind: Regnery/Gateway, 1953), chaps. 3–4.

CHAPTER FOUR

1. C. T. Driscoll et al., "Long-Term Trends in the Chemistry of Precipitation and Lake Water in the Adirondack Region of New York, USA," *Water, Air and Soil Pollution* 85 (1995): 583–588; G. E. Likens et al., "Long-Term Effects of Acid Rain: Response and Recovery of a Forest Ecosystem," *Science* 272 (1996): 244–246; C. T. Driscoll et al., "Recovery of Surface Waters in the Northeastern U.S. from Decreases in Atmospheric Deposition of Sulfur," *Water, Air and Soil Pollution* 105 (1998): 319–329; Alan Jenkins, "End of the Acid Reign?" *Nature* 401 (1999): 537–538.

2. Curtis Moore, "The Impracticality and Immorality of Cost-Benefit Analysis in Setting Health-Related Standards," *Tulane Environmental Law Journal* 11 (1998): 191; Robert N. Stavins, "What Can We Learn from the Grand Policy Experiment? Lessons from SO_2 Allowance Trading," *Journal of Economic Perspectives* 12, no. 3 (1998): 71. It is sometimes claimed that SO_2 is not worth

controlling because the National Acid Precipitation Assessment Program study found limited damages. But acid rain has indeed done considerable damage to the ecosystems of the Northeast; besides the sources in note 1, see John F. Sheehan, "Acid Rain: A Very Real Problem"; and Filson H. Ganz, "Acid Rain: A Very Real Problem," letters to the editor, *Christian Science Monitor*, June 10, 1996, p. 20. Moreover, NAPAP completely omitted any estimates of the two major areas of damages: to human health and to buildings; Darwin C. Hall, "Preliminary Estimates of Cumulative Private and External Costs of Energy," *Contemporary Policy Issues* 8, no. 3 (1990): 284.

3. Jane V. Hall et al., "Valuing the Health Benefits of Clean Air," *Science* 255 (1992): 812–817; George Tolley et al., "The Urban Ozone Abatement Problem," in *Cost Effective Control of Urban Smog*, ed. Richard F. Kosobud et al. (Chicago: Federal Reserve Bank, 1993), pp. 10–11, 14–15; Richard A. Wadden, "Discussion," in *Cost Effective Control of Urban Smog*, pp. 138–139; Luis A. Cifuentes and Lester B. Lave, "Economic Valuation of Air Pollution Abatement: Benefits from Health Effects," *Annual Review of Energy and the Environment* 18 (1993): 333; Craig N. Oren, "Getting Commuters Out of Their Cars: What Went Wrong?" *Stanford Environmental Law Journal* 17 (1998): 155–156. The problem here is ozone in the lower atmosphere. The ozone in the stratosphere is helpful, protecting all life on earth from harmful ultraviolet rays.

4. John D. Spengler, "Health Impacts of Ozone," in *Cost Effective Control of Urban Smog*, p. 122.

5. Tolley et al., "Urban Ozone Abatement Problem," p. 15; Oren, "Getting Commuters Out of Their Cars," p. 153; Moore, "Impracticality and Immorality of Cost-Benefit Analysis," pp. 195–196; Jocelyn Kaiser, "Evidence Mounts That Tiny Particles Can Kill," *Science* 289 (2000): 22–23; Jocelyn Kaiser, "Panel Backs EPA and Six Cities Study," *Science* 289 (2000): 711.

6. The EPA proposed new restrictions on particulates and ozone. The business community, as is far too typical, tried to avoid taking responsibility for the damage it causes to society and objected to the proposed rules. It claimed that the cost of complying would be as much as $150 billion per year. EPA director Carol Browner replied that businesses have a long history of wildly exaggerating costs of proposed new rules, and that the real costs of complying will be about $8.5 billion per year. The EPA studies calculate benefits to human health at up to $120 billion per year. See "Clean Air, Dirty Fight," *Economist*, March 15, 1997, pp. 29–30; Alexandra Marks, "Nation Holds Its Breath Over Clean-Air Rules," *Christian Science Monitor*, June 12, 1997, p. 1; Carl Pope, "A Good, Clean Fight," *Sierra*, March–April 1997, p. 12; "Clean Air Now!" *Planet*, March 1997, p. 1. The *Planet* is a newsletter produced by the Sierra Club. What the industries also failed to tell the American public is that the EPA's new rules are similar to standards California had adopted fifteen years earlier. The state of California, with a rigorous system for reviewing scientific studies, found that sufficient evidence existed long ago to prove that the federal standards were inadequate; James M. Lents, "A Review of National Ozone and Particulate Matter Air Quality Standards in Light of Long-Standing California Air Quality Standards," *Tulane Environmental Law Journal* 11 (1998): 415–424.The new standards were challenged in court, and as I write this the Supreme Court has just announced that it will hear the case; "News In Brief," *Christian Science Monitor*, May 23, 2000, p. 24. In any event, there is absolutely nothing in the theory of the free market or in the philosophy of conservatism that justifies telling those 64,000 people each year that we have decided to sacrifice them so that polluting industries can maximize their profits; Moore, "Impracticality and Immorality of Cost-Benefit Analysis," p. 209.

7. Oren, "Getting Commuters Out of Their Cars," pp. 151–152.

8. David M. Driesen, "Should Congress Direct the EPA to Allow Serious Harms to Public Health to Continue?: Cost-Benefit Tests and NAAQS Under the Clean Air Act," *Tulane Environmental Law Journal* 11 (1998): 233.

9. M. Jeff Hamond et al., *Tax Waste, Not Work* (San Francisco: Redefining Progress, 1997), p. 16.

10. See, for example, Frances Cairncross, "Costing the Earth (Survey)," *Economist*, September 2, 1989, p. 7; Robert Stavins and Thomas Grumbly, "The Greening of the Market: Making the Polluter Pay," in *Mandate for Change*, ed. Will Marshall and Martin Schram (New York: Berkley, 1993), pp. 197–198, 200–201; Robert Stavins and Bradley Whitehead, "Market-Based Environmental Policies," in *Thinking Ecologically: The Next Generation of Environmental Policy*, ed. Marian R. Chertow and Daniel C. Esty (New Haven: Yale University Press, 1997), p. 106.

11. Marshall J. Breger et al., "Providing Economic Incentives in Environmental Regulation," *Yale Journal on Regulation* 8 (1991): 476 [transcript of a symposium; E. Donald Elliott is speaking]; David Malin Roodman, "Harnessing the Market for the Environment," in *State of the World 1996,* ed. Linda Starke (New York: Norton, 1996), p. 175.

12. Stavins and Whitehead, "Market-Based Environmental Policies," p. 106; Robert N. Stavins, "Innovative Policies for Sustainable Development: The Role of Economic Incentives for Environmental Protection," *Harvard Public Policy Review* 7 (1990): 14; Tom H. Tietenberg, "Economic Incentive Policies and Sustainability," in *Towards an Ecologically Sustainable Economy,* ed. Britt Aniansson and Uno Svedin (Stockholm: Environmental Advisory Council, Swedish Council for Planning and Coordination of Research, 1990), pp. 71–72; Joe Alper, "Protecting the Environment with the Power of the Market," *Science* 260 (1993): 1884–1885.

13. Driesen, "Should Congress Direct the EPA to Allow Serious Harms?" pp. 220–223.

14. For a good summary of the theory behind pollution taxes, see Howard Gensler, "The Economics of Pollution Taxes," *Journal of Natural Resources and Environmental Law* 10 (1994–1995): 1–12.

15. Daniel J. Dudek et al., "Emissions Trading in Nonattainment Areas: Potential, Requirements, and Existing Programs," in *Market-Based Approaches to Environmental Policy,* ed. Richard F. Kosobud and Jennifer M. Zimmerman (New York: Van Nostrand Reinhold, 1997), p. 152.

16. Note that the cap-and-trade system does not require cost-benefit analysis to determine the cap. There are two separate policy decisions to be made. Air-quality standards should be determined on the basis of protecting human health and environmental amenities. Choosing the means to meet those standards is a separate policy decision and insofar as economic considerations are involved, they should be based on cost-effectiveness. Cap-and-trade policies often turn out to be very cost-effective.

17. Stavins and Grumbly, "Greening of the Market," p. 200.

18. T. H. Tietenberg, "Economic Instruments for Environmental Regulation," *Oxford Review of Economic Policy* 6, no. 1 (1990): 24.

19. Stavins, "Innovative Policies for Sustainable Development," p. 17; Catherine L. Kling, "Environmental Benefits from Marketable Discharge Permits or an Ecological vs. Economical Perspective on Marketable Permits," *Ecological Economics* 11 (1994): 58, 63.

20. Stavins, "Innovative Policies for Sustainable Development," pp. 17–18; Kelly Robinson, "The Regional Economic Impacts of Marketable Permit Programs: The Case of Los Angeles," in *Cost Effective Control of Urban Smog,* p. 170.

21. Robert W. Hahn and Robert N. Stavins, "Incentive-Based Environmental Regulation: A New Era from an Old Idea?" *Ecology Law Quarterly* 18 (1991): 16.

22. Robert W. Hahn and Gordon L. Hester, "Marketable Permits: Lessons for Theory and Practice," *Ecology Law Quarterly* 16 (1989): 380–387.

23. For a detailed description of the trading program, see General Accounting Office (GAO), *Air Pollution: Allowance Trading Offers an Opportunity to Reduce Emissions at Less Cost,* RCED-95-30 (Washington, D.C.: Government Printing Office, 1994).

24. Paul L. Joskow et al., "The Market for Sulfur Dioxide Emissions," *American Economic Review* 88 (1998): 673.

25. Douglas R. Bohi and Dallas Burtraw, "SO_2 Allowance Trading: How Do Expectations and Experience Measure Up?" *Electricity Journal,* August–September 1997, pp. 67–75; Dallas Burtraw, "The SO_2 Emissions Trading Program: Cost Savings Without Allowance Trades," *Contemporary Economic Policy* 14 (April 1996): 79–94; Klaus Conrad and Robert E. Kohn, "The U.S. Market for SO_2 Permits," *Energy Policy* 24 (1996): 1051–1059; GAO, *Air Pollution*; Richard A. Kerr, "Acid Rain Control: Success on the Cheap," *Science* 282 (1998): 1024–1027.

26. Barry D. Solomon, "New Directions in Emissions Trading: The Potential Contribution of New Institutional Economics," *Ecological Economics* 30 (1999): 376.

27. Stavins, "What Can We Learn from the Grand Policy Experiment?" p. 71.

28. GAO, *Air Pollution,* p. 37.

29. Joskow et al., "Market for Sulfur Dioxide Emissions," p. 684.

30. Jason S. Grumet, "Old West Justice: Federalism and Clean Air Regulation, 1970–1998," *Tulane Environmental Law Journal* 11 (1998): 378.

31. James M. Lents and Patricia Leyden, "RECLAIM: Los Angeles' New Market-Based Smog Cleanup Program," *Journal of the Air and Waste Management Association* 46 (1996): 195–206; Robinson, "Regional Economic Impacts," pp. 173–176; Solomon, "New Directions in Emissions Trading," pp. 374–375; Dudek et al., "Emissions Trading in Nonattainment Areas," pp. 166–168; Suellen Terrill Keiner, "State and Local Innovations in Air Quality Control," *Natural Resources and Environment* 13 (1998): 415–416.

32. Solomon, "New Directions in Emissions Trading," p. 379; Dudek et al., "Emissions Trading in Nonattainment Areas," pp. 168–169; Barry D. Solomon and Hugh S. Gorman, "State-Level Air Emissions Trading: The Michigan and Illinois Models," *Journal of the Air and Waste Management Association* 48 (1998): 1160–1162.

33. Charles Carter and Donald van der Vaart, "The Ozone Transport Dilemma: Is EPA's NO_x SIP Call the Solution?" *Natural Resources and Environment* 13 (1998): 399; Karl James Simon, "The Application and Adequacy of the Clean Air Act in Addressing Interstate Ozone Transport," *Environmental Lawyer* 5 (1998): 200.

34. Simon, "Application and Adequacy of the Clean Air Act," pp. 175–179.

35. Simon, "Application and Adequacy of the Clean Air Act," pp. 135–137, 151, 175.

36. Steve Twomey, "EPA Orders Plants to Cut Nitrogen Oxide Emissions," *Washington Post,* December 18, 1999, p. A6.

37. Dudek et al., "Emissions Trading in Nonattainment Areas," pp. 169–171; Carter and van der Vaart, "Ozone Transport Dilemma," p. 395; Solomon, "New Directions in Emissions Trading," p. 378.

38. Michael R. Miner, "A Market-Based Solution to Ozone Nonattainment: New Jersey's Nitrogen Oxides Budget Program," *Environmental Lawyer* 4 (1998): 896.

39. Daniel J. Dudek, "Incentives and the Car," in *Cost Effective Control of Urban Smog*, p. 103.

40. Keiner, "State and Local Innovations."

41. James D. Boyd, "Mobile Source Emissions Reduction Credits as a Cost Effective Measure for Controlling Urban Air Pollution," in *Cost Effective Control of Urban Smog*, p. 152.

42. Stefan Schaltegger and Tom Thomas, "Pollution Added Credit Trading (PACT): New Dimensions in Emissions Trading," *Ecological Economics* 19 (1996): 35–53.

43. For the details of this study, see Environmental Protection Agency, *Amoco-U.S. EPA Pollution Prevention Project, Yorktown, Virginia. Project Summary. Final Report.* PB92-228527/REB (Washington, D.C.: Environmental Protection Agency, 1992).

44. EPA, *Amoco-U.S. EPA Pollution Prevention Project*, p. 1–23.

45. EPA, *Amoco-U.S. EPA Pollution Prevention Project*, p. 1–5.

46. EPA, *Amoco-U.S. EPA Pollution Prevention Project*, pp. 1–13 to 1–14.

47. The EPA has a small pilot program called Project XL. Although most of the experiments—limited to just fifty—are mainly ways to cut red tape, the program may allow for experiments of the kind suggested here as a general policy. See *Project XL 1999 Comprehensive Report* <www.epa.gov/Project XL/file4.htm>.

48. Arnold W. Reitze Jr., "Transportation-Related Pollution and the Clean Air Act's Conformity Requirements," *Natural Resources and Environment* 13 (1998): 406.

49. Elmer W. Johnson, "The Coming Collision of Cities and Cars: Transportation Policy for the Twenty-First Century," in *Cost Effective Control of Urban Smog,* p. 118.

50. James N. Thurman and Todd Wilkinson, "EPA Takes Aim at the Exhaust Pipes of Trucks and Buses," *Christian Science Monitor,* May 18, 2000, p. 2.

51. Stuart P. Beaton et al., "On-Road Vehicle Emissions: Regulations, Costs, and Benefits," *Science* 268 (1995): 991–993; Gary A. Bishop and Donald H. Stedman, "Automobile Emissions—Control," in *The Wiley Encyclopedia of Energy and the Environment,* ed. Attilio Bisio and Sharon Boots (New York: Wiley, 1997), 1:234–243; J. G. Calvert et al., "Achieving Acceptable Air Quality: Some Reflections on Controlling Vehicle Emissions," *Science* 261 (1993): 37–45.

52. Bishop and Stedman, "Automobile Emissions"; Winston Harrington and Virginia D. McConnell, "Cost Effectiveness of Remote Sensing of Vehicle Emissions," in *Cost Effective Control of Urban Smog*, pp. 53–75.

53. Keiner, "State and Local Innovations," p. 414.

54. Dudek, "Incentives and the Car," pp. 101–103.

55. Beaton et al., "On-Road Vehicle Emissions," p. 992.

56. Wynn Van Bussmann, "Discussion," in *Cost Effective Control of Urban Smog*, p. 78.

57. Harrington and McConnell, "Cost Effectiveness of Remote Sensing," pp. 54–55, 69; Bishop and Stedman, "Automobile Emissions," p. 241.

58. Oren, "Getting Commuters Out of Their Cars," p. 160.

59. Oren, "Getting Commuters Out of Their Cars," p. 173; Jon Kessler and William Schroeer, "Meeting Mobility and Air Quality Goals: Strategies That Work," *Transportation* 22 (1995): 247.

60. Oren, "Getting Commuters Out of Their Cars," p. 163.

61. For the history of the legislation and its rapid repeal, see Oren, "Getting Commuters Out of Their Cars."

62. Reitze, "Transportation-Related Pollution."

63. Donald C. Shoup and Richard W. Willson, "Employer-Paid Parking: The Problem and Proposed Solutions," *Transportation Quarterly* 46 (1992): 169.

64. Kessler and Schroeer, "Meeting Mobility and Air Quality Goals," p. 261; see also Donald C. Shoup, "Evaluating the Effects of Cashing Out Employer-Paid Parking: Eight Case Studies," *Transport Policy* 4 (1997): 201–216.

65. Johnson, "Coming Collision of Cities and Cars," p. 117.

66. Kessler and Schroeder, "Meeting Mobility and Air Quality Goals," p. 251.

67. Johnson, "Coming Collision of Cities and Cars"; Kessler and Schroeer, "Meeting Mobility and Air Quality Goals," pp. 250–251.

68. Kessler and Schroeer, "Meeting Mobility and Air Quality Goals," p. 263.

69. Kessler and Schroeer, "Meeting Mobility and Air Quality Goals," pp. 255, 258; Oren, "Getting Commuters Out of Their Cars," p. 207.

70. Stephen R. Crutchfield et al., "Feasibility of Point-Nonpoint Source Trading for Managing Agricultural Pollutant Loadings to Coastal Waters," *Water Resources Research* 30 (1994): 2827.

71. Arun S. Malik et al., "Economic Incentives for Agricultural Nonpoint Source Pollution Control," *Water Resources Bulletin* 30 (1994): 474; John P. Almeida, "Nonpoint Source Pollution and Chesapeake Bay Pfiesteria Blooms: The Chickens Come Home to Roost," *Georgia Law Review* 32 (1998): 1200–1201.

72. Malik et al., "Economic Incentives for Agricultural Nonpoint Source Pollution Control," p. 472.

73. Michael M. Wenig, "How 'Total' Are 'Total Maximum Daily Loads'?—Legal Issues Regarding the Scope of Watershed-Based Pollution Control Under the Clean Water Act," *Tulane Environmental Law Journal* 12 (1998): 105. But see note 82.

74. Chris Carrel, "The Clean Water Clause," *Sierra*, May–June 1999, p. 58.

75. David Malakoff, "Death by Suffocation in the Gulf of Mexico," *Science* 281 (1998): 190–192.

76. John Charles Kluge, "Farming by the Foot: How Site-Specific Agriculture Can Reduce Nonpoint Source Water Pollution," *Columbia Journal of Environmental Law* 23 (1998): 103.

77. John C. Hall and Ciannat M. Howett, *Guide to Establishing a Point/Nonpoint Source Pollution Reduction Trading System for Basinwide Water Quality Management: The Tar-Pamlico River Basin Experience*, EPA–904-R–95–900 (North Carolina Department of Environment, Health, and Natural Resources for the EPA, 1995), p. P-1.

78. Crutchfield et al., "Feasibility of Point-Nonpoint Source Trading," p. 2825.

79. Kurt Stephenson et al., "Toward an Effective Watershed-Based Effluent Allowance Trading System: Identifying the Statutory and Regulatory Barriers to Implementation," *Environmental Lawyer* 5 (1999): 791; Almeida, "Nonpoint Source Pollution," p. 1197.

80. Esther Bartfeld, "Point-Nonpoint Source Trading: Looking Beyond Potential Cost Savings," *Environmental Law* 23 (1993): 48–49.

81. Wenig, "How 'Total' Are 'Total Maximum Daily Loads'?" p. 89; Carrel, "Clean Water Clause," pp. 58–59.

82. Wenig, "How 'Total' Are 'Total Maximum Daily Loads'?" p. 113. As I write this, the EPA has just issued a regulation requiring the states to make a comprehensive survey of water bod-

ies and develop plans to clean them up, including nonpoint pollution, over the next fifteen years; "EPA: Clean Up Waterways Now," *Denver Post,* July 12, 2000, p. 5A.

83. Wenig, "How 'Total' Are 'Total Maximum Daily Loads'?" p. 89.

84. Wenig, "How 'Total' Are 'Total Maximum Daily Loads'?" p. 115.

85. Carrel, "Clean Water Clause," p. 59.

86. Bartfeld, "Point-Nonpoint Source Trading," pp. 83–85.

87. For the details of the Tar-Pamlico trading system, see Hall and Howett, *Guide to Establishing a Point/Nonpoint Source Pollution Reduction Trading System;* and David W. Riggs, *Market Incentives for Water Quality: A Case Study of the Tar-Pamlico River Basin, North Carolina* (Clemson, S.C.: Clemson University Center for Policy Studies, 1993).

88. Environmental Protection Agency, *Draft Framework for Watershed-Based Trading,* EPA 800-R-96-001 (Washington, D.C.: Environmental Protection Agency, 1996).

89. Bartfeld, "Point-Nonpoint Source Trading," pp. 76–78; Stephenson et al., "Toward an Effective Watershed-Based Effluent Allowance Trading System," pp. 797–803; Elise M. Fulstone, "Effluent Trading: Legal Constraints on the Implementation of Market-Based Effluent Trading Programs Under the Clean Water Act," *Environmental Lawyer* 1 (1995): 478–485.

90. Crutchfield et al., "Feasibility of Point-Nonpoint Source Trading," p. 2826; Malik et al., "Economic Incentives for Agricultural Nonpoint Source Pollution Control," p. 477; David Letson, "Point/Nonpoint Source Pollution Reduction Trading: An Interpretive Survey," *Natural Resources Journal* 32 (1992): 221.

91. Kurt Stephenson et al., "Watershed-Based Effluent Trading: The Nonpoint Source Challenge," *Contemporary Economic Policy* 16 (October 1998): 413.

92. Malik et al., "Economic Incentives for Agricultural Nonpoint Source Pollution Control," p. 472.

93. Malik et al., "Economic Incentives for Agricultural Nonpoint Source Pollution Control," p. 473.

94. Michelle Jarvie and Barry Solomon, "Point-Nonpoint Effluent Trading in Watersheds: A Review and Critique," *Environmental Impact Assessment Review* 18 (1998): 139.

95. Bartfeld, "Point-Nonpoint Source Trading," pp. 76–77; Jarvie and Solomon, "Point-Nonpoint Effluent Trading," p. 141.

96. Malik et al., "Economic Incentives for Agricultural Nonpoint Source Pollution Control," p. 474.

97. Wenig, "How 'Total' Are 'Total Maximum Daily Loads'?" pp. 116–117.

98. Terry F. Young and Chelsea H. Congdon, *Plowing New Ground: Using Economic Incentives to Control Water Pollution from Agriculture* (Oakland, Calif.: Environmental Defense Fund, 1994).

99. Stephenson et al., "Watershed-Based Effluent Trading," p. 416.

100. Bartfeld, "Point-Nonpoint Source Trading," p. 64; Stephenson et al., "Watershed-Based Effluent Trading," pp. 417–420.

101. Stephenson et al., "Watershed-Based Effluent Trading," p. 419.

102. Stephenson et al., "Watershed-Based Effluent Trading," p. 418.

103. Bartfeld, "Point-Nonpoint Source Trading," pp. 66–68; Letson, "Point/Nonpoint Source Pollution Reduction Trading," pp. 222–227.

104. Hall and Howett, *Guide to Establishing a Point/Nonpoint Source Pollution Reduction Trading System,* p. 44.

105. Stephenson et al., "Toward an Effective Watershed-Based Effluent Allowance Trading System," pp. 811–814.

106. James J. Opaluch and Richard M. Kashmanian, "Assessing the Viability of Marketable Permit Systems: An Application in Hazardous Waste Management," *Land Economics* 61 (1985): 263–271.

107. M. Luttner, *President Clinton's Clean Water Act Initiative: Costs and Benefits,* quoted in Sean Blacklocke, *Effluent Trading in South Carolina: Pee Dee River Case Study* (Clemson, S.C.: Clemson University Center for Policy and Legal Studies, 1997), p. 5.

108. Jarvie and Solomon, "Point-Nonpoint Effluent Trading in Watersheds," p. 153.

109. Letson, "Point/Nonpoint Source Pollution Reduction Trading," pp. 226, 228; Bartfeld, "Point-Nonpoint Source Trading," p. 63.

110. Bartfeld, "Point-Nonpoint Source Trading," p. 69.

111. Malik et al., "Economic Incentives for Agricultural Nonpoint Source Pollution Control," p. 478.

112. Hans Th. A. Bressers and Jeannette Schuddeboom, "A Survey of Effluent Charges and Other Economic Instruments in Dutch Environmental Policy," in *OECD Documents: Applying Economic Instruments to Environmental Policies in OECD and Dynamic Non-Member Economies* (Paris: OECD, 1994), p. 158.

113. Chris Chung, "Summary of Discussions," in *OECD Documents: Applying Economic Instruments*, p. 10.

114. Malik et al., "Economic Incentives for Agricultural Nonpoint Source Pollution Control," p. 472.

115. Terry L. Anderson and Donald R. Leal, *Free Market Environmentalism* (San Francisco: Pacific Research Institute for Public Policy, 1991), pp. 138–140.

116. If you look at the history of toxic release policies, you will see many concrete illustrations of why principled conservatives cannot simply support whatever the business community wants. The fundamental irresponsibility of an unfortunately large portion of that community is vividly clear in their lack of concern about the dangers they impose on the public, as long as they can make money, and in their continuous attempts to block any actions to protect the public. For a couple of examples, see Gordon K. Durnil, *The Making of a Conservative Environmentalist* (Bloomington: Indiana University Press, 1995), pp. 116–117 for the chemical industry deliberately spreading falsehoods far and wide; and Clifford Rechtschaffen, "How to Reduce Lead Exposures with One Simple Statute: The Experience of Proposition 65," *Environmental Law Reporter* 29 (October 1999): 10585 for an industry task force insisting that the facts about lead leaching into drinking water must be kept secret.

117. Clifford Rechtschaffen, "The Warning Game: Evaluating Warnings Under California's Proposition 65," *Ecology Law Quarterly* 23 (1996): 352.

118. Environmental Protection Agency, "Environmental News," press release, May 11, 2000. <www.epa.gov/tri>.

119. Rechtschaffen, "Warning Game," p. 353.

120. Rechtschaffen, "Warning Game," p. 305.

121. Cal/EPA, Office of Environmental Health Hazard Assessment, "Proposition 65 in Plain Language!" updated March 1, 1999. <www.oehha.org/prop65/p65plain.htm>.

122. David Roe, "Barking Up the Right Tree: Recent Progress in Focusing the Toxics Issue," *Columbia Journal of Environmental Law* 13 (1988): 279.

123. Environmental Defense Fund, "Fact Sheet on Regulatory Productivity," February 1997, quoting the Cal/EPA Proposition 65 Review Panel *Report,* February 20, 1992.

124. Rechtschaffen, "Warning Game," p. 311.

125. Rechtschaffen, "How to Reduce Lead Exposures," p. 10586.

126. Rechtschaffen, "How to Reduce Lead Exposures," pp. 10584–10585.

127. See, e.g., Rechtschaffen, "How to Reduce Lead Exposures," pp. 10583–10588.

128. Quoted in Rechtschaffen, "Warning Game," p. 341.

129. Rechtschaffen, "How to Reduce Lead Exposures," p. 10589.

130. David Roe and William S. Pease, "Toxic Ignorance," *Environmental Forum,* May–June 1998, pp. 31–32.

131. Roe and Pease, "Toxic Ignorance," p. 32.

132. Roe and Pease, "Toxic Ignorance," p. 33.

133. Quoted in Leslie Roberts, "A Corrosive Fight over California's Toxics Law," *Science* 243 (1989): 307.

134. Quoted in Roberts, "Corrosive Fight over California's Toxics Law," p. 307.

135. Richard A. Lovett, "Prop 65's Non-Toxic Legacy," *Sacramento Bee,* November 30, 1997, Forum 6.

136. William S. Pease, "Chemical Hazards and the Public's Right to Know: How Effective Is California's Proposition 65?" *Environment* 33, no. 10 (December 1991): 14.

137. Quoted in Roberts, "Corrosive Fight over California's Toxics Law," p. 308.

138. Rechtschaffen, "Warning Game," p. 320; David Roe, "An Incentive-Conscious Approach to Toxic Chemical Controls," *Economic Development Quarterly* 3 (1989): 182.

139. In extending the Proposition 65 concept to national policy, we should be more explicit about the kinds of warnings companies must provide to the public. In California, many companies have weaseled out of their responsibility to provide clear warnings by using ambiguous language, burying the warnings in inconspicuous places, embedding them in texts full of self-praise about how great the companies are, etc.; Rechtschaffen, "Warning Game," pp. 320–358.

140. James M. Strock, "Wizards of Ooze," *Policy Review* 67 (Winter 1994): 42–45.

141. Steven A. Herman, "A Fundamentally Different Superfund Program," *Natural Resources and Environment* 12 (1998): 196–199, 227.

142. Quoted in Herman, "Fundamentally Different Superfund Program," p. 199.

143. Clifford S. Russell, "Economic Incentives in the Management of Hazardous Wastes," *Columbia Journal of Environmental Law* 13 (1988): 262.

144. Hilary A. Sigman, "A Comparison of Public Policies for Lead Recycling," *RAND Journal of Economics* 26 (1995): 454–456; Jeremy B. Hockenstein et al., "Crafting the Next Generation of Market-Based Environmental Tools," *Environment* 39, no. 4 (1997): 31; Robert N. Stavins, director, Senators Timothy E. Wirth and John Heinz, sponsors, *Project 88—Round II* (Washington, D.C., May 1991), pp. 57–58.

145. Stavins, *Project 88—Round II*, p. 64.

146. Russell, "Economic Incentives in the Management of Hazardous Wastes," pp. 265–266.

147. Russell, "Economic Incentives in the Management of Hazardous Wastes," p. 267; Stavins, *Project 88—Round II*, pp. 61–64; Hockenstein et al., "Crafting the Next Generation of Market-Based Environmental Tools," pp. 31–32; Sigman, "Comparison of Public Policies for Lead Recycling," pp. 469–470.

148. Stavins, *Project 88—Round II*, p. 62; Hockenstein et al., "Crafting the Next Generation of Market-Based Environmental Tools," pp. 31–32.

CHAPTER FIVE

1. See Randal O'Toole, "The Forest Service Has Already Been Reinvented," *Different Drummer: Reinventing the Forest Service* 2, no. 2 (1995): 39–53, for an excellent history of these changes in direction and types of personnel.

2. Majority Staff Report of the Subcommittee on Oversight and Investigations of the Committee on Natural Resources of the U.S. House of Representatives, *Taking from the Taxpayer: Public Subsidies for Natural Resource Development*, Committee Print no. 8, 103d Congress, 2d sess. (Washington, D.C.: Government Printing Office, 1994), p. 72; hereafter cited as *Taking from the Taxpayer*.

3. "Savaging the Forests," *New Voices* 4, no. 8 (1995): 1–2; Forest Water Alliance, *Our Forests and Our Future* (1997). *New Voices* is a newsletter published by the Wilderness Society.

4. Forest Water Alliance, *Our Forests and Our Future*.

5. *Seattle Audubon Society v. Evans*, 771 F. Supp. 1081 (W.D. Wash. 1991) at 1089–1090.

6. Claudia Goetz Phillips and John Randolph, "Has Ecosystem Management Really Changed Practices on the National Forests?" *Journal of Forestry*, May 1998, pp. 40–45.

7. Jenny Coyle, "Dombeck on Record," *Planet*, June 1999, p. 1. The *Planet* is a newsletter published by the Sierra Club.

8. For the gory details of the salvage rider and the damage it caused, see Patti A. Goldman and Kristen L. Boyles, "Forsaking the Rule of Law: The 1995 Logging Without Laws Rider and Its Legacy," *Environmental Law* 27 (1997): 1035–1089.

9. Paul W. Hirt, *A Conspiracy of Optimism: Management of the National Forests Since World War Two* (Lincoln: University of Nebraska Press, 1994), pp. 282–287. A survey of Forest Service

employees found that a majority believe the agency has changed for the better, but the changes have not gone far enough; Paul Mohai and Pamela Jakes, "The Forest Service in the 1990s: Is It Headed in the Right Direction?" *Journal of Forestry,* January 1996, pp. 31–37.

10. Quoted in Jim Lynch and J. Todd Foster, "Our Failing Forests," *Spokane Spokesman-Review,* November 21, 25, 28, 1993, A7. This is a reprint of a three-part series on the national forests, focusing primarily on the ones in eastern Washington, northern Idaho, and western Montana.

11. Randal O'Toole, *Reforming the Forest Service* (Washington, D.C.: Island, 1988), p. 157.

12. O'Toole, *Reforming the Forest Service*, p. 21; Hirt, *Conspiracy of Optimism*, p. xxvii.

13. Brad Knickerbocker, "A Road Rage over Future of U.S. Forests," *Christian Science Monitor,* July 19, 2000, p. 1.

14. Chris Maser, *The Redesigned Forest* (San Pedro: Miles, 1988), p. 159.

15. "Congress Continues Forest Service Harassment," *Outdoor News Bulletin*, April 17, 1998, p. 1. (The *Outdoor News Bulletin* is a newsletter published by the Wildlife Management Institute.) The $10.5 billion figure was frequently used in the debate on the floor of the House of Representatives on the Forest Recovery and Protection Act of 1998, *Congressional Record,* March 27, 1998, H1651-H1682.

16. Richard E. Rice, *The Uncounted Costs of Logging*, vol. 5 of *National Forests: Policies for the Future* (Washington, D.C.: Wilderness Society, 1989), p. 8.

17. W. F. Megahan and W. J. Kidd, "Effects of Logging and Logging Roads on Erosion and Sediment Deposition from Steep Terrain," *Journal of Forestry,* March 1972, pp. 136–141.

18. F. J. Swanson and C. T. Dyrness, "Impact of Clearcutting and Road Construction on Soil Erosion by Landslides in the Western Cascade Range, Oregon," *Geology* 3 (July 1975): 393–396.

19. Gordon Robinson, *The Forest and the Trees: A Guide to Excellent Forestry* (Washington, D.C.: Island, 1988), p. 147.

20. M. J. Furniss et al., "Road Construction and Maintenance," in *Influences of Forest and Rangeland Management on Salmonid Fishes and Their Habitats,* ed. William R. Meehan, American Fisheries Society Special Publication 19 (Bethesda, Md.: American Fishers Society, 1991), pp. 297–323. For eastern forests, see Andrew F. Egan, "Forest Roads: Where Soil and Water Don't Mix," *Journal of Forestry,* August 1999, pp. 18–21.

21. Megahan and Kidd, "Effects of Logging and Logging Roads."

22. Swanson and Dyrness, "Impact of Clearcutting and Road Construction."

23. Robinson, *The Forest and the Trees*, p. 147.

24. T. H. Watkins, "The Conundrum of the Forest," *Wilderness,* Spring 1986, p. 16.

25. Patrick Mazza, "The Mud Next Time," *Sierra,* May–June 1997, pp. 22–24.

26. Hirt, *Conspiracy of Optimism*, p. 149; Rodney J. Kennan and J. P. Kimmins, "The Ecological Effects of Clearcutting," *Environmental Reviews* 1 (1993): 129.

27. Robinson, *The Forest and the Trees*, p. 68.

28. F. H. Bormann et al., "Nutrient Loss Accelerated by Clearcutting of a Forest Ecosystem," *Science* 159 (1968): 882–884; Robinson, *The Forest and the Trees*, pp. 64, 194.

29. J. P. Kimmins, "Importance of Soil and Role of Ecosystem Disturbance for Sustained Productivity of Cool Temperate and Boreal Forests," *Soil Science Society of America Journal* 60 (1996): 1644.

30. *Stewardship or Stumps? National Forests at the Crossroads* (San Francisco: Sierra Club, 1997), p. 5.

31. *Stewardship or Stumps?*, p. 5.

32. See Kathie Durbin and Paul Koberstein, "Forests in Distress," *Oregonian,* October 15, 1990, p. 13, for the plight of a small town at the mercy of the Forest Service, pleading in vain for it to protect its water supply. This publication is a remarkable special report, twenty-eight pages long, detailing the problems of the forests, primarily in Oregon.

33. B. J. Hicks et al., "Responses of Salmonids to Habitat Changes," in *Influences of Forest and Rangeland Management on Salmonid Fishes and Their Habitats*, pp. 483–518.

34. Robinson, *The Forest and the Trees*, pp. 145, 215; for statistics from a number of studies, see Hicks et al., "Responses of Salmonids to Habitat Changes."

35. David S. Wilcove, "Turning Conservation Goals into Tangible Results: The Case of the Spotted Owl and Old-Growth Forests," in *Large-Scale Ecology and Conservation Biology,* ed. P. J. Edwards et al. (Oxford: Blackwell, 1994), p. 325.

36. Rebecca W. Rimel and Karin P. Sheldon, "Broad Approach to Old-Growth Forests," *Christian Science Monitor,* April 8, 1993, p. 18.

37. Durbin and Koberstein, "Forests in Distress," p. 15.

38. Rice, *Uncounted Costs of Logging,* p. 18.

39. Jason F. Shogren, "Economics and the Endangered Species Act," *Endangered Species UPDATE* 14, no. 1–2 (1997): 5.

40. *Taking from the Taxpayer,* p. 71.

41. Robinson, *The Forest and the Trees,* p. 18.

42. Hirt, *Conspiracy of Optimism.*

43. Robinson, *The Forest and the Trees,* pp. 18, 128.

44. Catherine Caufield, "The Ancient Forest," *New Yorker,* May 14, 1990, pp. 67–68.

45. Caufield, "The Ancient Forest," p. 68.

46. Quoted in Caufield, "The Ancient Forest," p. 68.

47. Staff of the Committee on Interior and Insular Affairs, U.S. House of Representatives, *Management of Federal Timber Resources: The Loss of Accountability,* June 15, 1992.

48. *Management of Federal Timber Resources,* pp. 3–4.

49. Caufield, "The Ancient Forest," p. 68.

50. Maser, *Redesigned Forest,* pp. xviii, 46.

51. Jerry F. Franklin, "Lessons from Old-Growth," *Journal of Forestry,* December 1993,Z Zp. 11.The claim is sometimes made that clearcutting mimics nature. Fires and windstorms take out all of the trees in a plot and clearcutting is similar to these natural disturbances that have affected the forests for millennia. This is, of course, simply not the case. At the most obvious level, very seldom do fires or windstorms build roads to get into the forests. But neither do they affect the forest as a clearcut does. Thomas Spies and Jerry Franklin explain that fires remove little of the coarse woody debris that is important for regenerating the forest. A clearcut removes almost all of it and the rest is burned. Fires and wind also leave some standing dead trees that are critical habitat for many species. Clearcutting leaves none of these. It also leaves smaller, more uniform patch sizes, with abrupt edges and much less structural diversity than do natural disturbances. Major fires in the Northwest occur infrequently, typically once every 350 to 450 years. Clearcutting on short rotations, by contrast, never allows the forest to reach later successional stages. Overall, natural disturbances leave important "biological legacies" that are important for sustaining the forest, whereas clearcutting removes most of them. Clearcutting does not mimic nature nor is it an even-aged monoculture tree farm like a natural forest. See, for example, Jerry F. Franklin et al., "Modifying Douglas-Fir Management Regimes for Nontimber Objectives," in *Douglas-Fir: Stand Management for the Future,* ed. Chadwick Dearing Oliver et al. (Seattle: University of Washington Institute of Forest Resources, 1986), p. 373–379; Thomas A. Spies and Jerry F. Franklin, "Old Growth and Forest Dynamics in the Douglas-Fir Region of Western Oregon and Washington," *Natural Areas Journal* 8 (1988): 190-201; Jerry Franklin, "Toward a New Forestry," *American Forests,* November–December 1989, pp. 37–44.

52. Franklin et al., "Modifying Douglas-Fir Management Regimes," p. 378.

53. Franklin et al., "Modifying Douglas-Fir Management Regimes," p. 375.

54. Jerry F. Franklin et al., "Importance of Ecological Diversity in Maintaining Long-Term Site Productivity," in *Maintaining the Long-Term Productivity of Pacific Northwest Forest Ecosystems,* ed. D. A. Perry et al. (Portland: Timber, 1989), p. 85.

55. According to the best and most recent analysis of the extent of old-growth forests in the Pacific Northwest, of the original 25 million acres only 2 million acres (8 percent) were left; Peter H. Morrison et al., *Ancient Forests in the Pacific Northwest* (Washington, D.C.: The Wilderness Society, 1991). There is less than that today. The occasional larger estimates are not based on consistently applied ecological definitions of old growth.

56. Spies and Franklin, "Old Growth and Forest Dynamics," p. 194.

57. Robinson, *The Forest and the Trees*, p. 202; R. Dennis Harr, "Effects of Clearcutting on Rain-on-Snow Runoff in Western Oregon: A New Look at Old Studies," *Water Resources Research* 22 (1986): 1095–1100; Steven N. Berris and R. Dennis Harr, "Comparative Snow Accumulation and Melt During Rainfall in Forested and Clearcut Plots in the Western Cascades of Oregon," *Water Resources Research* 23 (1987): 135–152; J. A. Jones and G. E. Grant, "Peak Flow Responses to Clear-Cutting and Roads in Small and Large Basins, Western Cascades, Oregon," *Water Resources Research* 32 (1996): 959–974.

58. Lynch and Foster, "Our Failing Forests," p. H4.

59. Rice, *Uncounted Costs of Logging*, p. 37.

60. Rice, *Uncounted Costs of Logging*, p. 45.

61. Robinson, *The Forest and the Trees*, p. 66.

62. Robinson, *The Forest and the Trees*, p. 46.

63. Quoted in Hirt, *Conspiracy of Optimism*, p. 272.

64. Durbin and Koberstein, "Forests in Distress," pp. 8–9, 11, 27; Hirt, *Conspiracy of Optimism*, pp. xxxviii–xxxix.

65. Franklin, "Toward a New Forestry." For an excellent summary of the scientific basis for "New Forestry," see Jerry F. Franklin, "Scientific Basis for New Perspectives in Forests and Streams," in *Watershed Management: Balancing Sustainability and Environmental Change*, ed. Robert J. Naiman (New York: Springer-Verlag, 1992), pp. 25–72.

66. "History of National Forest Conflicts," *Different Drummer: Taking the Forest Service into Its Second Century* 14 (1998): 28.

67. "History of National Forest Conflicts," p. 26.

68. For the roll call vote, see *Congressional Record*, March 27, 1998, H1681-H1682.

69. Quoted in B. J. Bergman, "Lay of the Land," *Sierra*, March–April 1998, p. 35.

70. Robert N. Stavins, director, and Senators Timothy Wirth and John Heinz, sponsors, *Project 88* (Washington, D.C., 1988, 1991), 2:80.

71. Stavins, *Project 88*, 1:60.

72. General Accounting Office, *Forest Service: Distribution of Timber Sales Receipts, Fiscal Years 1992–1994*, GAO/RCED-95-237FS (Washington, D.C.: Government Printing Office, 1995).

73. *Economic Report of the President, 1997* (Washington, D.C.: Government Printing Office, 1997), p. 217.

74. General Accounting Office, *Forest Service: Distribution of Timber Sales Receipts, Fiscal Years 1995 Through 1997*, GAO/RCED–99–24 (Washington, D.C.: Government Printing Office, 1998).

75. "National Forest Timber Sales and the Forest Service Budget," *Different Drummer: Taking the Forest Service into Its Second Century* 14 (1998): S-1, S-4.

76. Quoted in *The Citizens' Guide to the Forest Service Budget, Forest Watch Special Issue* 12, no. 9 (April 1992): 3.

77. O'Toole, *Reforming the Forest Service*. See also Andy Stahl, "The New Question," *Journal of Forestry*, May 1999, p. 6.

78. *Citizens' Guide to the Forest Service Budget*, pp. 12–14.

79. *Citizens' Guide to the Forest Service Budget*, p. 14.

80. Lynch and Foster, "Our Failing Forests," p. H4.

81. Ted Williams, "The Unkindest Cuts," *Audubon*, January–February 1998, p. 30.

82. *Citizens' Guide to the Forest Service Budget*, p. 14.

83. *Citizens' Guide to the Forest Service Budget*, p. 26.

84. *Taking from the Taxpayer*, p. 73; O'Toole, *Reforming the Forest Service*, p. 113.

85. O'Toole, *Reforming the Forest Service*, p. 33; see also *Taking from the Taxpayer*, pp. 76–77.

86. *Citizens' Guide to the Forest Service Budget*, p. 7.

87. Quoted in "Recreation and Wildlife Are Big Business," *Outdoor News Bulletin*, September 27, 1996, p. 4. The *Outdoor News Bulletin* is a newsletter published by the Wildlife Management Institute. See also *Economic Report of the President, 1997* (Washington, D.C.: Government Printing Office, 1997), pp. 222–223.

88. "The Forests' Future," editorial, *Christian Science Monitor,* February 23, 1998, p. 12.

89. Quoted in "Recreation and Wildlife Are Big Business," p. 4.

90. Quoted in *Citizens' Guide to the Forest Service Budget,* p. 20.

91. *Citizens' Guide to the Forest Service Budget,* p. 20.

92. Quoted in Brad Knickerbocker, "Critics Whipsaw the Forest Service," *Christian Science Monitor,* December 22, 1998, p. 3.

93. Steven W. Selin et al., "Has Collaborative Planning Taken Root in the National Forests?" *Journal of Forestry,* May 1997, pp. 25–28.

94. *Different Drummer: State Lands and Resources* 2, no. 3 (1995).

95. For recent practices, see Durbin and Koberstein, "Forests in Distress." For the irresponsibility of timber companies when the salvage rider let them do whatever they wanted, see Goldman and Boyles, "Forsaking the Rule of Law," pp. 1060, 1070–1078.

96. Mazza, "The Mud Next Time."

97. Carl Pope, "Unnatural Causes," *Sierra,* January–February 1998, p. 13.

98. Russell Kirk, "America the Beautiful? Not Without Some Countryside," *Detroit News,* March 22, 1973, p. B15.

99. John C. Sawhill, "Intact Landscapes in a Fragmentary World," *Nature Conservancy,* July–August 1997, p. 5.

100. Rice, *Uncounted Costs of Logging,* p. 56; Stavins, *Project 88,* 2:77–86; K. Norman Johnson et al., "Sustaining the People's Lands," *Journal of Forestry,* May 1999, pp. 8, 11; O'Toole, *Reforming the Forest Service,* pt. 3.

101. See Thomas Michael Power and Paul Rauber, "The Price of Everything," *Sierra,* November–December 1993, p. 94; Todd Wilkinson, "Crush of Off-Road Vehicles Plies West's Public Lands," *Christian Science Monitor,* October 5, 1999, p. 3. For the problems industrial recreation causes on a national forest, and an attempt to control it to protect the natural resources, see Allen Best, "A National Forest Tries to Rein in Recreation," *High Country News,* January 17, 2000, pp. 1, 6–11. Texas state parks now must charge fees to support themselves. To protect the parks from money-raising activities that would damage their natural resources, regional managers must approve all customer services proposed by field personnel; Donald R. Leal and Holly Lippke Fretwell, *Back to the Future to Save Our Parks,* PERC Policy Series PS-10 (Bozeman, Mont.: PERC, 1997), p. 17. Perhaps a similar system, or an independent review board, could be used for federal lands.

102. "The Reinvented Budget: Still the Wrong Incentives," *Different Drummer: Reinventing the Forest Service* 2, no. 2 (1995): 18.

103. O'Toole, *Reforming the Forest Service,* p. 199.

104. "True Reinvention: Run Forests Like Businesses," *Different Drummer: Reinventing the Forest Service* 2, no. 2 (1995): 58.

105. "True Reinvention," p. 58.

106. Careful harvesting of trees using selection logging can preserve the recreational values of the forest; Robinson, *The Forest and the Trees,* p. 81.

107. The public generally favors the idea of charging higher fees to support recreation on federal lands. Before the fee demonstration program, a national survey by Colorado State University found that 80 percent of the public supports higher fees for the national parks if all of the revenue goes to the parks; Human Dimensions in Natural Resources Unit, Colorado State University, *National Public Opinion Survey on the National Park System* (Fort Collins: Colorado State University/National Parks and Conservation Association, 1995).The fee demonstration program has now been in effect long enough for some preliminary results. So far, the National Park Service has been much more successful than the other three agencies in raising money from increased fees. But the Park Service has had fee collection systems in place for years, and the other agencies are learning how best to collect them.Two journals recently devoted an entire issue each to studies of the program: *Journal of Park and Recreation Administration* 17, no. 3 (1999); and *Journal of Leisure Research* 31, no. 3 (1999); see also the main article, "Land of the Fee," by Hal Clifford, in *High Country News,* February 14, 2000, pp. 1, 6–10. Some important themes on public acceptance run through the studies and are reflected in the newspaper

report. They find general support for charging fees for recreation, but several things qualify that support. There is a widely expressed concern that fees may exclude low-income people from our public lands. Empirical results, so far at least, indicate that this has not happened. After all, the entry fees we are dealing with are roughly comparable to lunch at a fast-food restaurant and buying an annual pass brings the cost per visit way below that. For the vast majority of tourists, the fees are a tiny fraction of the cost of a vacation trip.The most important factor that affects public acceptance of the fees, however, is perceived "fairness." People seem to accept paying higher fees if they know that (1) all other people and companies who use the resources are also paying their fair share and (2) the money will actually be spent where it is collected, to protect the natural area and to provide the services used. As long as loggers and ranchers are subsidized on federal land, campers and hikers will resent paying fees. If the money collected just goes to support bureaucrats in Washington, acceptance of fees will be low. The keys, on the basis of this research, seem to be communication along with the fee collection, to inform people about the costs of trail construction and maintenance, trash collection, campground operating expenses, and the like, and to show them "your fees at work." The fees should be set relative to the level of services provided (e.g., a developed campground should cost more than a primitive campsite). And, of course, subsidies for the extractive industries should be eliminated. Another theme from the research is that people have a "reference point" from which they evaluate fees, which is the amount they last paid. If the last time someone hiked a trail it was free, paying a fee for the next hike will be less acceptable. This suggests that acceptance should increase over time, as people become accustomed to paying for recreational use of federal land. They will also become more accustomed to paying for recreation because many state park systems are increasing the fees they charge. Sixteen states get half or more of their park operating budgets from user fees. New Hampshire and Vermont fund their parks entirely from fees, and Nebraska's fees now cover expenses. Texas is in the process of making its state parks independent of general revenues. For details on state park systems, see *Different Drummer: State Lands and Resources* 2, no. 3 (1995): 18–33.

108. Richard Conniff, "Once the Secret Domain of Miners and Ranchers, the BLM Is Going Public," *Smithsonian,* September 1990, p. 34.

109. The BLM has explicit authority to protect its lands by reducing or discontinuing grazing temporarily or permanently; Joseph M. Feller, "What Is Wrong with the BLM's Management of Livestock Grazing on the Public Lands," *Idaho Law Review* 30 (1993–1994): 566–568.

110. *Different Drummer: Reforming the Western Range* 1, no. 2 (1994): 24. The preface notes that the unsigned sections in this issue were written by Karl Hess Jr. and Randal O'Toole. See also Andy Kerr, "Expanding the Market for Grazing Permits," *Different Drummer: Inside the Bureau of Land Management* 13 (1998): 53.

111. *Different Drummer: Reforming the Western Range*, pp. 26, 28.

112. Bruce M. Pendery, "Reforming Livestock Grazing on the Public Domain: Ecosystem Management-Based Standards and Guidelines Blaze a New Path for Range Management," *Environmental Law* 27 (1997): 523; see also *Taking from the Taxpayer*, pp. 90–91.

113. Pendery, "Reforming Livestock Grazing," p. 523.

114. Ed Chaney et al., *Livestock Grazing on Western Riparian Areas* (Washington, D.C.: Government Printing Office, 1993), p. 2.

115. For summaries of the research on grazing impacts, with extensive lists of references, see Thomas L. Fleischner, "Ecological Costs of Livestock Grazing in Western North America," *Conservation Biology* 8 (1994): 629–644; and A. J. Belsky et al., "Survey of Livestock Influences on Stream and Riparian Ecosystems in the Western United States," *Journal of Soil and Water Conservation* 54 (1999): 419–431.

116. David N. Cole, "Trampling Disturbance and Recovery of Cryptogamic Soil Crusts in Grand Canyon National Park," *Great Basin Naturalist* 50 (1990): 324.

117. Jeffrey R. Johansen and Larry L. St. Clair, "Cryptogamic Soil Crusts: Recovery from Grazing near Camp Floyd State Park, Utah, USA," *Great Basin Naturalist* 46 (1986): 632–640.

118. Charles F. Wilkinson, *Crossing the Next Meridian* (Washington, D.C.: Island, 1992), p. 80.

119. Chaney et al., *Livestock Grazing on Western Riparian Areas*, p. 5.

120. J. Boone Kauffman and W. C. Krueger, "Livestock Impacts on Riparian Ecosystems and Streamside Management Implications: A Review," *Journal of Range Management* 37 (1984): 434; Chaney et al., *Livestock Grazing on Western Riparian Areas*, p. 33.

121. Pendery, "Reforming Livestock Grazing," p. 541.

122. Pendery, "Reforming Livestock Grazing," p. 529.

123. *Different Drummer: Reforming the Western Range*, pp. 14–15.

124. For details, see National Research Council, *Rangeland Health* (Washington, D.C.: National Academy Press, 1994), chap. 3.

125. Pendery, "Reforming Livestock Grazing," p. 609. This article is a detailed analysis of the 1995 regulations and standards. The BLM in large part adopted the criteria for evaluating rangeland ecosystems recommended by the National Research Council in *Rangeland Health*.

126. Wayne Elmore, "Riparian Responses to Grazing Practices," in *Watershed Management*, ed. Robert J. Naiman (New York: Springer-Verlag, 1992), p. 442.

127. Chaney et al., *Livestock Grazing on Western Riparian Areas*, p. 5.

128. Chaney et al., *Livestock Grazing on Western Riparian Areas*, p. 42, quoting the Office of Technology Assessment.

129. Joseph M. Feller, "'Til the Cows Come Home: The Fatal Flaw in the Clinton Administration's Public Lands Grazing Policy," *Environmental Law* 25 (1995): 712; Karl Hess Jr., "A New Approach to Range Reform," *Different Drummer: A New Environmental Agenda for the 105th Congress* 3, no. 4 (1997): 49.; Todd Oppenheimer, "The Rancher Subsidy," *Atlantic Monthly*, January 1996, p. 26; Wilkinson, *Crossing the Next Meridian*, p. 81.

130. *Different Drummer: Reforming the Western Range*, p. 39.

131. *Different Drummer: Reforming the Western Range*, p. 44.

132. *Public Lands Council v. U.S. Department of Interior* 929 F. Supp. 1436 (D.Wyo. 1996); *Public Lands Council v. Babbitt* 154 F. 3d 1160 (10th Cir. 1998); *Public Lands Council v. Babbitt* 167 F. 3d 1287 (10th Cir. 1999); *Public Lands Council v. Babbitt* 120 S. Ct. 1815 (2000) at 1826.

133. *Different Drummer: Reforming the Western Range*, p. 43.

134. Quoted in George Wuerthner, "How the West Was Eaten," *Wilderness*, Spring 1991, p. 36.

135. William E. Riebsame, "Ending the Range Wars?" *Environment* 38, no. 4 (1996): 27.

136. Feller, "What Is Wrong," pp. 576–583.

137. Feller, "What Is Wrong," pp. 577–581; Jerry L. Holechek and Karl Hess Jr., "Free Market Policy for Public Land Grazing," *Rangelands* 16, no. 2 (1994): 65.

138. Pendery, "Reforming Livestock Grazing," pp. 517, 586, 590–591.

139. Holechek and Hess, "Free Market Policy," p. 64.

140. Wilkinson, *Crossing the Next Meridian*, pp. 107–108.

141. Karl Hess Jr. and Jerry L. Holechek, "Babbitt Inherited a Mess; His Plan Will Make It Worse," *High Country News*, November 1, 1993, p. 16; Karl Hess Jr., "Storm over the Rockies," *Reason*, June 1995, p. 23.

142. See, for example, *Different Drummer: Reforming the Western Range*, pp. 46–47, concerning a notorious program on the Vale District in Oregon.

143. Marc Reisner and Sarah Bates, *Overtapped Oasis: Reform or Revolution for Western Water* (Washington, D.C.: Island, 1990), pp. 27–34.

144. Karl Hess Jr., "Reforming Public Range Lands," *Different Drummer: Congress and the Federal Lands* 2, no. 4 (1995): 49.

145. Brad Knickerbocker, "New Call to End Regulation on Range," *Christian Science Monitor*, October 24, 1995, p. 4.

146. Hess and Holechek, "Babbitt Inherited a Mess," p. 16; Holechek and Hess, "Free Market Policy," p. 63; Riebsame, "Ending the Range Wars?" p. 9.

147. *Different Drummer: Reforming the Western Range*, pp. 28–29; Mark N. Salvo, "The Declining Importance of Public Lands Ranching in the West," *Public Land and Resources Law Review* 19 (1998): 106; *Taking from the Taxpayer*, p. 91.

148. *Different Drummer: Reforming the Western Range*, p. 29; Holechek and Hess, "Free Market Policy," p. 63.

149. Holechek and Hess, "Free Market Policy," p. 63.

150. Feller, "What Is Wrong," p. 559.
151. Feller, "What Is Wrong," p. 559.
152. Feller, "What Is Wrong," p. 560.
153. Feller, "What Is Wrong," pp. 586–591.
154. Feller, "What Is Wrong," p. 591.
155. *Different Drummer: Reforming the Western Range*, p. 40.
156. Karl Hess Jr., "A New Approach to Range Reform," *Different Drummer: A New Environmental Agenda for the 105th Congress* 3, no. 4 (1997): 50.
157. Feller, "What Is Wrong," pp. 570–573.
158. Karl Hess Jr., "Public Rangelands: Twenty-Five Years of Stalemate," *Different Drummer: The State of the Environmental Movement* 3, no. 3 (1996): 56–57.
159. Oppenheimer, "Rancher Subsidy," p. 37.
160. Wilkinson, *Crossing the Next Meridian*, p. 109; *Different Drummer: Reforming the Western Range*, pp. 36–37. For numerous examples of the dramatic ability of riparian zones to recover, complete with before-and-after photographs, see Chaney et al., *Livestock Grazing on Western Riparian Areas.*
161. Chaney et al., *Livestock Grazing on Western Riparian Areas*, p. 42; Oppenheimer, "Rancher Subsidy," p. 38.
162. Hess, "Storm over the Rockies," p. 20; Kerr, "Expanding the Market for Grazing Permits," p. 50.
163. Jerry L. Holechek, "Policy Changes on Federal Rangelands: A Perspective," *Journal of Soil and Water Conservation* 48 (1993): 172. There is a fear that if fees are raised, some of these marginal operations will quit raising cattle and sell out to developers who will subdivide the property for ranchettes; Wilkinson, *Crossing the Next Meridian*, p. 106. That may happen anyway, of course. Giving these owners the chance to get cash for their permits may help them keep their property intact. See also George Wuerthner, "Subdivisions Versus Agriculture," *Conservation Biology* 8 (1994): 905–908, contending that some ranchettes are environmentally better than cows everywhere.
164. For further development of similar proposals, see Hess, "New Approach to Range Reform," pp. 51–52; *Different Drummer: Reforming the Western Range*, p. 54; Holechek and Hess, "Free Market Policy," pp. 65–66.
165. "The Wayward West," *High Country News*, January 17, 2000, p. 3.
166. See, for example, *Different Drummer: Reforming the Western Range*, p. 58; *Different Drummer: Inside the Bureau of Land Management*, pp. 61–62.
167. *Taking from the Taxpayer*, p. 14.
168. "The True Price of Gold" [fact sheet] (Washington, D.C.: Mineral Policy Center, n.d.), p. 1.
169. *Taking from the Taxpayer*, p. 14.
170. *Taking from the Taxpayer*, p. 17.
171. Brad Knickerbocker, "Call to Ban Mining Pits New West Against Old," *Christian Science Monitor*, February 5, 1999, p. 3.
172. Thomas J. Hilliard, *Golden Patents, Empty Pockets* (Washington, D.C.: Mineral Policy Center, 1994), pp. 4, 25, 29.
173. *Taking from the Taxpayer*, p. 15.
174. Thomas J. Aley and Wilgus B. Creath, "Mining and Hydrology," in *Golden Dreams, Poisoned Streams*, ed. Carlos D. Da Rosa and James S. Lyon (Washington, D.C.: Mineral Policy Center, 1997), p. 128. (The first four chapters of this book were written by Da Rosa and Lyon; the remaining chapters are by others and will be cited by author and chapter title.)
175. Da Rosa and Lyon, *Golden Dreams*, p. 44.
176. "True Price of Gold," pp. 1–2.
177. Johnnie Moore and Samual L. Luoma, "Impacts of Water Pollution from Mining: A Case Study," in *Golden Dreams*, p. 170.
178. Da Rosa and Lyon, *Golden Dreams*, p. 43.
179. Da Rosa and Lyon, *Golden Dreams*, pp. 64–72.
180. *Taking from the Taxpayer*, p. 13.

181. Susan R. Poulter, "Cleanup and Restoration: Who Should Pay?" *Journal of Land, Resources, and Environmental Law* 18 (1998): 77.

182. "True Price of Gold," p. 2.

183. *Taking from the Taxpayer*, p. 19.

184. Da Rosa and Lyon, *Golden Dreams*, p. 82.

185. Thomas W. Sonandres, "Desert Ecology," *MPC News*, Spring 1999, p. 10. *MPC News* is a newsletter published by the Mineral Policy Center.

186. Moore and Luoma, "Impacts of Water Pollution," pp. 183–185.

187. Da Rosa and Lyon, *Golden Dreams*, chap. 4. For an example of a well-designed mine, see the description of the McLaughlin gold mine in California, pp. 110–111.

188. James S. Lyon et al., *Burden of Gilt* (Washington, D.C.: Mineral Policy Center, 1993), pp. 22–23; L. Thomas Galloway and Karen L. Perry, "Mining Regulatory Problems and Fixes," in *Golden Dreams*, p. 213; Cathy Carlson, "Money Pits," *MPC News*, Spring 2000, p. 8; Mark H. Hunter, "Colorado Considers a Mining Ban," *High Country News*, June 19, 2000, p. 6.

189. Eric Whitney, "South Dakota Tells a Mine to Stay Put," *High Country News*, February 1, 1999. p. 4.

190. *Six Mines, Six Mishaps* (Washington, D.C.: Mineral Policy Center, 1999), pp. 5–8.

191. Knickerbocker, "Call to Ban Mining," p. 3.

192. Carlson, "Money Pits," p. 8.

193. "A Pickaxe Too Far," *Economist*, April 25, 1992, p. 27.

194. Galloway and Perry, "Mining Regulatory Problems," pp. 194–199.

195. Galloway and Perry, "Mining Regulatory Problems," pp. 199–200; *Taking from the Taxpayer*, p. 35.

196. Wilkinson, *Crossing the Next Meridian*, p. 66; see also Daphne Werth, "Where Regulation and Property Rights Collide: Reforming the Hardrock Act of 1872," *University of Colorado Law Review* 65 (1994): 430–434, 438–439. The BLM is currently in the process of rewriting its regulations, so we can hope for at least some improvement within the next year.

197. Wilkinson, *Crossing the Next Meridian*, p. 58.

198. "Pickaxe Too Far," p. 27; "The Last American Dinosaur: The 1872 Mining Law" [fact sheet] (Washington, D.C.: Mineral Policy Center, n.d.); Werth, "Where Regulation and Property Rights Collide," pp. 433, 442.

199. Galloway and Perry, "Mining Regulatory Problems," pp. 207–215; Thomas J. Hilliard, *States' Rights, Miners' Wrongs* (Washington, D.C.: Mineral Policy Center/American Fisheries Society/American Rivers/Trout Unlimited, 1994); Werth, "Where Regulation and Property Rights Collide," p. 446.

200. Da Rosa and Lyon, *Golden Dreams*, pp. 15–16; "Last American Dinosaur," p. 4.

201. Dan Randolph et al., "It's Time to Protect Special Places from Mining," *MPC News*, Spring 1999, p. 1. Recently the Interior Department issued a legal opinion that the BLM does have the right to veto a mine under certain circumstances. At issue is a proposed gold mine in California that would destroy Native American sacred sites and habitat for endangered species. But if the BLM does reject the mine, it will take years of litigation to determine if Interior's new interpretation will stand; Michelle Nijhuis, "Mine Proposal Stumbles," *High Country News*, March 27, 2000, p. 5.

202. Knickerbocker, "Call to Ban Mining," p. 3.

203. Michael Graf, "Application of Takings Law to the Regulation of Unpatented Mining Claims," *Ecology Law Quarterly* 24 (1997): 72.

204. Graf, "Application of Takings Law," pp. 120–121.

205. Thomas Michael Power, *Not All That Glitters* (Washington, D.C.: Mineral Policy Center/National Wildlife Federation, 1993), pp. 9, 13.

206. Wilkinson, *Crossing the Next Meridian*, p. 60.

207. Lyon et al., *Burden of Gilt*, pp. 6–7.

208. Power, *Not All That Glitters*, p. 1.

209. "Last American Dinosaur," p. 4.

210. *Taking from the Taxpayer*, pp. 14–15.

211. Graf, "Application of Takings Law," p. 60. The original law required $100 worth of work to be done on the claim each year. Since that requirement was impossible to enforce, in 1993 Congress replaced it with a $100 annual fee.

212. Lyon et al., *Burden of Gilt*, p. 9.

213. "Pickaxe Too Far," p. 27.

214. Power, *Not All That Glitters*, pp. 29–30.

215. The industry's arguments in this paragraph are typical ones from past years; see Power, *Not All That Glitters.*

216. Wilkinson, *Crossing the Next Meridian*, p. 260.

217. Here I am only concerned with surface water and the system for allocating it. Groundwater is also a problem in the West, but the laws governing rights to it are different from those for surface water. In many places groundwater is being pumped far faster than the aquifers can recharge and water tables are being lowered dramatically. This can cause land to subside, springs and wetlands to dry up, and stream flow to be reduced. A complete environmental and policy analysis would have to include a look at groundwater as well.

218. Western Water Policy Review Advisory Commission, *Water in the West: Challenge for the Next Century,* June 1998, pp. 5-4, 5-11, 6-26. This commission was created by Congress to review water policy. Hereafter cited as Commission, *Water in the West.*

219. Robert Gottlieb, *A Life of Its Own: The Politics and Power of Water* (San Diego: Harcourt Brace Jovanovich, 1988), pp. 47–48.

220. Federal water projects originally had a social goal as well, to promote family farms. Federal water was to go only to small-acreage farms. This, of course, suited the water industry's corner of the iron triangle not at all, and the limitations were widely abused from the start. Much of the water now goes to large agribusinesses—but that too is a separate story.

221. Commission, *Water in the West*, p. 2-24.

222. Terry L. Anderson and Pamela S. Snyder, *Priming the Invisible Pump,* PERC Policy Series PS-9 (Bozeman, Mont.: PERC, 1997), p. 12.

223. *Taking from the Taxpayer*, p. 43.

224. *Taking from the Taxpayer*, pp. 43–46, 49–50.

225. *Taking from the Taxpayer*, p. 50.

226. Anderson and Snyder, *Priming the Invisible Pump*, pp. 9–10.

227. *Taking from the Taxpayer*, p. 50.

228. Michael C. Blumm, "Public Choice Theory and the Public Lands: Why 'Multiple Use' Failed," *Harvard Environmental Law Review* 18 (1994): 411.

229. Blumm, "Public Choice Theory," p. 411.

230. David Pimentel et al., "Water Resources: Agriculture, the Environment, and Society," *BioScience* 47 (1997): 102.

231. Commission, *Water in the West*, p. 2-24.

232. *Taking from the Taxpayer*, p. 53.

233. Donald Worster, *Rivers of Empire* (New York: Pantheon, 1985), p. 278.

234. Anderson and Snyder, *Priming the Invisible Pump*, p. 11.

235. Pimentel et al., "Water Resources," p. 103.

236. Anderson and Snyder, *Priming the Invisible Pump*, p. 6. On the high plains of Texas, irrigators were forced to adopt efficient means of watering their crops because the aquifer was being depleted. They cut water use by up to 50 percent. This was not federal water, but it shows the potential for conservation while maintaining production; Reisner and Bates, *Overtapped Oasis*, p. 119. Improving water efficiency can, however, have environmental consequences that would have to be considered. In some places water that leaks out of unlined canals helps recharge aquifers or serves as a source of water for wetlands; Commission, *Water in the West*, pp. 2–25, 3–14, 6–34.

237. Thomas J. Graff and David Yardas, "Reforming Western Water Policy: Markets and Regulation," *Natural Resources and Environment* 12 (1998): 165.

238. Wilkinson, *Crossing the Next Meridian*, chap. 5.

239. Michael Collier et al., *Dams and Rivers: Primer on the Downstream Effects of Dams*, U.S. Geological Survey Circular 1126 (Tucson, June 1996).

240. Terry L. Anderson and Donald R. Leal, *Free Market Environmentalism* (San Francisco: Pacific Research Institute for Public Policy, 1991), p. 111.

241. Pimentel et al., "Water Resources," p. 104.

242. Reisner and Bates, *Overtapped Oasis*, p. 58.

243. Commission, *Water in the West*, pp. 2-30 to 2-32.

244. Wilkinson, *Crossing the Next Meridian*, p. 263; Harrison C. Dunning, "Confronting the Environmental Legacy of Irrigated Agriculture in the West: The Case of the Central Valley Project," *Environmental Law* 23 (1993): 953, 959.

245. Dunning, "Confronting the Environmental Legacy," pp. 953–954.

246. Dunning, "Confronting the Environmental Legacy," p. 954.

247. Reisner and Bates, *Overtapped Oasis*, p. 60; Commission, *Water in the West*, pp. 3–15.

248. Commission, *Water in the West*, p. 5-5. Unfortunately, the agencies are still engaged in many smaller projects that are also losers, financially and environmentally; see, e.g., Gawain Kripke et al., eds., *Green Scissors 99* (Washington, D.C.: Friends of the Earth, 1999), pp. 51–67.

249. Wilkinson, *Crossing the Next Meridian*, pp. 286–287.

250. Reisner and Bates, *Overtapped Oasis*, p. 32.

251. Quoted in Anderson and Leal, *Free Market Environmentalism*, p. 117.

252. Commission, *Water in the West*, pp. 3–16, 4–23.

253. Anderson and Snyder, *Priming the Invisible Pump*, pp. 12–13.

254. Wilkinson, *Crossing the Next Meridian*, p. 289; Commission, *Water in the West*, pp. 6–28.

255. Graff and Yardas, "Reforming Western Water Policy," p. 169.

256. Charles W. Howe, "Increasing Efficiency in Water Markets: Examples from the Western United States," in *Water Marketing—The Next Generation*, ed. Terry L. Anderson and Peter J. Hill (Lanham, Md.: Rowman & Littlefield, 1997), p. 93.

257. For some specific recommendations that deal with many of the complexities involved, see Reisner and Bates, *Overtapped Oasis*, pp. 122–143. Many water transfers have significant third party effects that would have to be taken into account; Anderson and Snyder, *Priming the Invisible Pump*, pp. 5–6.

258. One of the many issues that will be involved in taking significant steps toward expanding water marketing concerns windfall profits. It would hardly be fair to let a farmer buy highly subsidized water for a few dollars per acre-foot and then sell it to a city for hundreds or even thousands of dollars. On the other hand, irrigators should be able to profit from selling water or there will be no incentive to conserve it. One possibility would be, while subsidies are being phased out, to charge the farmer full cost for any federal water sold. Another policy that could be considered is a "tax" to enter the market with federally subsidized water, requiring that a portion of the conserved water be dedicated to in-stream flows. That way the public would benefit as well. Oregon may be a model here. It allows 75 percent of conserved water to be sold, with 25 percent left in the streams; Reisner and Bates, *Overtapped Oasis*, pp. 79, 124.

259. Barton H. Thompson Jr., "Water Markets and the Problem of Shifting Paradigms," in *Water Marketing: The Next Generation*, p. 2.

260. Howe, "Increasing Efficiency in Water Markets," pp. 85–86.

261. Wilkinson, *Crossing the Next Meridian*, p. 285.

262. Graff and Yardas, "Reforming Western Water Policy," p. 168.

263. Wilkinson, *Crossing the Next Meridian*, p. 286.

264. This list of some of the barriers to marketing water is taken from Anderson and Snyder, *Priming the Invisible Pump*; Commission, *Water in the West*; Reisner and Bates, *Overtapped Oasis*; Thompson, "Water Markets"; and Wilkinson, *Crossing the Next Meridian*.

265. Few of the water rights of Indian tribes have been finally adjudicated or negotiated. These rights are both large and very senior. "The existence of unquantified tribal rights adds great uncertainty to all other rights holders in a given basin." Commission, *Water in the West*, p. 5-3.

266. The notorious case in the 1920s in which Los Angeles grabbed the water of the Owens Valley still casts a large shadow over interbasin transfers; Thompson, "Water Markets," p. 6; Reis-

ner and Bates, *Overtapped Oasis*, p. 71. The legitimate concerns of local communities will have to be taken into consideration, perhaps by limiting the percentage of water that can be transferred out of a watershed or by requiring buyers to pay to mitigate damages; Thompson, "Water Markets," p. 24; Reisner and Bates, *Overtapped Oasis*, p. 135.

267. Graff and Yardas, "Reforming Western Water Policy," pp. 168–169; Thompson, "Water Markets," pp. 17–18.

268. Thompson, "Water Markets," p. 1.

269. Terry Anderson and Pamela Snyder, "Markets and Western Water," *Different Drummer: An Environmental Agenda for the 105th Congress* 3, no. 4 (1997): 25.

270. Commission, *Water in the West*, pp. 5–9.

271. Commission, *Water in the West*, pp. 5–39.

272. Commission, *Water in the West*, pp. 5–24. The program is strictly voluntary. The federal government pays 65 percent of the costs and the state and local irrigators split the rest.

273. Commission, *Water in the West*, p. 5-38.

274. Alexandra Ravinet, "Rivers Get Over the Dam," *Christian Science Monitor*, July 8, 1999, p. 14.

275. Commission, *Water in the West*, pp. 5-33 and 5-35.

276. John L. Hammond, "Wilderness and Heritage Values," *Environmental Ethics* 7 (1985): 165.

277. Quoted in Robert G. Athearn, *The Mythic West in Twentieth-Century America* (Lawrence: University Press of Kansas, 1986), p. 222.

278. Athearn, *Mythic West*, p. 221.

CHAPTER SIX

1. Robert Solo, "Problems of Modern Technology," *Journal of Economic Issues* 8 (1974): 863.

2. "Fewer than a dozen scientists, many of them on the payroll of coal and energy companies, say not to worry. On the evening news, both sides get equal time." Paul Rauber, "The Uncertainty Principle," *Sierra*, September–October 1996, p. 20. See also Jeremy Leggett, "Energy and the New Politics of the Environment," *Energy Policy* 19 (1991): 165; Meredith Wadman, "U.S. 'Wastes Vital Time' as Climate-Change Minority Sows Confusion," *Nature* 400 (1999): 5.

3. Ross Gelbspan, "A Good Climate for Investment," *Atlantic Monthly*, June 1998, p. 26; Brad Knickerbocker, "The Global Warming Debate Cuts Two Ways," *Christian Science Monitor*, May 11, 1998, p. 3; Stephen J. DeCanio, *The Economics of Climate Change* (San Francisco: Redefining Progress, 1997), p. 15; "Climate Protection Happens," *Rocky Mountain Institute Newsletter*, Summer 1998, p. 5; Brad Knickerbocker, "Business Takes a 'Greener' Stand on Global Warming," *Christian Science Monitor*, January 24, 2000, pp. 2–3. Major companies that publicly recognize the seriousness of global warming include Boeing, 3M, AT&T, Allied Signal, Dow, DuPont, Eastman Kodak, Enron, GE, Lockheed Martin, Toyota, United Technologies, Weyerhauser, General Motors, Ford, and Johnson and Johnson, among others.

4. Quoted in Leggett, "Energy and the New Politics of the Environment," p. 166.

5. Paul R. Ehrlich and Anne H. Ehrlich, *Betrayal of Science and Reason* (Washington, D.C.: Island, 1996), p. 31; emphasis in the original.

6. Gordon K. Durnil, *The Making of a Conservative Environmentalist* (Bloomington: Indiana University Press, 1995), pp. 88–95.

7. John Houghton, *Global Warming: The Complete Briefing*, 2d ed. (Cambridge: Cambridge University Press, 1997). Houghton was head of the United Kingdom Meteorological Office until he retired in 1991. He is currently cochairman of the Science Assessment Working Group of the Intergovernmental Panel on Climate Change and chairman of the United Kingdom's Royal Commission on Environmental Pollution.

8. See J. D. Mahlman, "Uncertainties in Projections of Human-Caused Climate Warming," *Science* 278 (1997): 1416–1417, for a more extensive list, categorizing scientific findings on global warming by levels of certainty, uncertainty, and probability.

9. Houghton, *Global Warming*, p. 11.

10. Houghton, *Global Warming*, p. 164.

11. J. T. Houghton et al., eds., *Climate Change 1995: The Science of Climate Change* (Cambridge: Cambridge University Press, 1996), pp. 78–79. This volume is the contribution of Working Group I to the *Second Assessment Report* of the Intergovernmental Panel on Climate Change. Each chapter has multiple coauthors, so it will be easier to cite it hereafter as IPCC 1995, WGI. There are two other working groups in the IPCC as well: Working Group II on Impacts, Adaptations, and Mitigation of Climate Change and Working Group III on Economic and Social Dimensions of Climate Change. Citations from their reports will take the same form.For the problems in estimating tropical forest destruction, see R. A. Houghton, "The Worldwide Extent of Land-Use Change," *BioScience* 44 (1994): 309–311. A recent study indicates that less CO_2 is being released from tropical forests than previously thought; Richard Monastersky, "The Case of the Missing Carbon Dioxide," *Science News* 155 (1999): 383. It is sometimes confusing that some authors refer to the amount of carbon dioxide being emitted and others refer to the amount of carbon as carbon dioxide. One ton of carbon makes 3.67 tons of carbon dioxide.

12. The ocean takes up carbon dioxide from the atmosphere as it is dissolved into seawater. On top of that form of storage, there is the "biological pump," which sequesters carbon in the deep ocean. Phytoplankton in the sea use the dissolved CO_2 for photosynthesis. As phytoplankton take up $CO_{2,}$ the water can then absorb more from the atmosphere. The remains of these organisms, plus those of the zooplankton that feed on them, sink into the depths, removing the carbon from the carbon cycle for hundreds or thousands of years, or even longer; Houghton, *Global Warming*, p. 236.

13. Houghton, *Global Warming*, p. 26; IPCC 1995, WGI, p. 79.

14. Eric T. Sundquist, "The Global Carbon Dioxide Budget," *Science* 259 (1993): 937.

15. Houghton, *Global Warming*, p. 24; IPCC 1995, WGI, p. 78; J. R. Petit et al., "Climate and Atmospheric History of the Past 420,000 Years from the Vostok Ice Core, Antarctica," *Nature* 399 (1999): 429–436; C. D. Keeling and T. P. Whorf, "Atmospheric Carbon Dioxide Record from Mauna Loa." <http://cdiac.esd.ornl.gov/trends/emis/tre-usa.htm> The evidence for atmospheric CO_2 (and other greenhouse gases) for early years comes primarily from analysis of air bubbles trapped in ice in Greenland and Antarctica. Since 1957 precise measurements of CO_2 have been made at the Mauna Loa Observatory in Hawaii. A fair amount of publicity was given to estimates that worldwide carbon dioxide emissions in 1998 declined slightly. However, much of the decrease was a result of the severe recession in Southeast Asia. U.S. emissions increased, so we certainly have nothing to be proud about. But some of the decrease shows the promise of policies that will be discussed in the next chapter. If the statistics can be believed, CO_2 emissions in China declined while production increased. China had reduced subsidies on coal and banned coal burning in Beijing. The economies in transition in eastern Europe saw a decline of emissions as well. They have long been very inefficient in using energy, and as they make the transition to modern market economies their carbon intensity will go down. See "Better News—Perhaps—on Global Warming," *Nature* 400 (1999): 487; Tony Reichhardt, "Emissions Fall Despite Economic Growth," *Nature* 400 (1999): 494; Michael Jefferson, "Global Emissions Could Soon Start Rising Again," *Nature* 400 (1999): 810.

16. Houghton, *Global Warming*, p. 34. "Global warming potential" is an index that compares the radiative forcing power of the greenhouse gases. It is not, by any means, a simple calculation to make. The different gases have different lifetimes in the atmosphere, and they react in different ways with other gases. The IPCC in its latest report gives three different global warming potentials for each greenhouse gas for time horizons of 20, 100, and 500 years. Since conservatives are supposed to take the long view, I have given the numbers for the 500-year time horizon. For shorter horizons, the index numbers are much larger. For example, the three index numbers for CH_4 are 56, 21, and 6.5, respectively; IPCC 1995, WGI, pp. 22, 121. (Carbon dioxide is the reference gas, so its index number is 1.)

17. Houghton, *Global Warming*, pp. 33–34; IPCC 1995, WGI, p. 15; Petit et al., "Climate and Atmospheric History of the Past 420,000 Years."

18. L. P. Steele et al., "Slowing Down of the Global Accumulation of Atmospheric Methane During the 1980s," *Nature* 358 (1992): 313–316; E. J. Dlugokencky et al., "Continuing Decline in the Growth Rate of the Atmospheric Methane Burden," *Nature* 393 (1998): 447–450.

19. Markus Leuenberger and Ulrich Siegenthaler, "Ice-Age Atmospheric Concentration of Nitrous Oxide from an Antarctic Ice Core," *Nature* 360 (1992): 449–451.

20. Leuenberger and Siegenthaler, "Ice-Age Atmospheric Concentration of Nitrous Oxide," p. 450; Houghton, *Global Warming*, p. 35; IPCC 1995, WGI, p. 22.

21. Richard A. Kerr, "Ozone-Destroying Chlorine Tops Out," *Science* 271 (1996): 32; R. Monastersky, "Drop in Ozone Killers Means Global Gain," *Science News* 149 (1996): 151. For the greenhouse effect of ozone, see J. T. Kiehl et al., "Climate Forcing Due to Tropospheric and Stratospheric Ozone," *Journal of Geophysical Research* 104 (1999): 31239–31254. For discussions of the CFC-ozone reactions in relation to warming potential, see Houghton, *Global Warming*, pp. 35–38; IPCC 1995, WGI, pp. 109–111; V. Ramaswamy et al., "Radiative Forcing of Climate from Halocarbon-Induced Global Stratospheric Ozone Loss," *Nature* 355 (1992): 810–812; J. Hansen et al., "Radiative Forcing and Climate Response," *Journal of Geophysical Research* 102 (1997): 6831–6864; John S. Daniel et al., "On the Evaluation of Halocarbon Radiative Forcing and Global Warming Potentials," *Journal of Geophysical Research* 100 (1995): 1271–1285; Susan Solomon and John S. Daniel, "Impact of the Montreal Protocol and Its Amendments on the Rate of Change of Global Radiative Forcing," *Climatic Change* 32 (1996): 7–17. For the global warming potential of CFC replacements, see Vaishali Naik et al., "Consistent Sets of Atmospheric Lifetimes and Radiation Forcing on Climate for CFC Replacements: HCFC and HFC," *Journal of Geophysical Research* 105 (2000): 6903–6914.

22. Houghton, *Global Warming*, p. 22. A new study by James Hansen et al. concludes that CO_2 has been responsible for approximately 50 percent of greenhouse gas forcing to date, CH_4 for 24 percent, CFCs for 12 percent, ozone for 10 percent, and N_2O for 5 percent. The relative importance of CO_2, they conclude, will increase in the future; James Hansen et al., "Global Warning in the Twenty-First Century: An Alternative Scenario," *Proceedings of the National Academy of Sciences* 97 (2000): 9875–9880.

23. P. D. Jones et al., "Surface Air Temperature and Its Changes over the Past 150 Years," *Reviews of Geophysics* 37 (1999): 173–199; Phil Jones, "It Was the Best of Times, It Was the Worst of Times," *Science* 280 (1998): 544–545; P. D. Jones, "Recent Warming in Global Temperature Series," *Geophysical Research Letters* 21 (1994): 1149–1152.

24. Houghton, *Global Warming*, p. 3; Jones et al., "Surface Air Temperature," pp. 174, 176, 196; Michael E. Mann et al., "Northern Hemisphere Temperatures During the Past Millennium: Inferences, Uncertainties and Limitations," *Geophysical Research Letters* 26 (1999): 759–762. The latest analysis of surface temperature trends by NASA's Goddard Institute for Space Studies concludes that the global average increased by about .7°C since the late 1880s; J. Hansen et al., "GISS Analysis of Surface Temperature Change," *Journal of Geophysical Research* 104 (1999): 30997–31022. Since 1979, satellites have used microwave instruments to collect data on atmospheric temperatures at several altitudes. Data from the lower troposphere, at an altitude of about 3.5 kilometers, indicated a cooling trend. This is, of course, a different measurement than surface air temperatures, so their trends need not be the same. Nevertheless, as James Hansen and his colleagues observe, the satellite data "have been the principal refuge for those who deny the reality of global warming." "Global Climate Data and Models: A Reconciliation," *Science* 281 (1998): 932. However, the satellite time series is far too short to draw conclusions about climate. Later analyses discovered that, when corrected for events with short-term effects such as the Pinatubo eruption and El Niños, the satellite data are in better agreement with surface air temperature measurements. Moreover, on a longer time series (with earlier measurements from radiosondes), the surface air and lower tropospheric temperature trends both show warming.The satellite data require extremely complicated corrections to factor out noise from several sources and to combine readings from different satellites at different times. Some climate scientists contend that the attempts to make these corrections have not been fully successful. Recently it was discovered that the microwave data must also be corrected for "orbital decay" of the satellites, as they lose altitude over time. When this correction is made, the cooling trend disappears and the temperatures show a warming trend in the lower troposphere.The scientists who have been compiling and preparing the satellite measurements now have a new version that corrects for orbital decay and other factors. This version still shows a slight cooling trend from 1979 to 1997. (Problems using such a short time span are clearly shown here because merely adding the 1998

data turns the overall trend into a warming one.) There is a considerable effort now to discover an explanation, since the general expectation was that the lower troposphere and surface air temperatures would coincide fairly closely. Some forcings are apparently disconnecting the two levels. One recent study found that when greenhouse gases, tropospheric ozone, stratospheric ozone, aerosols, and the Mount Pinatubo eruption are all factored in, a climate model produces the diverging temperature trends, but other problems remain in the results. As I write this, the final word clearly is not yet in. There are still uncertainties about the effects of ozone at different altitudes, about the direct and indirect effects of aerosols, and about the satellite data.None of this should be taken as a "denial" of global warming. It calls for an explanation, to be sure. For a long time, surface air temperatures and lower troposphere temperatures coincided. Now it appears that they may react differently when subjected to current natural and anthropogenic forces. But to find that a temperature trend at one altitude is not what was expected, although trends at other levels coincide with expectations, does not "refute" global warming. It certainly does not deny the fact that the surface air temperatures have increased during the twentieth century at a rate far faster than natural and to a level far greater than natural. Consider a completely independent type of evidence that directly measures the earth's temperature at the ground. Temperature changes at the ground propagate slowly downward into the rocks beneath the surface, so subsurface temperatures record a history of surface temperature changes from the past. Many studies have been done measuring these temperatures from boreholes. They confirm that, worldwide, the earth has warmed about 0.5°C during the twentieth century, that the twentieth century was the warmest in 500 years, and that the rate of warming was the highest in at least five centuries. For details and analyses of the satellite data, see, among others, John R. Christy and Richard T. McNider, "Satellite Greenhouse Signal," *Nature* 367 (1994): 325; Stephen H. Schneider, "Detecting Climatic Change Signals: Are There Any 'Fingerprints'?" *Science* 263 (1994): 341–342; P. D. Jones et al., "Recent Warming in Global Temperature Series," pp. 1149–1152; John R. Christy and Roy W. Spencer, "Assessment of Precision in Temperatures for the Microwave Sounding Units," *Climatic Change* 30 (1995): pp. 97–102; James Hansen et al., "Satellite and Surface Temperature Data at Odds?" *Climatic Change* 30 (1995): 103–117; John R. Christy et al., "Reducing Noise in the MSU Daily Lower-Tropospheric Global Temperature Dataset," *Journal of Climate* 8 (1995): 888–896; IPCC 1995, WGI, pp. 137, 147–148; T. J. L. Wigley, "Climate Change Report," *Science* 271 (1996): 1461–1462; James W. Hurrell and Kevin E. Trenberth, "Satellite Versus Surface Estimates of Air Temperature Since 1979," *Journal of Climate* 9 (1996): 2222–2232; P. D. Jones et al., "Comparisons Between the Microwave Sounding Unit Temperature Record and the Surface Temperature Record from 1979 to 1996: Real Differences or Potential Discontinuities?" *Journal of Geophysical Research* 102 (1997): 30, 145; James W. Hurrell and Kevin E. Trenberth, "Spurious Trends in Satellite MSU Temperatures from Merging Different Satellite Records," *Nature* 386 (1997): 164–167; Kevin E. Trenberth and James W. Hurrell, "How Accurate Are Satellite Thermometers?" *Nature* 389 (1997): 342–343; Houghton, *Global Warming*, pp. 47–48; James W. Hurrell and Kevin E. Trenberth, "Difficulties in Obtaining Reliable Temperature Trends: Reconciling the Surface and Satellite Microwave Sounding Unit Records," *Journal of Climate* 11 (1998): 945–967; Frank J. Wentz and Matthias Schabel, "Effects of Orbital Decay on Satellite-Derived Lower-Tropospheric Temperature Trends," *Nature* 394 (1998): 661–664; C. Prabhakara et al., "Global Warming Deduced from MSU," *Geophysical Research Letters* 25 (1998): 1927–1930; Richard A. Kerr, "Among Global Thermometers, Warming Still Wins Out," *Science* 281 (1998): 1948–1949; Jones et al., "Surface Air Temperature," pp. 186–189; B. D. Santer et al., "Uncertainties in Observationally Based Estimates of Temperature Change in the Free Atmosphere," *Journal of Geophysical Research* 104 (1999): 6305–6333; Dian J. Gaffen et al., "Multidecadal Changes in the Vertical Temperature Structure of the Tropical Troposphere," *Science* 287 (2000): 1242–1245; David E. Parker, "Temperatures High and Low," *Science* 287 (2000): 1216–1217; B. D. Santer et al., "Interpreting Differential Temperature Trends at the Surface and in the Lower Troposphere," *Science* 287 (2000): 1227–1232; James W. Hurrell et al., "Comparison of Tropospheric Temperatures from Radiosondes and Satellites: 1979–1998," *Bulletin of the American Meteorological Society* 81 (2000): 2165–2177. For the latest version of the satellite data, showing a small warming trend in the lower troposphere through 1998, see John R. Christy et al., "MSU Tropospheric Temperatures: Dataset Construction and Radiosonde Comparisons," *Journal of Atmospheric and Oceanic Technology* 17 (2000): 1153–1170. For a good analysis of the problems involved in interpreting satellite data, see National Research Council, *Reconciling Observations of Global Temperature Change* (Washington, D.C.: National Academy Press, 2000). For evidence from boreholes, see John Sass, "Climate Plumbs the Depths," *Nature* 349 (1991): 458; Henry N. Pollack and David S. Chapman, "Underground Records of Changing Climate," *Scientific American,* June 1993, pp. 44–50; David Deming, "Climatic Warming in North America: Analysis of Borehole Temperatures," *Science* 268 (1995): 1576–1577; Robert N. Harris and David S. Chapman, "Borehole Temperatures and a Baseline for 20th-Century Global Warming Estimates," *Science* 275 (1997): 1618–1621; Henry N. Pollack et al., "Climate Change Record in Subsurface Temperatures: A Global Perspective," *Science* 282 (1998): 279–281; Jonathan T. Overpeck, "The Hole Record," *Nature* 403 (2000): 714–715; Shaopeng Huang et al., "Temperature Trends over the Past Five Centuries Reconstructed from Borehole Temperatures," *Nature* 403 (2000): 756–758.

25. Houghton, *Global Warming*, p. 8.

26. See, for example, Richard A. Kerr, "Model Gets It Right—Without Fudge Factors," *Science* 276 (1997): 1041. For a description of climate modeling, see Houghton, *Global Warming*, chap. 5; and IPCC 1995, WGI, chap. 5.

27. Jorge L. Sarmiento et al., "Limiting Future Atmospheric Carbon Dioxide," *Global Biogeochemical Cycles* 9 (1995): 121–137.

28. Houghton, *Global Warming*, pp. 76–77.

29. Nicholas R. Bates et al., "Contribution of Hurricanes to Local and Global Estimates of Air-Sea Exchange of CO_2," *Nature* 395 (1998): 58–61.

30. M. A. K. Khalil and R. A. Rasmussen, "Climate-Induced Feedbacks for the Global Cycles of Methane and Nitrous Oxide," *Tellus* 41B (1989): 554–559.

31. Walter C. Oechel et al., "Recent Change of Arctic Tundra Ecosystems from a Net Carbon Dioxide Sink to a Source," *Nature* 361 (1993): 520–523; M. L. Goulden et al., "Sensitivity of Boreal Forest Carbon Balance to Soil Thaw," *Science* 279 (1998): 214–217.

32. Houghton, *Global Warming*, pp. 74, 77–78; G. Ramstein et al., "Cloud Processes Associated with Past and Future Climate Changes," *Climate Dynamics* 14 (1998): 233–247; David Rind, "Just Add Water Vapor," *Science* 281 (1998): 1152–1153; Brian J. Soden, "Enlightening Water Vapour," *Nature* 406 (2000): 247–248.

33. Houghton, *Global Warming*, p. 125.

34. Jon D. Erickson, "From Ecology to Economics: The Case Against CO_2 Fertilization," *Ecological Economics* 8 (1993): 165. See also Fakhri A. Bazzaz and Eric D. Fajer, "Plant Life in a CO_2-Rich World," *Scientific American* (January 1992): 68–74.

35. H. A. Mooney et al., "Predicting Ecosystem Responses to Elevated CO_2 Concentrations," *BioScience* 41 (1991): 96–104; Evan H. DeLucia et al., "Net Primary Production of a Forest Ecosystem with Experimental CO_2 Enrichment," *Science* 284 (1999): 1177–1179; Christian Körner, "CO_2 Fertilization: The Great Uncertainty in Future Vegetation Development," in *Vegetation Dynamics and Global Change,* ed. Allen M. Solomon and Herman H. Shugart (New York: Chapman & Hall, 1993), pp. 53–70; Erickson, "From Ecology to Economics," pp. 168–171.

36. Mingkui Cao and F. Ian Woodward, "Dynamic Responses of Terrestrial Ecosystem Carbon Cycling to Global Climate Change," *Nature* 393 (1998): 249–252; P. J. Sellers et al., "Comparison of Radiative and Physiological Effects of Doubled Atmospheric CO_2 on Climate," *Science* 271 (1996): 1402–1406; Allen M. Solomon and Wolfgang Cramer, "Biospheric Implications of Global Environmental Change," in *Vegetation Dynamics and Global Change*, pp. 25–52.

37. George M. Woodwell, "Biotic Feedbacks from the Warming of the Earth," in *Biotic Feedbacks in the Global Climatic System: Will the Warming Feed the Warming?* ed. George M. Woodwell and Fred T. MacKenzie (New York: Oxford University Press, 1995), p. 15.

38. George M. Woodwell et al., "Will the Warming Speed the Warming?" in *Biotic Feedbacks in the Global Climatic System*, p. 394; see also D. S. Jenkinson et al., "Model Estimates of CO_2 Emissions from Soil in Response to Global Warming," *Nature* 351 (1991): 304–306. For grasslands, see Bruce A. Hungate et al., "The Fate of Carbon in Grasslands Under Carbon Dioxide

Enrichment," *Nature* 388 (1997): 576–579. See also a study of an artificial tropical ecosystem which found that elevated CO_2 levels did not stimulate any increase in biomass but tripled carbon losses from the soil; Christian Körner and John A. Arnone III, "Responses to Elevated Carbon Dioxide in Artificial Tropical Ecosystems," *Science* 257 (1992): 1672–1675.

39. G. M. Woodwell et al., "Biotic Feedbacks in the Warming of the Earth," *Climatic Change* 40 (1998): 497.

40. T. M. Smith and H. H. Shugart, "The Transient Response of Terrestrial Carbon Storage to a Perturbed Climate," *Nature* 361 (1993): 523. See also Peter M. Cox et al., "Acceleration of Global Warming Due to Carbon-cycle Feedbacks in a Coupled Climate Model," *Nature* 408 (2000): 184–187.

41. Woodwell et al., "Will the Warming Speed the Warming?" p. 395.

42. Woodwell et al., "Will the Warming Speed the Warming?" p. 404; Woodwell et al., "Biotic Feedbacks," p. 495.

43. R. J. Charlson et al., "Climate Forcing by Anthropogenic Aerosols," *Science* 255 (1992): 423–430; J. T. Kiehl and B. P. Briegleb, "The Relative Roles of Sulfate Aerosols and Greenhouse Gases in Climate Forcing," *Science* 260 (1993): 311–314; Robert J. Charlson and Tom M. L. Wigley, "Sulfate Aerosol and Climate Change," *Scientific American,* February 1994, pp. 48–57; J. F. B. Mitchell et al., "Climate Response to Increasing Levels of Greenhouse Gases and Sulphate Aerosols," *Nature* 376 (1995): 501–504; IPCC 1995, WGI, pp. 112–118; J. Hansen et al., "The Missing Climate Forcing," *Philosophical Transactions of the Royal Society of London* B 352 (1997): 231–240; Hansen et al., "Radiative Forcing and Climate Response."The warming of the earth during the twentieth century has not been even. From about 1940 to the mid-1970s, global average temperature was fairly stable. That, of course, was the period of rapid industrial growth that preceded general environmental awareness and the adoption of numerous pollution control laws in OECD countries. The heightened levels of aerosols from air pollution during that time may be the cause of the temporary stabilization of temperatures; Charlson and Wigley, "Sulfate Aerosol and Climate Change," p. 53. Recent studies show that black carbon aerosols (soot) in air pollution add to global warming; see, e.g., J. M. Haywood and V. Ramaswamy, "Global Sensitivity Studies of the Direct Radiative Forcing Due to Anthropogenic Sulfate and Black Carbon Aerosols," *Journal of Geophysical Research* 103 (1998): 6043–6058.

44. IPCC 1995, WGI, pp. 5–6; Houghton, *Global Warming*, pp. 42, 94–95; T. M. L. Wigley and S. C. B. Raper, "Implications for Climate and Sea Level of Revised IPCC Emissions Scenarios," *Nature* 357 (1992): 293–300. Wigley and Raper give an excellent short summary of all the different scenarios and the ranges of temperature projections. I have given only the most likely one. A new IPCC report is coming out shortly, which apparently will project even higher temperatures by 2100; "News in Brief," *Christian Science Monitor,* October 27, 2000, p. 24.

45. Houghton, *Global Warming*, p. 95.

46. For a report on the completed Vostok core, see Petit et al., "Climate and Atmospheric History of the Past 420,000 Years." For earlier studies, see C. Lorius et al., "A 150,000-Year Climatic Record from Antarctic Ice," *Nature* 316 (1985): 593–596; A. Neftel et al., "Evidence from Polar Ice Cores for the Increase in Atmospheric CO_2 in the Past Two Centuries," *Nature* 315 (1985): 45–47; J. M. Barnola et al., "Vostok Ice Core Provides 160,000-Year Record of Atmospheric CO_2," *Nature* 329 (1987): 408–414; C. Genthon et al., "Vostok Ice Core: Climatic Response to CO_2 and Orbital Forcing Changes over the Last Climatic Cycle," *Nature* 329 (1987): 414–418; J. Jouzel et al., "Vostok Ice Core: A Continuous Isotope Temperature Record Over the Last Climatic Cycle (160,000 Years)," *Nature* 329 (1987): 403–408; A. Neftel et al., "CO_2 Record in the Byrd Ice Core 50,000–5,000 Years BP," *Nature* 331 (1988): 609–611; C. Lorius et al., "The Ice-Core Record: Climate Sensitivity and Future Greenhouse Warming," *Nature* 347 (1990): 139–145; J. M. Barnola et al., "CO_2-Climate Relationships as Deduced from the Vostok Ice Core: A Re-examination Based on New Measurements and on a Re-evaluation of the Air Dating," *Tellus* 43B (1991): 83–90; Greenland Ice-Core Project (GRIP) Members, "Climate Instability During the Last Interglacial Period Recorded in the GRIP Ice Core," *Nature* 364 (1993): 203–207; J. Jouzel et al., "Extending the Vostok Ice-Core Record of Palaeoclimate to the Penultimate Glacial Period," *Nature* 364 (1993): 407–412; D. Raynaud, "Ice Core Records as a Key to Understanding the History of Atmospheric Trace Gases," in *Biogeochemistry of Global Change,* ed. R. S. Oremland (New York: Chapman & Hall, 1993), pp. 29–45; D. Ray-

naud et al., "The Ice Record of Greenhouse Gases," *Science* 259 (1993): 926–934; Michael Bender et al., "Climate Correlations Between Greenland and Antarctica During the Past 100,000 Years," *Nature* 372 (1994): 663–666; Hubertus Fischer et al., "Ice Core Records of Atmospheric CO_2 Around the Last Three Glacial Terminations," *Science* 283 (1999): 1712–1714.

47. Michael E. Mann et al., "Global-Scale Temperature Patterns and Climate Forcing Over the Past Six Centuries," *Nature* 392 (1998): 783–784.

48. Mann et al., "Global-Scale Temperature Patterns," p. 779. It is sometimes asserted that the current warming may be part of the natural recovery from the Little Ice Age of the fifteenth to the early nineteenth centuries, but this cannot be the explanation; see T. Wigley et al., "The Observed Global Warming Record: What Does It Tell Us?" *Proceedings of the National Academy of Sciences, USA* 94 (1997): 8218; Melissa Free and Alan Robock, "Global Warming in the Context of the Little Ice Age," *Journal of Geophysical Research* 104 (1999): 19057–19070.

49. Gabriele Hegerl, "The Past as Guide to the Future," *Nature* 392 (1998): 759.

50. Thomas J. Crowley, "Causes of Climate Change Over the Past 1000 Years," *Science* 289 (2000): 270.

51. Houghton, *Global Warming*, p. 158.

52. Houghton, *Global Warming*, p. 159.

53. Houghton, *Global Warming*, p. 158.

54. IPCC 1995, WGI, pp. 4–6. Since this report was completed, dozens of additional studies have been published that support these conclusions. A new IPCC report should appear within the next year, which will update the summaries of studies. "The draft of the year 2000 report now circulating concludes that 'there has been a discernable human influence on global climate.' That's stronger than the panel's 1995 assessment"; Robert C. Cowen, "Human Influence on Global Weather Now a Fact," *Christian Science Monitor,* June 8, 2000, p. 16.

55. Klaus Hasselmann, "Climate-Change Research After Kyoto," *Nature* 390 (1997): 225.

56. The last interglacial period (Eemian), 125,000 to 115,000 years ago, includes a period of some 5,000 years that were about 2°C warmer than the present; Greenland Ice Core Project Members, "Climate Instability During the Last Interglacial Period," p. 205; Jouzel et al., "Vostok Ice Core," p. 406.

57. Wallace S. Broecker, "Unpleasant Surprises in the Greenhouse?" *Nature* 328 (1987): 123.

58. Woodwell et al., "Will the Warming Speed the Warming?" p. 395.

59. Scott Lehman, "Sudden End of an Interglacial," *Nature* 390 (1997): 117. Even previous interglacial ages have had unstable climates; see Greenland Ice Core Project Members, "Climate Instability During the Last Interglacial Period"; Andrew J. Weaver and Tertia M. C. Hughes, "Rapid Interglacial Climate Fluctuations Driven by North Atlantic Ocean Circulation," *Nature* 367 (1994): 447–450.

60. Wigley and Raper, "Implications for Climate and Sea Level," p. 300.

61. Thomas R. Karl et al., "The Coming Climate," *Scientific American,* May 1997, p. 80.

62. Houghton, *Global Warming*, p. 132.

63. L. S. Kalkstein and K. E. Smoyer, "The Impact of Climate Change on Human Health: Some International Implications," *Experientia* 49 (1993): 969–979; Richard Stone, "If the Mercury Soars, So May Health Hazards," *Science* 267 (1995): 957–958; James P. Bruce et al., eds., *Climate Change 1995: Economic and Social Dimensions of Climate Change* (Cambridge: Cambridge University Press, 1996), p. 198. (This is the Contribution of Working Group III to the Second Assessment Report of the Intergovernmental Panel on Climate Change, hereafter cited as IPCC 1995, WGIII.) Dian J. Gaffen and Rebecca J. Ross, "Increased Summertime Heat Stress in the U.S.," *Nature* 396 (1998): 529–530.

64. Andrew Dobson and Robin Carper, "Global Warming and Potential Changes in Host-Parasite and Disease-Vector Relationships," in *Global Warming and Biological Diversity,* ed. Robert L. Peters and Thomas E. Lovejoy (New Haven: Yale University Press, 1992), p. 215.

65. Stone, "If the Mercury Soars," pp. 957–958; Jocelyn Kaiser, "Helping Those Most at Risk," *Science* 278 (1997): 217; Houghton, *Global Warming*, p. 132; W. J. M. Martens et al., "Climate Change and Vector-Borne Diseases," *Global Environmental Change* 5 (1995): 195–209; IPCC 1995, WGIII, p. 198; Paul R. Epstein, "Is Global Warming Harmful to Health?" *Scientific American,* August 2000, pp. 50–57; Paul R. Epstein et al., "Biological and Physical Signs of Climate

Change: Focus on Mosquito-Borne Diseases," *Bulletin of the American Meteorological Society* 79 (1998); 409–417.

66. Dobson and Carper, "Global Warming and Potential Changes," p. 201.

67. Houghton, *Global Warming*, pp. 109–110.

68. Houghton, *Global Warming*, pp. 111–113.

69. "World," *Christian Science Monitor*, June 14, 1999, p. 6.

70. Houghton, *Global Warming*, pp. 113, 115; IPCC 1995, WGIII, p. 192.

71. Kaiser, "Helping Those Most at Risk," p. 217; IPCC 1995, WGIII, p. 193.

72. Houghton, *Global Warming*, p. 115.

73. C. S. M. Doake and D. G. Vaughan, "Rapid Disintegration of the Wordie Ice Shelf in Response to Atmospheric Warming," *Nature* 350 (1991): 328–330.

74. Helmut Rott et al., "Rapid Collapse of Northern Larsen Ice Shelf, Antarctica," *Science* 271 (1996): 788–792; C. S. M. Doake et al., "Breakup and Conditions for Stability of the Northern Larsen Ice Shelf, Antarctica," *Nature* 391 (1998): 778–780.

75. D. G. Vaughan and C. S. M. Doake, "Recent Atmospheric Warming and Retreat of Ice Shelves on the Antarctic Peninsula," *Nature* 379 (1996): 328–330. On the decline in Antarctic sea ice, see Eugene Murphy and John King, "Icy Message from the Antarctic," *Nature* 389 (1997): 20–21; William K. de la Mare, "Abrupt Mid-Twentieth Century Decline in Antarctic Sea Ice Extent from Whaling Records," *Nature* 389 (1997): 57–60.

76. "News in Brief," *Christian Science Monitor*, April 9, 1999, p. 24.

77. D. A. Rothrock et al., "Thinning of the Arctic Sea-Ice Cover," *Geophysical Research Letters* 26 (1999): 3469–3472; Richard A. Kerr, "Will the Arctic Ocean Lose All Its Ice?" *Science* 286 (1999): 1828; Konstantin Y. Vinnikov et al., "Global Warming and Northern Hemisphere Sea Ice Extent," *Science* 286 (1999): 1934–1937; Ola M. Johannessen et al., "Satellite Evidence for an Arctic Sea Ice Cover in Transformation," *Science* 286 (1999): 1937–1939.

78. See, among others, J. H. Mercer, "West Antarctic Ice Sheet and CO_2 Greenhouse Effect: A Threat of Disaster," *Nature* 271 (1978): 321–325; Robert H. Thomas et al., "Effect of Climatic Warming on the West Antarctic Ice Sheet," *Nature* 277 (1979): 355–358; Philippe Huybrechts and Johannes Oerlemans, "Response of the Antarctic Ice Sheet to Future Greenhouse Warming," *Climate Dynamics* 5 (1990): 93–102; R. B. Alley and I. M. Whillans, "Changes in the West Antarctic Ice Sheet," *Science* 254 (1991): 959–963; Douglas R. MacAyeal, "Irregular Oscillations of the West Antarctic Ice Sheet," *Nature* 359 (1992): 29–32; Donald Blankenship et al., "Active Volcanism Beneath the West Antarctic Ice Sheet and Implications for Ice-Sheet Stability," *Nature* 361 (1993): 526–529; Richard B. Alley and Douglas R. MacAyeal, "West Antarctic Ice Sheet Collapse: Chimera or Clear Danger?" *Antarctic Journal-Review* 28 (1993): 59–60; Steven M. Hodge and Sheila K. Doppelhammer, "Satellite Imagery of the Onset of Streaming Flow of Ice Streams C and D, West Antarctica," *Journal of Geophysical Research* 101 (1996): 6669–6677; Charles R. Bentley, "Rapid Sea-Level Rise Soon from West Antarctic Ice Sheet Collapse?" *Science* 275 (1997): 1077–1078; Robert Bindschadler, "Actively Surging West Antarctic Ice Streams and Their Response Characteristics," *Annals of Glaciology* 24 (1997): 409–414; the exchange of letters between Robert Bindschadler and Charles E. Bentley in *Science* 276 (1997): 62–64; Robert Bindschadler and Patricia Vornberger, "Changes in the West Antarctic Ice Sheet Since 1963 from Declassified Satellite Photography," *Science* 279 (1998): 689–692; Michael Oppenheimer, "Global Warming and the Stability of the West Antarctic Ice Sheet," *Nature* 393 (1998): 325–332; John VanDecar, "On the Shelf," *Nature* 391 (1998): 747; Richard A. Kerr, "West Antarctic's Weak Underbelly Giving Way?" *Science* 281 (1998): 499–500; E. J. Rignot, "Fast Recession of a West Antarctic Glacier," *Science* 281 (1998): 549–551; Reed P. Scherer et al., "Pleistocene Collapse of the West Antarctic Ice Sheet," *Science* 281 (1998): 82–85; R. E. Bell et al., "Influence of Subglacial Geology on the Onset of a West Antarctic Ice Stream from Aerogeophysical Observations," *Nature* 394 (1998): 58–62; S. Anandakrishnan et al., "Influence of Subglacial Geology on the Position of a West Antarctic Ice Stream from Seismic Observations," *Nature* 394 (1998): 62–65; H. Conway et al., "Past and Future Grounding-Line Retreat of the West Antarctic Ice Sheet," *Science* 286 (1999): 280–283. For the IPCC's conclusions, see IPCC 1995, WGI, p. 389.Recent studies concluded that parts of the Greenland ice sheet are rapidly thinning. Since they are on land, they

would contribute to a rising sea level; W. Krabil et al., "Rapid Thinning of Parts of the Southern Greenland Ice Sheet," *Science* 283 (1999): 1522–1524; W. Krabil et al., "Greenland Ice Sheet: High-Elevation Balance and Peripheral Thinning," *Science* 289 (2000): 428–430.

79. Broecker, "Unpleasant Surprises in the Greenhouse?" p. 123.

80. H. B. Gordon et al., "Simulated Changes in Daily Rainfall Intensity Due to the Enhanced Greenhouse Effect: Implications for Extreme Rainfall Events," *Climate Dynamics* 8 (1992): 83–102; Thomas R. Karl et al., "Trends in High-Frequency Climate Variability in the Twentieth Century," *Nature* 377 (1995): 217–220; A. M. Fowler and K. J. Hennessy, "Potential Impacts of Global Warming on the Frequency and Magnitude of Heavy Precipitation," *Natural Hazards* 11 (1995): 283–303; Karl et al., "The Coming Climate," p. 82; K. J. Hennessy et al., "Changes in Daily Precipitation Under Enhanced Greenhouse Conditions," *Climate Dynamics* 13 (1997): 667–680; Kevin E. Trenberth, "Conceptual Framework for Changes of Extremes of the Hydrological Cycle with Climate Change," *Climatic Change* 42 (1999): 327–339.

81. IPCC 1995, WGIII, p. 193; Roger R. Revelle and Paul E. Waggoner, "Effects of Climatic Change on Water Supplies in the Western United States," in *The Challenge of Global Warming*, ed. Dean Edwin Abrahamson (Washington, D.C.: Island, 1989), pp. 151–160; David Pimentel et al., "Water Resources: Agriculture, the Environment, and Society," *BioScience* 47 (1997): 99; Western Water Policy Review Advisory Commission, *Water in the West: Challenge for the Next Century*, June 1998, pp. 2-1 to 2-3.

82. See, for example, Kerry A. Emanuel, "The Dependence of Hurricane Intensity on Climate," *Nature* 326 (1987): 483–485; S. B. Idso et al., "Carbon Dioxide and Hurricanes: Implications of Northern Hemispheric Warming for Atlantic/Caribbean Storms," *Meteorology and Atmospheric Physics* 42 (1990): 259–263; A. J. Broccoli and S. Manabe, "Can Existing Climate Models Be Used to Study Anthropogenic Changes in Tropical Cyclone Climate?" *Geophysical Research Letters* 17 (1990): 1917–1920; Brian F. Ryan et al., "Tropical Cyclone Frequencies Inferred from Gray's Yearly Genesis Parameter: Validation of GCM Tropical Climates," *Geophysical Research Letters* 19 (1992): 1831–1834; R. J. Haarsma et al., "Tropical Disturbances in a GCM," *Climate Dynamics* 8 (1993): 247–257; James Lighthill et al., "Global Climate Change and Tropical Cyclones," *Bulletin of the American Meteorological Society* 75 (1994): 2147–2157; Kerry A. Emanuel, "Comments on 'Global Climate Change and Tropical Cyclones': Part I," *Bulletin of the American Meteorological Society* 76 (1995): 2241–2243; L. Bengtsson et al., "Will Greenhouse Gas-Induced Warming over the Next 50 Years Lead to Higher Frequency and Greater Intensity of Hurricanes?" *Tellus* 48A (1996): 57–73; Karl et al., "The Coming Climate," pp. 82–83; Mahlman, "Uncertainties in Projections," p. 1417; Bates et al., "Contribution of Hurricanes," p. 61; Michael Tucker, "Climate Change and the Insurance Industry: The Cost of Increased Risk and the Impetus for Action," *Ecological Economics* 22 (1997): 85–96; J. F. Royer et al., "A GCM Study of the Impact of Greenhouse Gas Increase on the Frequency of Occurrence of Tropical Cyclones," *Climatic Change* 38 (1998): 307–343; Thomas R. Knutson et al., "Simulated Increase of Hurricane Intensities in a CO_2 Warmed Climate," *Science* 279 (1998): 1018–1020; A. Henderson-Sellers et al., "Tropical Cyclones and Global Climate Change: A Post-IPCC Assessment," *Bulletin of the American Meteorological Association* 79 (1998): 19–38. For the IPCC's conclusions, see IPCC 1995, WGI, p. 334.

83. Christopher Flavin, "Storm Warnings: Climate Change Hits the Insurance Industry," *World Watch*, November–December 1994, pp. 13–14; Kevin E. Trenberth and Timothy J. Hoar, "The 1990–1995 El Niño–Southern Oscillation Event: Longest on Record," *Geophysical Research Letters* 23 (1996): 57–60; Houghton, *Global Warming*, p. 4; Kevin E. Trenberth and Timothy J. Hoar, "El Niño and Climate Change," *Geophysical Research Letters* 24 (1997): 3057–3060; Gretchen Vogel and Andrew Lawler, "Hot Year, But Cool Response in Congress," *Science* 280 (1998): 1684; Richard A. Kerr, "Big El Niños Ride the Back of Slower Climate Change," *Science* 283 (1999): 1108–1109; A. Timmermann et al., "Increased El Niño Frequency in a Climate Model Forced by Future Greenhouse Warming," *Nature* 398 (1999): 694–696; Franklin W. Nutter, "Global Climate Change: Why U.S. Insurers Care," *Climatic Change* 42 (1999): 45–49.

84. *Rocky Mountain Institute Newsletter*, Summer 1998, p. 4.

85. Tucker, "Climate Change and the Insurance Industry"; Flavin, "Storm Warnings," p. 14; IPCC 1995, WGIII, pp. 194–195.

86. Cynthia Rosenzweig and Martin L. Parry, "Potential Impact of Climate Change on World Food Supply," *Nature* 367 (1994): 133–138; Richard M. Adams et al., "Global Climate Change and U.S. Agriculture," *Nature* 345 (1990): 219–224; Richard M. Adams et al., "A Reassessment of the Economic Effects of Global Climate Change on U.S. Agriculture," *Climatic Change* 30 (1995): 147–167; Kaiser, "Helping Those Most at Risk," p. 217; Houghton, *Global Warming*, pp. 123–124.

87. Erickson, "From Ecology to Economics: The Case Against CO_2 Fertilization"; Bazzaz and Fajer, "Plant Life in a CO_2-Rich World."

88. Adams et al., "Global Climate Change and U.S. Agriculture," p. 222; Adams et al., "Reassessment," pp. 156, 159; Rosenzweig and Parry, "Potential Impact of Climate Change on World Food Supply," p. 136.

89. R. T. Wetherald and S. Manabe, "The Mechanisms of Summer Dryness Induced by Greenhouse Warming," *Journal of Climate* 8 (1995): 3096–3108; S. Manabe and R. T. Wetherald, "Reduction in Summer Soil Wetness Induced by an Increase in Atmospheric Carbon Dioxide," *Science* 232 (1986): 626–628.

90. Robert C. Cowen, "Evidence Mounts: Warming Trend Changes Climate," *Christian Science Monitor*, June 24, 1999, p. 12; Johannes Oerlemans, "Quantifying Global Warming from the Retreat of Glaciers," *Science* 264 (1994): 243–245; Houghton, *Global Warming*, p. 118.

91. Houghton, *Global Warming*, pp. 124–125; Richard M. Adams et al., *Agriculture and Global Climate Change* (Arlington, Va.: Pew Center on Global Climate Change, 1999), pp. 4, 13, 20.

92. Rosenzweig and Parry, "Potential Impact of Climate Change on World Food Supply," p. 138; see also IPCC 1995, WGIII, p. 190.

93. Robert L. Peters, "Conservation of Biological Diversity in the Face of Climate Change," in *Global Warming and Biological Diversity*, p. 15; IPCC 1995, WGIII, p. 200.

94. Robert L. Peters, "Effects of Global Warming on Biological Diversity," in *The Challenge of Global Warming*, p. 87.

95. Margaret B. Davis and Catherine Zabinski, "Changes in Geographical Range Resulting from Greenhouse Warming: Effects on Biodiversity in Forests," in *Global Warming and Biological Diversity*, p. 306.

96. Camille Parmesan, "Climate and Species' Range," *Nature* 382 (1996): 765–766; Chris D. Thomas and Jack L. Lennon, "Birds Extend Their Ranges Northwards," *Nature* 399 (1999): 213; Camille Parmesan et al., "Poleward Shifts in Geographical Ranges of Butterfly Species Associated with Regional Warming," *Nature* 399 (1999): 579–583; J. P. Barry et al., "Climate-Related, Long-Term Faunal Changes in a California Rocky Intertidal Community," *Science* 267 (1995): 672–675.

97. Robert L. Peters and Joan D. S. Darling, "The Greenhouse Effect and Nature Reserves," *BioScience* 35 (1985): 707–717.

98. Allen M. Solomon and Wolfgang Cramer, "Biospheric Implications of Global Environmental Change," in *Vegetation Dynamics and Global Change*, p. 42.

99. Davis and Zabinski, "Changes in Geographical Range," p. 304; Woodwell, "Biotic Feedbacks from the Warming of the Earth," p. 12.

100. Leslie Roberts, "How Fast Can Trees Migrate?" *Science* 243 (1989): 737; Solomon and Cramer, "Biospheric Implications," pp. 40–41, 44; Norman Myers, "The World's Forests and Their Ecosystem Services," in *Nature's Services*, ed. Gretchen C. Daily (Washington, D.C.: Island, 1997), p. 223.

101. Houghton, *Global Warming*, p. 129; Robert L. Peters, "Effects of Global Warming on Forests," *Forest Ecology and Management* 35 (1990): 16; Myers, "The World's Forests and Their Ecosystem Services," p. 223; M. D. Flannigan and L. E. Van Wagner, "Climate Change and Wildlife in Canada," *Canadian Journal of Forest Research* 21 (1991): 66–72; IPCC 1995, WGIII, p. 192.

102. Jerry F. Franklin et al., "Effects of Global Climatic Change on Forests in Northwestern North America," in *Global Warming and Biological Diversity*, pp. 244, 246–247, 250.

103. IPCC 1995, WGIII, p. 192.

104. Woodwell et al., "Biotic Feedbacks in the Warming of the Earth," pp. 510–511.

105. Woodwell, "Biotic Feedbacks from the Warming of the Earth," p. 15.

106. Roberts, "How Fast Can Trees Migrate?" pp. 736–737; see also Myers, "The World's Forests and Their Ecosystem Services," pp. 222–223; Ronald P. Neilson, "Vegetation Redistribution: A Possible Biosphere Source of CO_2 During Climatic Change," *Water, Air, and Soil Pollution* 70 (1993): 659–673.

107. Chris Bright, "Tracking the Ecology of Climate Change," in *State of the World 1997,* ed. Lester R. Brown et al. (New York: Norton, 1997), p. 92; Elizabeth Pennisi, "New Threat Seen from Carbon Dioxide," *Science* 279 (1998): 989; R. Monastersky, "Carbon Dioxide Buildup Harms Coral Reefs," *Science News* 155 (1999): 214; Peter Pockley, "Global Warming 'Could Kill Most Coral Reefs by 2100,'" *Nature* 400 (1999): 98; Thomas J. Goreau and M. Pecheux, "Weighing Up the Threat to the World's Corals," letters, *Nature* 402 (1999): 457; Dennis Normile, "Warmer Waters More Deadly to Coral Reefs Than Pollution," *Science* 290 (2000): 682–683; IPCC 1995, WGIII, p. 200.

108. Houghton, *Global Warming,* p. 134. See Norman Myers, "Environmental Refugees in a Globally Warmed World," *BioScience* 43(1993): 752–761.

109. For the worldwide circulation system of surface and deep water, see the map in Houghton, *Global Warming,* p. 81.

110. Jodie E. Smith et al., "Rapid Climate Change in the North Atlantic During the Younger Dryas Recorded by Deep-Sea Corals," *Nature* 386 (1997): 818–820; Scott J. Lehman and Lloyd D. Keigwin, "Sudden Changes in North Atlantic Circulation During the Last Deglaciation," *Nature* 356 (1992): 757–762; R. B. Alley et al., "Abrupt Increase in Greenland Snow Accumulation at the End of the Younger Dryas Event," *Nature* 362 (1993): 527–529; W. Dansgaard et al., "The Abrupt Termination of the Younger Dryas Climate Event," *Nature* 339 (1989): 532–534.

111. Zicheng Yu and Ulrich Eiches, "Abrupt Climate Oscillations During the Last Deglaciation in Central North America," *Science* 282 (1998): 2235–2238; F. Alayne Street-Perrott and R. Alan Perrott, "Abrupt Climate Fluctuations in the Tropics: The Influence of Atlantic Ocean Circulation," *Nature* 343 (1990): 607–612; L. G. Thompson et al., "Late Glacial Stage and Holocene Tropical Ice Core Records from Huascarán, Peru," *Science* 269 (1995): 46–50; Konrad A. Hughen et al., "Rapid Climate Changes in the Tropical Atlantic Region During the Last Deglaciation," *Nature* 380 (1996): 51–54. Some studies conclude that the thermohaline circulation did not shut down completely; rather, the North Atlantic water sank only to midlevels and thus North Atlantic Deep Water was no longer formed and the sites of convection shifted; see Stefan Rahmstorf, "Rapid Climate Transitions in a Coupled Ocean-Atmosphere Model," *Nature* 372 (1994): 82–85; Wallace S. Broecker, "Massive Iceberg Discharges as Triggers for Global Climate Change," *Nature* 372 (1994): 421–424. Whichever is the case, the impact on climate was significant.

112. Syukuro Manabe and Ronald J. Stouffer, "Century-Scale Effects of Increased Atmospheric CO_2 on the Ocean-Atmosphere System," *Nature* 364 (1993): 215–218. Another study indicates that the rate at which greenhouse gases increase is as important as their total concentration in the atmosphere. Continuing today's growth rate of CO_2 in the atmosphere for a century results in a complete shutdown of the thermohaline circulation. If that same final concentration is reached more slowly, the circulation system slows down but does not stop; Thomas F. Stocker and Andreas Schmittner, "Influence of CO_2 Emission Rates on the Stability of the Thermohaline Circulation," *Nature* 388 (1997): 862–865. A recent advance in modeling the circulation system indicates that it could be significantly altered in the next few decades, under business as usual; Stefan Rahmstorf, "Shifting Seas in the Greenhouse?" *Nature* 399 (1999): 523–524; Richard A. Wood et al., "Changing Spatial Structure of the Thermohaline Circulation in Response to Atmospheric CO_2 Forcing in a Climate Model," *Nature* 399 (1999): 572–575.

113. Peter N. Spotts, "Ice Station Pieces Together Arctic's Global Sway," *Christian Science Monitor,* May 20, 1999, p. 17; Richard Monastersky, "Sea Changes in the Arctic," *Science News* 155 (1999): 104–106; Miles G. McPhee et al., "Freshening of the Upper Ocean in the Arctic: Is Perennial Sea Ice Disappearing?" *Geophysical Research Letters* 25 (1998): 1729–1732.

114. Jorge L. Sarmiento and Corinne Le Quéré, "Oceanic Carbon Dioxide Uptake in a Model of Century-Scale Global Warming," *Science* 274 (1996): 1346–1350.

115. Stefan Rahmstorf, "Bifurcations of the Atlantic Thermohaline Circulation in Response to Changes in the Hydrological Cycle," *Nature* 378 (1995): 145–149; see also Richard A. Kerr, "Warming's Unpleasant Surprise: Shivering in the Greenhouse?" *Science* 281 (1998): 156–158.

116. Stefan Rahmstorf, "Risk of Sea-Change in the Atlantic," *Nature* 388 (1997): 826. See also Wallace S. Broecker, "Thermohaline Circulation, the Achilles Heel of Our Climate System: Will Man-Made CO_2 Upset the Current Balance?" *Science* 278 (1997): 1582–1588.

117. Franklin et al., "Effects of Global Climatic Change on Forests in Northwest America," p. 250; see also Stephen H. Schneider and Terry L. Root, "Ecological Implications of Climate Change Will Include Surprises," *Biodiversity and Conservation* 5 (1996): 1109–1119.

118. Christian Azar and Henning Rodhe, "Targets for Stabilization of Atmospheric CO_2," *Science* 276 (1997): 1818.

119. William D. Nordhaus, "The Ghosts of Climates Past and the Specters of Climate Change Future," *Energy Policy* 23 (1995): 276.

120. IPCC 1995, WGIII, p. 183.

121. Richard S. J. Tol, "The Damage Costs of Climate Change Toward More Comprehensive Calculations," *Environmental and Resource Economics* 5 (1995): 360.

122. William D. Nordhaus, "A Sketch of the Economics of the Greenhouse Effect," *American Economic Review*, AEA Papers and Proceedings 81 (1991): 146–150; William D. Nordhaus, "To Slow or Not to Slow: The Economics of the Greenhouse Effect," *Economic Journal* 101 (1991): 920–937; William D. Nordhaus, "An Optimal Transition Path for Controlling Greenhouse Gases," *Science* 258 (1992): 1315–1319; William D. Nordhaus, "Optimal Greenhouse-Gas Reductions and Tax Policy in the 'DICE' Model," *American Economic Review*, AEA Papers and Proceedings 83 (1993): 313–317; William D. Nordhaus, "Reflections on the Economics of Climate Change," *Journal of Economic Perspectives* 7 (1993): 11–25; William D. Nordhaus, "Rolling the 'DICE': An Optimal Transition Path for Controlling Greenhouse Gases," *Resource and Energy Economics* 15 (1993): 27–50; Nordhaus, "Ghosts of Climates Past."

123. Nordhaus, "Reflections on the Economics of Climate Change," p. 22; Nordhaus, "Ghosts of Climates Past," p. 280.

124. Nordhaus, "Optimal Transition Path for Controlling Greenhouse Gases," p. 1318.

125. William R. Cline, *The Economics of Global Warming* (Washington, D.C.: Institute for International Economics, 1992); see also William R. Cline, "Give Greenhouse Abatement a Fair Chance," *Finance and Development*, March 1993, pp. 3–5; William R. Cline, "Costs and Benefits of Greenhouse Abatement: A Guide to Policy Analysis," in *The Economics of Climate Change* (Paris: OECD, 1994), pp. 87–105.

126. Samuel Fankhauser, "The Social Costs of Greenhouse Gas Emissions: An Expected Value Approach," *Energy Journal* 15 (1994): 157–184; Samuel Fankhauser, "The Economic Costs of Global Warming Damage: A Survey," *Global Environmental Change* 4 (1994): 301–309; Samuel Fankhauser and David W. Pearce, "The Social Costs of Greenhouse Gas Emissions," in *The Economics of Climate Change*, pp. 71–86; Samuel Fankhauser and Richard S. J. Tol, "Climate Change Costs," *Energy Policy* 24 (1996): 665–673; Samuel Fankhauser et al., "The Aggregation of Climate Change Damages: A Welfare Theoretic Approach," *Environmental and Resource Economics* 10 (1997): 249–266.

127. Richard S. J. Tol, "The Damage Costs of Climate Change: A Note on Tangibles and Intangibles, Applied to DICE," *Energy Policy* 22 (1994): 436–438; Tol, "The Damage Costs of Climate Change Toward More Comprehensive Calculations"; Richard S. J. Tol, "The Damage Costs of Climate Change Towards a Dynamic Representation," *Ecological Economics* 19 (1996): 67–90; Richard S. J. Tol, "On the Optimal Control of Carbon Dioxide Emissions: An Application of FUND," *Environmental Modeling and Assessment* 2 (1997): 151–163; Richard S. J. Tol, "On the Difference in Impact of Two Almost Identical Climate Change Scenarios," *Energy Policy* 26 (1998): 13–20; Richard S. J. Tol, "The Marginal Costs of Greenhouse Gas Emissions," *Energy Journal* 20 (1999): 61–81; Richard S. J. Tol, "Time Discounting and Optimal Emission Reduction: An Application of *FUND*," *Climatic Change* 41 (1999): 351–362.

128. William D. Nordhaus, "Expert Opinion on Climatic Change," *American Scientist* 82 (January–February 1994): 45–51.

129. A couple of quotations that Nordhaus gives from his survey responses suggest pretty strongly that this is, indeed, the attitude of some economists.

130. "IPCC Second Assessment Synthesis of Scientific-Technical Information Relevant to Interpreting Article 2 of the UN Framework Convention on Climate Change," in *IPCC Second*

Assessment: Climate Change 1995 (New York: World Meteorological Organization/United Nations Environment Program, 1996), p. 6. See also Jerry M. Melillo, "Warm, Warm on the Range," *Science* 283 (1999): 183–184.

131. IPCC 1995, WGIII, pp. 188–189.

132. Fankhauser, "Economic Costs of Global Warming Damage," p. 307. Cline briefly discusses the risks of catastrophes, which add support to the conclusion he had already reached, that a significant reduction in greenhouse gases is warranted on economic grounds; *Economics of Global Warming*, pp. 303–305.

133. Fankhauser and Tol, "Climate Change Costs," p. 669. Tol's model now includes, for example, damages from heat waves, malaria, and river floods.

134. Fankhauser et al., "Aggregation of Climate Change Damages," p. 255.

135. Clive L. Spash, "Double CO_2 and Beyond: Benefits, Costs and Compensation," *Ecological Economics* 10 (1994): 30.

136. Nordhaus, "Ghosts of Climates Past," pp. 280–281.

137. Tol, "Damage Costs of Climate Change Toward More Comprehensive Calculations," p. 366.

138. Fankhauser and Tol, "Climate Change Costs," p. 668.

139. IPCC 1995, WGIII, pp. 186–187.

140. For example, Nordhaus, "Ghosts of Climates Past," p. 281.

141. Tol, "Marginal Costs of Greenhouse Gas Emissions," p. 61.

142. Christian Azar, "Are Optimal CO_2 Emissions Really Optimal?" *Environmental and Resource Economics* 11 (1998): 305.

143. Paul Ekins, "Rethinking the Costs Related to Global Warming: A Survey of the Issues," *Environmental and Resource Economics* 6 (1995): 235.

144. Quoted in Ekins, "Rethinking the Costs," p. 235, from an unpublished working paper.

145. IPCC 1995, WGIII, p. 205.

146. Christian Azar and Thomas Sterner, "Discounting and Distributional Considerations in the Context of Global Warming," *Ecological Economics* 19 (1996): 181.

147. Tol, "Marginal Costs of Greenhouse Gas Emissions," p. 61.

148. Daniel A. Farber and Paul A. Hemmersbaugh, "The Shadow of the Future: Discount Rates, Later Generations, and the Environment," *Vanderbilt Law Review* 46 (1993): 277; emphasis in the original.

149. Clive L. Spash, "Economics, Ethics, and Long-Term Environmental Damages," *Environmental Ethics* 15 (1993): 120.

150. Ralph C. d'Arge et al., "Carbon Dioxide and Intergenerational Choice," *American Economic Review* 72 (1982): 251.

151. Quoted in Geoffrey Heal, "Discounting and Climate Change," *Climatic Change* 37 (1997): 335.

152. Heal, "Discounting and Climate Change," pp. 335–336.

153. M. Granger Morgan et al., "Why Conventional Tools for Policy Analysis Are Often Inadequate for Problems of Global Change," *Climatic Change* 41 (1999): 271. Ari Rabl has an interesting way of demonstrating that normal discounted cost-benefit analysis reduces to absurdity when applied to the very long time scales we are dealing with here. As he points out, if one uses a typical discount rate and applies it to a typical project, over a very long time scale the supposed benefits of the project become larger than the entire gross domestic product; "Discounting of Long-Term Costs: What Would Future Generations Prefer Us to Do?" *Ecological Economics* 17 (1996): 137–145.

154. Rabl, "Discounting of Long-Term Costs," p. 138.

155. Nordhaus, "To Slow or Not to Slow?" p. 936.

156. Peter A. Schultz and James F. Kasting, "Optimal Reductions in CO_2 Emissions," *Energy Policy* 25 (1997): 491. For problems with Nordhaus's assumptions about future accumulation of CO_2 in the atmosphere, see Robert K. Kaufman, "Assessing the DICE Model: Uncertainty Associated with the Emission and Retention of Greenhouse Gases," *Climatic Change* 35 (1997): 435–448.

157. This study by Cline is summarized in Frankhauser and Tol, "Climate Change Costs," p. 670. We saw earlier that on the basis of his own original estimates Cline takes global warm-

ing much more seriously than do other economists. The reasons are that he uses a low discount rate (1.5 percent) and he factors in risk aversion because of the possibility of very high damage scenarios. He also adopts a very long time horizon. He realizes that business as usual does not stop at a doubling of greenhouse gases, but it keeps on going up. The result is that farther in the future the global average temperatures become much larger: between 10°C and 18°C by the year 2300 and the damages escalate rapidly; *Economics of Global Warming*; "Give Greenhouse Abatement a Fair Chance"; "Costs and Benefits of Greenhouse Abatement."

158. Richard B. Howarth, "Climate Change and Overlapping Generations," *Contemporary Economic Policy* 14 (October 1996): 108.

159. Fankhauser, "Social Costs of Greenhouse Gas Emissions," p. 178.

160. Tol, "Marginal Costs of Greenhouse Gas Emissions," p. 69.

161. Azar and Sterner, "Discounting and Distributional Considerations," p. 169, 182.

162. Derek Parfit, "Energy Policy and the Further Future: The Social Discount Rate," in *Energy and the Future*, ed. Douglas MacLean and Peter G. Brown (Totowa, N.J.: Rowman & Littlefield, 1983), p. 31; emphasis in the original.

163. Herman E. Daly and John B. Cobb Jr., *For the Common Good* (Boston: Beacon, 1989); Clifford W. Cobb and John B. Cobb Jr., *The Green National Product: A Proposed Index of Sustainable Economic Welfare* (Lanham, Md.: University Press of America, 1994); Clifford Cobb et al., *The Genuine Progress Indicator* (San Francisco: Redefining Progress, 1995); Clifford Cobb et al., *Why Bigger Isn't Better: The Genuine Progress Indicator—1999 Update* (San Francisco: Redefining Progress, 1999).

164. Tol, "Damage Costs of Climate Change: A Note on Tangibles and Intangibles," p. 437.

165. See Eric Neumayer, "Global Warming: Discounting Is Not the Issue, but Substitutability Is," *Energy Policy* 27 (1999): 33–43.

166. Cline, "Give Greenhouse Abatement a Fair Chance," p. 5.

167. Robert C. Lind, "Intergenerational Equity, Discounting, and the Role of Cost-Benefit Analysis in Evaluating Global Climate Policy," *Energy Policy* 23 (1995): 384.

168. Lind, "Intergenerational Equity," p. 382.

169. William D. Nordhaus, "Discounting in Economics and Climate Change," *Climatic Change* 37 (1997): 320, 327; emphasis in the original.

170. Schultz and Kasting, "Optimal Reductions in CO_2 Emissions," p. 498; Azar, "Are Optimal CO_2 Emissions Really Optimal?" pp. 309–310; Milind Kandlikar, "Indices for Comparing Greenhouse Gas Emissions: Integrating Science and Economics," *Energy Economics* 18 (1996): 271–272.

CHAPTER SEVEN

1. Edward S. Rubin et al., "Realistic Mitigation Options for Global Warming," *Science* 257 (1992): 148.

2. Organization for Economic Cooperation and Development (OECD), *Environmental Performance Reviews: United States* (Paris: OECD, 1996), p. 217.

3. Quoted in John Houghton, *Global Warming: The Complete Briefing*, 2d ed. (Cambridge: Cambridge University Press, 1997), p. 175.

4. Houghton, *Global Warming*, p. 202; Richard Monastersky, "Good-bye to a Greenhouse Gas," *Science News*, June 19, 1999, pp. 392–394.

5. Joyce Kaiser, "A Way to Make CO_2 Go Away: Deep-Six It," *Science* 281 (1998): 505; E. A. Parson and D. W. Keith, "Fossil Fuels Without CO_2 Emissions," *Science* 282 (1998): 1053–1054; Monastersky, "Good-bye to a Greenhouse Gas;" Peter G. Brewer et al., "Direct Experiments on the Ocean Disposal of Fossil Fuel CO_2," *Science* 284 (1999): 943–945; Rex Dalton, "U.S. Warms to Carbon Sequestration Research," *Nature* 401 (1999): 315.

6. F. Joos et al., "Estimates of the Effect of Southern Ocean Iron Fertilization on Atmospheric CO_2 Concentrations," *Nature* 349 (1991): 772–775; T. H. Peng and W. S. Broecker, "Dynamical Limitations on the Antarctic Iron Fertilization Strategy," *Nature* 349 (1991): 227–229; U. Siegenthaler and J. L. Sarmiento, "Atmospheric Carbon Dioxide and the Ocean," *Nature* 365 (1993):

124; N. Kumar et al., "Increased Biological Productivity and Export Production in the Glacial Southern Ocean," *Nature* 378 (1995): 675–680; Kenneth H. Coale et al., "A Massive Phytoplankton Bloom Induced by an Ecosystem-Scale Iron Fertilization Experiment in the Equatorial Pacific Ocean," *Nature* 383 (1996): 495–501; D. J. Cooper et al., "Large Decrease in Ocean-Surface CO_2 Fugacity in Response to *in Situ* Iron Fertilization," *Nature* 383 (1996): 511–513; Houghton, *Global Warming*, p. 28. Sallie W. Chisholm, "Stirring Times in the Southern Ocean," *Nature* 407 (2000): 685–687; Philip W. Boyd et al., "A Mesoscale Phytoplankton Bloom in the Polar Southern Ocean Stimulated by Iron Fertilization," *Nature* 407 (2000): 695–702.

7. Roger A. Sedjo, "Forests to Offset the Greenhouse Effect," *Journal of Forestry*, July 1989, p. 13.

8. R. K. Dixon et al., "Carbon Pools and Flux of Global Forest Ecosystems," *Science* 263 (1994): 189.

9. Roger A. Sedjo et al., "The Economics of Managing Carbon Via Forestry: Assessment of Existing Studies," *Environmental and Resource Economics* 6 (1995): 146.

10. Daniel J. Dudek and Alice LeBlanc, "Offsetting New CO_2 Emissions: A Rational First Greenhouse Policy Step," *Contemporary Policy Issues* 8 (July 1990): 32; Peter J. Parks and Ian W. Hardie, "Least-Cost Forest Carbon Reserves: Cost-Effective Subsidies to Convert Marginal Agricultural Land to Forests," *Land Economics* 71 (1995): 122–136.

11. Sedjo et al., "Economics of Managing Carbon Via Forestry," p. 156.

12. Sedjo et al., "Economics of Managing Carbon Via Forestry," p. 159.

13. Matt Schwartz, "Used Pallets No Longer on the Skids," *Christian Science Monitor*, August 13, 1996, p. 10.

14. Mark E. Harmon et al., "Effects on Carbon Storage of Conversion of Old-Growth Forests to Young Forests," *Science* 247 (1990): 699–702; see also L. L. Wright and E. E. Hughes, "U.S. Carbon Offset Potential Using Biomass Energy Systems," *Water, Air, and Soil Pollution* 70 (1993): 491.

15. Another possible sink for carbon, although not nearly as large as forests, is agricultural land. Conservation tillage lets carbon build up in the soil, while still producing crops; R. Lal et al., "Managing U.S. Cropland to Sequester Carbon in Soil," *Journal of Soil and Water Conservation* 54 (1999): 374–381; James P. Bruce et al., "Carbon Sequestration in Soils," *Journal of Soil and Water Conservation* 54 (1999): 382–389. Already, a consortium of Canadian utilities is paying Iowa farmers to sequester carbon in order to offset emissions from their power plants; Shereen El Feki, "Agriculture and Technology (Survey)," *Economist*, March 25, 2000, p. 12.

16. Dudek and LeBlanc, "Offsetting New CO_2 Emissions," p. 38.

17. Houghton, *Global Warming*, p. 201.

18. "Power to the People," *Economist*, February 20, 1999, p. 60.

19. Evan Mills et al., "Getting Started: No-Regrets Strategies for Reducing Greenhouse Gas Emissions," *Energy Policy* 19 (1991): 528; Wolfgang Rüdig, "Energy Conservation and Electricity Utilities: A Comparative Analysis of Organizational Obstacles to CHP/DH," *Energy Policy* 14 (1986): 105; Joseph J. Romm, *Cool Companies* (Washington, D.C.: Island, 1999), pp. 119–127.

20. Amory B. Lovins and L. Hunter Lovins, *Climate: Making Sense and Making Money* (Old Snowmass, Colo.: Rocky Mountain Institute, 1997), p. 8.

21. Tim Jackson, "Renewable Energy," *Energy Policy* 20 (1992): 868.

22. Amory B. Lovins, "Four Revolutions in Electric Efficiency," *Contemporary Policy Issues* 8 (July 1990): 132.

23. Jackson, "Renewable Energy," p. 868.

24. Jackson, "Renewable Energy," p. 870.

25. Ross Gelbspan, "A Good Climate for Investment," *Atlantic Monthly*, June 1998, p. 27; Gelbspan, "How Green Is Browne?" *Economist*, April 17, 1999, p. 74; Gelbspan, "When Virtue Pays a Premium," *Economist*, April 18, 1998, p. 57–58. The U.S. government is getting actively involved once again. One program encourages installing solar PV panels on a million roofs by 2010 and includes solarizing 20,000 federal buildings; Dashka Slater, "Sunny Prospects," *Sierra*, May–June 1998, p. 28.

26. Jackson, "Renewable Energy," p. 870.

27. Penny Street and Ian Miles, "Transition to Alternative Energy Supply Technologies: The Case of Windpower," *Energy Policy* 24 (1996): 415; Francis McGowan, "Controlling the Greenhouse Effect," *Energy Policy* 19 (1991): 116.

28. Colin Woodard, "Wind Power Takes Off as Major Energy Source," *Christian Science Monitor,* June 30, 1999, p. 6.

29. Street and Miles, "Transition to Alternative Energy Supply Technologies," p. 415.

30. Woodard, "Wind Power Takes Off," p. 1.

31. Tom Gray, "More Than Hot Air," *Christian Science Monitor,* August 30, 1999, p. 8.

32. D. O. Hall et al., "Cooling the Greenhouse with Bioenergy," *Nature* 353 (1991): 11–12; D. O. Hall, "Biomass Energy," *Energy Policy* 24 (1991): 711–737; Gregg Marland and Scott Marland, "Should We Store Carbon in Trees?" *Water, Air, and Soil Pollution* 64 (1992): 181–195; R. Neil Sampson et al., "Biomass Management and Energy," *Water, Air, and Soil Pollution* 70 (1993): 139–159.

33. Wright and Hughes, "U.S. Carbon Offset Potential," p. 486.

34. D. O. Hall, "Biomass Energy," p. 711.

35. Wright and Hughes, "U.S. Carbon Offset Potential," p. 488.

36. Lee R. Lynd, "Overview and Evaluation of Fuel Ethanol from Cellulosic Biomass," *Annual Review of Energy and the Environment* 21 (1996): 403–465.

37. "Fuel Cells Meet Big Business," *Economist,* July 24, 1999, pp. 59–60.

38. "Intel on Wheels," *Economist,* October 31, 1998, p. 69; "Fuel Cells Hit the Road," *Economist,* Apil 24, 1999, p. 77.

39. Kevin A. Wilson, "Sound System," *AutoWeek,* April 12, 1999, pp. 30–31.

40. Kevin A. Wilson, "Clean Politics," *AutoWeek,* March 29, 1999, p. 4.

41. Wilson, "Sound System," p. 31.

42. "Selling Fuel Cells," *Economist,* July 1, 2000, p. 83. The source of hydrogen makes a big difference in the amount of greenhouse gases emitted. Taking into account the entire fuel cycle, from well to wheel, using gasoline as the source of hydrogen saves little in emissions, methanol saves a significant amount, and natural gas would save a lot; "How Green Is Your Hydrogen?" *Economist,* April 1, 2000, p. 74. Presumably best of all would be to use a renewable source of electricity, such as wind power, to get hydrogen from water by electrolysis. Fuel cell developments are coming rapidly and they include new ways to store hydrogen on a vehicle, rather than having a "reformer" to extract it on each car or truck. Large reformers could be located at filling stations and the hydrogen could be stored on board the vehicle in hydrides; "How Green Is Your Hydrogen?" p. 74. Another way of storing hydrogen on a vehicle would use carbon "nanotubes" and "nanofibers" that act like sponges for hydrogen; "Space Age Soot," *Economist,* December 11, 1999, pp. 73–74.

43. T. B. Johansson et al., "Options for Reducing CO_2 Emissions from the Energy Supply Sector," *Energy Policy* 24 (1996): 989. Shell Corporation also makes this claim; see, e.g., their ad in the *Economist,* July 10, 1999, p. 35.

44. Jackson, "Renewable Energy," pp. 873–874; Johansson et al., "Options for Reducing CO_2 Emissions," p. 999; Amory B. Lovins and L. Hunter Lovins, "Least-Cost Climatic Stabilization," *Annual Review of Energy and the Environment* 16 (1991): 461; Alliance to Save Energy et al., *Energy Innovations: A Prosperous Path to a Clean Environment* (Washington, D.C.: Alliance to Save Energy/American Council for an Energy-Efficient Economy/Natural Resources Defense Council/Tellus Institute/Union of Concerned Scientists, 1997), p. 37.

45. "Mining Landfills," *PERC Reports,* March 1998, p. 10. *PERC Reports* is a newsletter published by the Political Economy Research Center, Bozeman, Montana.

46. Kathleen B. Hogan et al., "Methane on the Greenhouse Agenda," *Nature* 354 (1991): 181; see also Rubin et al., "Realistic Mitigation Options for Global Warming, p. 263.

47. Hogan et al., "Methane on the Greenhouse Agenda," p. 181.

48. Hogan et al., "Methane on the Greenhouse Agenda," p. 182; R. L. Sass et al., "Methane Emission from Rice Fields: The Effect of Floodwater Management," *Global Biogeochemical Cycles* 6 (1992): 249–262.

49. Lovins and Lovins, *Climate,* pp. i, 2; emphasis in the original.

50. Lovins and Lovins, *Climate*, p. i.

51. Amory B. Lovins, letter to the editor, *Science* 251 (1991): 1297.

52. Amory B. Lovins, "The Great Negawatts Debate," *Electricity Journal*, May 1994, p. 47.

53. "Climate Protection for Fun and Profit," *Rocky Mountain Institute Newsletter* 13 (Fall–Winter 1997): 3.

54. Lovins and Lovins, *Climate*, p. 6.

55. Lovins, "Great Negawatts Debate," pp. 35, 38.

56. Lovins, letter to the editor, p. 1297.

57. L. Hunter Lovins et al., "Energy Policy," in *Changing America: Blueprints for the New Administration*, ed. Mark Green (New York: Newmarket, 1992), pp. 676–677.

58. Ian Brown, editor's introduction, *Energy Policy* 19 (1991): 195.

59. Clark W. Gellings et al., "Potential Energy Savings from Efficient Electric Technologies," *Energy Policy* 19 (1991): 220–225. This article is based on the EPRI study.

60. Summarized in Neha Khanna and Duane Chapman, "Time Preference, Abatement Costs, and International Climate Policy: An Appraisal of IPCC 1995," *Contemporary Economic Policy* 14 (April 1996): 61; see also Florentin Krause, "The Costs of Mitigating Carbon Emissions," *Energy Policy* 24 (1996): 899–915.

61. Arthur Rosenfeld et al., "Conserved Energy Supply Curves for U.S. Buildings," *Contemporary Policy Issues* 11 (January 1993): 45.

62. National Academy of Sciences, *Policy Implications of Greenhouse Warming* (Washington, D.C.: National Academy Press, 1992), p. 489.

63. Mark D. Levine et al., "Energy Efficiency Policy and Market Failures," *Annual Review of Energy and the Environment* 20 (1995): 536.

64. Intergovernmental Panel on Climate Change, *IPCC Second Assessment: Climate Change 1995* (New York: United Nations/World Meteorological Organization, 1995), p. 54. This publication contains a synthesis of scientific information, plus a summary for policymakers from each of the three working groups. The calculation is from WGIII.

65. Thomas C. Schelling, "Some Economics of Global Warming," *American Economic Review* 82 (1992): 11.

66. William D. Nordhaus, "The Cost of Slowing Climate Change: A Survey," *Energy Journal* 12 (1991): 48–49. Yet he coauthored the recent "Economists' Statement on Climate Change" (see below).

67. Robert Repetto, *Jobs, Competitiveness, and Environmental Regulation: What Are the Real Issues?* (Washington, D.C.: World Resources Institute, 1995), p. 12.

68. Lovins, "Great Negawatts Debate," pp. 331–332. For numerous case studies, see Romm, *Cool Companies*.

69. For example, in James R. Udall, "Amory Lovins: Walking the Soft Path," *Sierra*, January–February 1990, p. 128.

70. Alan S. Manne and Richard G. Richels, "CO_2 Emission Limits: An Economic Cost Analysis for the USA," *Energy Journal* 11 (1990): 68.

71. William D. Nordhaus, "Rolling the 'DICE': An Optimal Transition Path for Controlling Greenhouse Gas," *Resource and Energy Economics* 15 (1993): 41. For a summary of other top-down models, see John P. Weyant, "Costs of Reducing Global Carbon Emissions," *Journal of Economic Perspectives* 7 (1993): 36.

72. Stephen J. DeCanio, *The Economics of Climate Change* (San Francisco: Redefining Progress, 1997), p. 16; emphasis in the original. The models assume that the economy is operating on its "production possibilities frontier"; see Michael Grubb et al., "The Costs of Limiting Fossil-Fuel CO_2 Emissions: A Survey and Analysis," *Annual Review of Energy and the Environment* 18 (1993): 433–437; Stephen J. DeCanio, "Economic Modeling and the False Tradeoff Between Environmental Protection and Economic Growth," *Contemporary Economic Policy* 15 (October 1997): 12–17. The bottom-up studies clearly prove that it is not. Some top-down models now factor in a bottom-up component, but it is often very small, still based on the assumption that our economy is efficient now. Top-down models also contain other assumptions that bias the results on the extreme high side. For example, some assume that there is absolutely no alternative

to fossil fuel energy, completely ignoring even renewables that are proven and nearly competitive today. Others assume that the only alternatives are extremely expensive synfuels. For a detailed analysis of the different kinds of models with their assumptions, see Grubb et al., "Costs of Limiting Fossil-Fuel CO_2 Emissions"; Deborah Wilson and Joel Swisher, "Exploring the Gap: Top-Down Versus Bottom-Up Analyses of the Cost of Mitigating Global Warming," *Energy Policy* 21 (1993): 249–263; James P. Bruce et al., eds., *Climate Change 1995: Economic and Social Dimensions of Climate Change* (Cambridge: Cambridge University Press, 1996), pp. 282–289, hereafter cited as IPCC 1995, WGIII; Robert Repetto and Duncan Austin, *The Costs of Climate Protection: A Guide for the Perplexed* (Washington, D.C.: World Resources Institute, 1997).

73. Lovins and Lovins, *Climate*, p. 1.

74. DeCanio, "Economic Modeling and the False Tradeoff," p. 23.

75. See, for example, Eberhard Jochem and Edelgard Gruber, "Obstacles to Rational Electricity Use and Measures to Alleviate Them," *Energy Policy* 18 (1990): 340–350; Adam B. Jaffe and Robert N. Stavins, "The Energy-Efficiency Gap: What Does It Mean?" *Energy Policy* 22 (1994): 804–810; Richard B. Howarth and Alan H. Sanstad, "Discount Rates and Energy Efficiency," *Contemporary Economic Policy* 13 (July 1995): 101–109; William H. Golove and Joseph H. Eto, *Market Barriers to Energy Efficiency: A Critical Reappraisal of the Rationale for Public Policies to Promote Energy Efficiency* (Berkeley: Lawrence Berkeley National Laboratory, 1996); Lovins and Lovins, *Climate*, pp. 11–20.

76. Willett Kempton and Laura Montgomery, "Folk Quantification of Energy," *Energy* 7 (1982): 817–827.

77. Ted Flanigan and June Weintraub, "The Most Successful DSM Programs in North America," *Electricity Journal*, May 1993, p. 56.

78. Harry Chernoff, "Individual Purchase Criteria for Energy-Related Durables: The Misuse of Life Cycle Cost," *Energy Journal* 4 (1983): 82; Kenneth Train, "Discount Rates in Consumers' Energy-Related Decisions: A Review of the Literature," *Energy* 10 (1985): 1246, 1248; Raymond S. Hartman and Michael J. Doane, "Household Discount Rates Revisited," *Energy Journal* 7 (1986): 139; Henry Ruderman et al., "The Behavior of the Market for Energy Efficiency in Residential Appliances Including Heating and Cooling Equipment," *Energy Journal* 8 (1987): 101–124; Jerry A. Hausman, "Individual Discount Rates and the Purchase and Utilization of Energy-Using Durables," *Bell Journal of Economics* 10 (1979): 33–54; Dermot Gately, "Individual Discount Rates and the Purchase and Utilization of Energy-Using Durables: Comment," *Bell Journal of Economics* 11 (1980): 373–374.

79. "Money to Burn," *Economist*, January 6, 1990, p. 65.

80. Ronald J. Sutherland, "Market Barriers to Energy-Efficiency Investments," *Energy Journal* 12 (1991): 19.

81. Gilbert E. Metcalf, "Economics and Rational Conservation Policy," *Energy Policy* 22 (1994): 819–825.

82. Paul C. Stern, "Blind Spots in Policy Analysis: What Economics Doesn't Say About Energy Use," *Journal of Policy Analysis and Management* 5 (1986): 216.

83. See John Conlisk, "Why Bounded Rationality?" *Journal of Economic Literature* 34 (1996): 669–700, for a survey of the literature and issues.

84. Ruderman et al., "Behavior of the Market for Energy Efficiency," p. 117.

85. Quoted in Stephen J. DeCanio, "Barriers Within Firms to Energy-Efficient Investments," *Energy Policy* 21 (1993): 907.

86. DeCanio, "Barriers Within Firms," p. 912.

87. On organizational barriers, see DeCanio, "Barriers Within Firms," pp. 908–910; Flanigan and Weintraub, "The Most Successful DSM Programs," p. 56; Peter B. Cebon, "'Twixt Cup and Lip: Organizational Behaviour, Technical Prediction and Conservation Practice," *Energy Policy* 20 (1992): 802–814.

88. Sutherland, "Market Barriers to Energy-Efficiency Investments," pp. 31–32.

89. Richard S. J. Tol, "The Damage Costs of Climate Change Toward More Comprehensive Calculations," *Environmental and Resource Economics* 5 (1995): 361.

90. C. Joly, "Climate Change, Insurance and Investment Management," in *Climate Change: Mobilising Global Effort* (Paris: OECD, 1997), pp. 51–52.

91. International Energy Agency, *Voluntary Actions for Energy-Related CO_2 Abatement* (Paris: OECD/IEA, 1997), pp. 148–149. Ford and General Motors have finally promised to make some improvements in the efficiency of their trucks and SUVs, but they will still be gas guzzlers; "News in Brief," *Christian Science Monitor,* August 4, 2000, p. 24.

92. Houghton, *Global Warming,* p. 166.

93. Wallace S. Broecker, "Greenhouse Surprises," in *The Challenge of Global Warming,* ed. Dean Edwin Abrahamson (Washington, D.C.: Island, 1989), p. 205.

94. Christian Azar and Henning Rodhe, "Targets for Stabilization of Atmospheric CO_2," *Science* 276 (1997): 1818.

95. See the trajectories in Azar and Rodhe, "Targets for Stabilization," p. 1818, fig. 1.

96. Alan Manne and Richard Richels, "The Greenhouse Debate: Economic Efficiency, Burden Sharing and Hedging Strategies," *Energy Journal* 16 (1995): 3.

97. T. M. L. Wigley et al., "Economic and Environmental Choices in the Stabilization of Atmospheric CO_2 Concentrations," *Nature* 379 (1996): 240–243.

98. Azar and Rodhe, "Targets for Stabilization," p. 1819.

99. Weyant, "Costs of Reducing Global Carbon Emissions," p. 36. Some companies are already negotiating carbon offset trades and the World Bank is organizing a precompliance market; "Seeing Green," *Economist,* October 30, 1999, p. 73; "Cost Free," *Economist,* January 22, 2000, pp. 64–65.

100. OECD, *Reforming Energy and Transport Subsidies: Environmental and Economic Implications* (Paris: OECD, 1997), p. 3.

101. *IPCC Second Assessment,* p. 55.

102. Howard Geller and Scott McGaraghan, "Successful Government-Industry Partnership: The U.S. Department of Energy's Role in Advancing Energy-Efficient Technologies," *Energy Policy* 26 (1998): 167–177. Unfortunately, the energy sector of our economy invests very little in R&D, and government research funds are shrinking; J. J. Dooley, "Unintended Consequences: Energy R&D in a Deregulated Energy Market," *Energy Policy* 26 (1998): 547–555; Robert M. Margolis and Daniel M. Kammen, "Underinvestment: The Energy Technology and R&D Policy Challenge," *Science* 285 (1999): 690–692; Robert M. Margolis and Daniel M. Kammen, "Evidence of Under-Investment in Energy R&D in the United States and the Impact of Federal Policy," *Energy Policy* 27 (1999): 575–584.

103. Gretchen Vogel and Andrew Lawler, "Hot Year, But Cool Response in Congress," *Science* 280 (1998): 1684.

104. Quoted in "Fuel Cells Meet Big Business," p. 59.

105. Thomas B. Johansson et al., eds., *Renewable Energy: Sources for Fuels and Electricity* (Washington, D.C.: Island, 1993), p. 51.

106. Some earlier studies claimed that recycling revenues from a carbon tax would reduce the distortions that income taxes and payroll taxes impose on the economy. They concluded that this would produce a substantial economic benefit—a "double dividend"—that would also greatly increase the "optimal" level of CO_2 abatement. For studies finding a double dividend, see William D. Nordhaus, "Optimal Greenhouse-Gas Reductions and Tax Policy in the 'DICE' Model," *American Economic Review* 83 (1993): 316; Repetto and Austin, *Costs of Climate Protection,* pp. 24–26; and the summary in Grubb et al., "Costs of Limiting Fossil-Fuel CO_2 Emissions," pp. 461–462. Some recent analyses argue that these studies ignored the interaction effects of a carbon tax with other distortionary taxation. When this effect is taken into account, a carbon tax still distorts the economy and impacts its efficiency. These analyses all agree, however, that proper recycling of the revenue from a carbon tax can offset much of its impact. See A. Lans Bovenberg and Ruud A. de Mooij, "Environmental Levies and Distortionary Taxation," *American Economic Review* 84 (1994): 1085–1089; Lawrence H. Goulder, "Effects of Carbon Taxes in an Economy with Prior Tax Distortions: An Intertemporal General Equilibrium Analysis," *Journal of Environmental Economics and Management* 29 (1995): 271–297; Ian W. H. Parry, "Pollution Taxes and Revenue Recycling," *Journal of Environmental Economics and Management* 29 (1995): S64-S77; A. Lans Bovenberg and Lawrence H. Goulder, "Optimal Environmental Taxa-

tion in the Presence of Other Taxes: General-Equilibrium Analyses," *American Economic Review* 86 (1996): 985–1000; Ian W. H. Parry et al., "When Can Carbon Abatement Policies Increase Welfare? The Fundamental Role of Distorted Factor Markets," *Journal of Environmental Economics and Management* 37 (1999): 52–84. For a short summary of the issues, see Wallace E. Oates, "Green Taxes: Can We Protect the Environment and Improve the Tax System at the Same Time?" *Southern Economic Journal* 61 (1995): 915–922. For a detailed analysis of the issues, see Lawrence H. Goulder, "Environmental Taxation and the Double Dividend: A Reader's Guide," *International Tax and Finance* 2 (1995): 157–183. For a defense of the double dividend even with the tax interaction effect, see Repetto and Austin, *Costs of Climate Protection*, p. 26. I doubt that the final word has yet been said.

107. Ralph Cavanagh, "Energy-Efficiency Solutions: What Commodity Prices Can't Deliver," *Annual Review of Energy and the Environment* 20 (1995): 521–522.

108. Barry D. Solomon, "Global CO_2 Emissions Trading: Early Lessons from the U.S. Acid Rain Program," *Climatic Change* 30 (1995): 75–96. Some economic analyses indicate that the quotas should be auctioned and taxes should be reduced on a revenue-neutral basis; see note 106.

109. Stephen Bernow et al., "An Integrated Approach to Climate Policy in the U.S. Electric Power Sector," *Energy Policy* 26 (1998): 375.

110. Michael Shelby et al., "Climate Change Implications of Eliminating U.S. Energy Subsidies," in *Reforming Energy and Transport Subsidies* (Paris: OECD, 1997), pp. 79–81.

111. Joanne C. Burgess, "The Contribution of Efficient Energy Pricing to Reducing Carbon Dioxide Emissions," *Energy Policy* 18 (1990): 453–454.

112. Chris Flavin and Nicholas Lenssen, "Policies for a Solar Economy," *Energy Policy* 20 (1992): 248.

113. Paul Ekins, "The Secondary Benefits of CO_2 Abatement: How Much Emission Reduction Do They Justify?" *Ecological Economics* 16 (1996): 18, 20.

114. IPCC 1995, WGIII, p. 217.

115. Laurent Belsie, "No Pollution, and Power to Spare," *Christian Science Monitor,* August 23, 1999, p. 14. Toyota's operations in southern California get 100 percent of their electricity from renewables. This increases their utility bill by only 10–15 percent, which they will make up in efficiency improvements; Romm, *Cool Companies*, p. 131.

116. An alternative to the renewables portfolio standard is the system benefits charge. This policy lets the utilities generate power from whatever sources they want and levies a surcharge on their sales. The surcharge can then be spent by the regulatory agency on a variety of programs. The agency could subsidize renewables, it could fund energy efficiency programs such as insulating homes of low-income people or distributing compact fluorescent bulbs, and the like. From a conservative point of view, the renewables portfolio standard seems to be the better choice. The system benefits charge leaves a governmental agency in charge of selecting and administering projects. The renewables portfolio standard directly assures an increasing role for renewable energy—which is important to meet our long-term targets—and leaves it up to individual utilities and private generators, operating through the marketplace, to determine the best and most efficient way to meet the standard.For discussion of these two systems and their relative merits, see Alliance to Save Energy et al., *Energy Innovations*, pp. 38–40; Steve Bernow et al., "Quantifying the Impacts of a National, Tradable Renewables Portfolio Standard," *Electricity Journal,* May 1997, pp. 42–52; Bernow et al., "Integrated Approach to Climate Policy," pp. 380–381; Eric Hirst et al., "The Future of DSM in a Restructured U.S. Electricity Industry," *Energy Policy* 24 (1996): 311–312; Daniel Kirshner et al., "A Cost-Effective Renewables Policy Can Advance the Transition to Competition," *Electricity Journal,* January–February 1997, pp. 54–61; Nancy Rader and Richard Norgaard, "Efficiency and Sustainability in Restructured Electricity Markets: The Renewables Portfolio Standard," *Electricity Journal,* July 1996, pp. 37–49; Ryan Wiser et al., "Renewable Energy Policy and Electricity Restructuring: A California Case Study," *Energy Policy* 26 (1998): 465–475.

117. Lovins and Lovins, *Climate*, p. 12.

118. Arnold P. Fickett et al., "Efficient Use of Electricity," *Scientific American,* September 1990, p. 73.

119. Flanigan and Weintraub, "Most Successful DSM Programs"; see also Clark W. Gellings, "Then and Now: The Perspective of the Man Who Coined the Term 'DSM,'" *Energy Policy* 24 (1996): 286; Joseph Eto et al., "The Total Cost and Measured Performance of Utility-Sponsored Energy Efficiency Programs," *Energy Journal* 17 (1996): 31–51.

120. Eto et al., "Total Cost and Measured Performance," p. 49; Gellings, "Then and Now," p. 287; Hirst et al., "Future of DSM," p. 314; Kenneth M. Keating, "What Roles for Utility-Sponsored DSM in a Competitive Environment?" *Energy Policy* 24 (1996): 317–321; Lovins, "Great Negawatts Debate," p. 58.

121. International Energy Agency, *Voluntary Actions for Energy-Related CO_2 Abatement*, pp. 142–143.

122. DeCanio, "Barriers Within Firms," p. 911; Robin W. Gates, "Investing in Energy Conservation," *Energy Policy* 11 (1983): 71.

123. Steven M. Nadel et al., "A Review of U.S. and Canadian Lighting Programs for the Residential, Commercial, and Industrial Sectors," *Energy* 18 (1993): 153.

124. Mark D. Levine et al., "Energy Efficiency Policy and Market Failures," *Annual Review of Energy and the Environment* 20 (1995): 546.

125. Alliance to Save Energy et al., *Energy Innovations*, p. 108; Richard B. Howarth and Margrethe A. Winslow, "Energy Use and CO_2 Emissions Reduction: Integrating Pricing and Regulatory Policies," *Energy* 19 (1994): 860.

126. Alliance to Save Energy et al., *Energy Innovations*, p. 108. The charge is sometimes made that energy efficiency standards limit consumer choice. But, as Richard Howarth and Bo Andersson put it, "Well designed standards . . . would limit consumers' choices only in the trivial sense of eliminating options that no one would willingly exercise," if he or she had the necessary information and the ability to make the calculations; "Market Barriers to Energy Efficiency," *Energy Economics* 15 (1993): 271. Besides, renters and home buyers almost never have a choice.

127. Jack M. Hollander and Thomas R. Schneider, "Energy-Efficiency: Issues for the Decade," *Energy* 21 (1996): 282–284.

128. Steven Nadel and Howard Geller, "Utility DSM: What Have We Learned? Where Are We Going?" *Energy Policy* 24 (1996): 292.

129. Golove and Eto, *Market Barriers to Energy Efficiency*, p. 39.

130. "Designing Incentives," *Rocky Mountain Institute Newsletter*, Summer 1996, p. 4.

131. Howard Geller and Steven Nadel, "Market Transformation Strategies to Promote End-Use Efficiency," *Annual Review of Energy and the Environment* 19 (1994): 301–346; see p. 341 for a list of opportunities they have identified.

132. Geller and Nadel, "Market Transformation Strategies," p. 319.

133. DeCanio, "Barriers Within Firms," p. 906; L. Lovins et al., "Energy Policy," pp. 685–686.

134. World Wildlife Fund, information printed on the back of an envelope used in a mailing to members, September 1999.

135. William R. Moomaw, "Industrial Emissions of Greenhouse Gases," *Energy Policy* 24 (1996): 965.

136. Lovins and Lovins, *Climate*, p. 7.

137. Alliance to Save Energy et al., *Energy Innovations*, p. 95.

138. Lovins and Lovins, *Climate*, p. 25.

139. John DeCicco and Jason Mark, "Meeting the Energy and Climate Challenge for Transportation in the United States," *Energy Policy* 26 (1998): 395.

140. DeCicco and Mark, "Meeting the Energy and Climate Challenge," p. 396.

141. I also think that the current fad for ever bigger and more hideous trucks and grossly misnamed "sport utility vehicles" is the most ridiculous, most wasteful, and certainly the ugliest fad ever to plague our once fair land. (In case anyone wonders, my car is a 1995 Honda Civic EX coupe.)

142. DeCicco and Mark, "Meeting the Energy and Climate Challenge," p. 398.

143. Clifford W. Cobb, *The Roads Aren't Free: Estimating the Full Social Cost of Driving and the Effects of Accurate Pricing* (San Francisco: Redefining Progress, 1998), pp. 8–17. I have

subtracted his category for global warming because I will consider that externality separately.

144. James J. MacKenzie et al., *The Going Rate: What It Really Costs to Drive* (Washington, D.C.: World Resources Institute, 1992). Again, I have subtracted their estimate for greenhouse gas damages because I am considering that externality separately.

145. Cobb, *The Roads Aren't Free*, p. 25.

146. Carmen Difiglio, "Using Advanced Technologies to Reduce Motor Vehicle Greenhouse Gas Emissions," *Energy Policy* 25 (1997): 1177.

147. Laurie Michaelis and Ogunlade Davidson, "GHG Mitigation in the Transport Sector," *Energy Policy* 24 (1996): 980.

148. Eric C. Evarts, "Car Buyers Put Fuel Economy in Back Seat," *Christian Science Monitor*, October 8, 1999, p. 2. A higher CAFE standard would, of course, reduce emissions below a baseline drawn from current trends, so it would have some impact.

149. Difiglio, "Using Advanced Technologies," p. 1177; see also U.S. Department of Energy, Office of Policy, *Effects of Feebates on Vehicle Fuel Economy, Carbon Emissions, and Consumer Surplus*, DOE/PO-0031 (Washington, D.C.: Department of Energy, 1995).

150. Robert T. Watson et al., eds., *Climate Change 1995: Impacts, Adaptations and Mitigation of Climate Change: Scientific-Technical Analyses* (Cambridge: Cambridge University Press, 1996), p. 14. This is the report of Working Group II.

151. Geller and Nadel, "Market Transformation Strategies," p. 330.

152. Based on 1996 sales figures, *Statistical Abstract of the United States, 1998* (Washington, D.C.: Department of Commerce, 1998), p. 644, no. 1053. Of course, gasoline sales would go down significantly under full-cost pricing, but so would greenhouse gas emissions.

153. A study in Sacramento found that buildings with light-colored roofs use 40 percent less energy for air conditioning than those with darker roofs; Laura Gatland, "Chicago Rooftops: From Gravel and Tar to Greenery," *Christian Science Monitor*, April 15, 1999, p. 2; see also Arthur H. Rosenfeld et al., "Painting the Town White—and Green," *Technology Review*, February–March 1997, pp. 52–59.

154. International Dark-Sky Association, "Help Us Save the Stars" (brochure); <www.darksky.org>.

155. International Dark-Sky Association, "Summary of the City of San Diego Street Light Conversion Program," Information Sheet 13 (October 1996); <www.darksky.org>.

156. For many possibilities, see International Energy Agency, *Voluntary Actions for Energy-Related CO_2 Abatement.*

157. Geller and Nadel, "Market Transformation Strategies," p. 332.

158. Michel Colombier and Philippe Menanteau, "From Energy Labelling to Performance Standards: Some Methods of Stimulating Technical Change to Obtain Greater Energy Efficiency," *Energy Policy* 25 (1997): 432.

159. Hollander and Schneider, "Energy-Efficiency," p. 281.

160. Some economists have argued that if people reduce their energy consumption in one place, say, by buying more efficient appliances, they will use the savings to increase their energy consumption elsewhere. If a car gets better gas mileage, the owner will just drive it more. This is known as a "rebound effect" and the result will be little or no net reduction in total energy use. Empirical studies, however, have found that this effect is very small and that energy efficiency improvements do indeed result in a reduction of total energy use. See Amory B. Lovins, "Energy Saving Resulting from the Adoption of More Efficient Appliances: Another View," *Energy Journal* 9 (1988): 155–162; John Henley et al., "Energy Saving Resulting from the Adoption of More Efficient Appliances: A Follow-up," *Energy Journal* 9 (1988): 163–170; M. J. Grubb, "Energy Efficiency and Economic Fallacies," *Energy Policy* 18 (1990): 783–785; David L. Greene, "Vehicle Use and Fuel Economy: How Big Is the 'Rebound Effect'?" *Energy Journal* 13 (1992): 117–143; Michael Grubb, reply to Brookes, *Energy Policy* 20 (1992): 392–393; Richard B. Howarth, "Energy Efficiency and Economic Growth," *Contemporary Economic Policy* 15 (October 1997): 1–9; John A. Laitner, "Energy Efficiency: Rebounding to a Sound Analytical Perspective," *Energy Policy* 28 (2000): 471–475.

161. "Economists' Statement on Climate Change," c.1997, reprinted with permission, Redefining Progress; 1904 Franklin Street, 6th Floor, Oakland, CA 94612; (510) 444–3041; <www.rprogress.org>.

162. Manne and Richels, "CO_2 Emission Limits," p. 68; Nordhaus, "Rolling the 'DICE,'" p. 41.

163. Grubb et al., "Costs of Limiting Fossil-Fuel CO_2 Emissions," p. 423.

164. Schelling, "Some Economics of Global Warming," p. 8.

165. Nordhaus, "Rolling the 'DICE,'" pp. 41, 47–48.

166. Grubb et al., "Costs of Limiting Fossil-Fuel CO_2 Emissions," p. 471; see also DeCanio, *Economics of Climate Change*, p. 18; Howarth and Winslow, "Energy Use and CO_2 Emissions Reduction," p. 858; Darius W. Gaskins Jr. and John P. Weyant, "Model Comparisons of the Costs of Reducing CO_2 Emissions," *American Economic Review* 83 (1993): 320.

167. Wilson and Swisher, "Exploring the Gap," p. 251; see also IPCC 1995, WGIII, pp. 287–288; Grubb et al., "Costs of Limiting Fossil-Fuel CO_2 Emissions," pp. 448–454; Jean-Charles Hourcade and John Robinson, "Mitigating Factors: Assessing the Costs of Reducing GHG Emissions," *Energy Policy* 24 (1996): 866; Repetto and Austin, *Costs of Climate Protection*, p. 19.

168. Grubb et al., "Costs of Limiting Fossil-Fuel CO_2 Emissions," p. 456.

169. DeCanio, *Economics of Climate Change*, p. 25; Manne and Richels, "CO_2 Emission Limits," p. 73.

170. Lovins and Lovins, *Climate*, p. 25; emphasis in the original.

171. J. W. C. White, "Don't Touch That Dial," *Nature* 364 (1993): 186.

172. Lovins and Lovins, *Climate*, p. 2.

CHAPTER EIGHT

1. Stuart L. Pimm et al., "The Future of Biodiversity," *Science* 269 (1995): 347.

2. Paul R. Ehrlich and Edward O. Wilson, "Biodiversity Studies: Science and Policy," *Science* 253 (1991): 758.

3. Pimm et al., "Future of Biodiversity," p. 347.

4. Michael E. Soulé and M. A. Sanjayan, "Conservation Targets: Do They Help?" *Science* 279 (1998): 2060.

5. "Q&A with Norman Myers," *Focus*, March–April 1996, p. 6. *Focus* is a newsletter published by the World Wildlife Fund.

6. Brad Knickerbocker, "Noah's Ark or Nuisance?" *Christian Science Monitor*, September 26, 1995, p. 4.

7. Fish and Wildlife Service, "Box Score," <www.fws.gov>.

8. William Stoltzenburg, "Greener Acres," *Nature Conservancy*, November–December 1999, p. 8.

9. David S. Wilcove et al., "Quantifying Threats to Imperiled Species in the United States," *BioScience* 48 (1998): 608.

10. Wilcove et al., "Quantifying Threats," p. 612.

11. A. P. Dobson et al., "Geographic Distribution of Endangered Species in the United States," *Science* 275 (1997): 550–553.

12. Richard M. Weaver, *Ideas Have Consequences* (Chicago: University of Chicago Press, 1948), pp. 171–172.

13. Psalm 24:1.

14. Genesis 6:19–21.

15. Genesis 9:8–17.

16. Patricia Byrnes, "Wild Medicine," *Wilderness*, Fall 1995, p. 28. This article is based on an interview with Thomas Eisner (see notes 20–21).

17. Byrnes, "Wild Medicine," p. 30.

18. The Wilderness Society, *The Endangered Species Act: A Commitment Worth Keeping* (Washington, D.C.: Wilderness Society, n.d.), p. 10.

19. Wilderness Society, *Endangered Species Act*, p. 10.

20. Thomas Eisner, "The Hidden Value of Species Diversity," *BioScience* 42 (1992): 578.

21. Thomas Eisner et al., "Building a Scientifically Sound Policy for Protecting Endangered Species," *Science* 268 (1995): 1231.

22. Richard C. Bishop, "Endangered Species and Uncertainty: The Economics of a Safe Minimum Standard," *American Journal of Agricultural Economics* 60 (1978): 11.

23. Charles Perrings et al., "The Ecology and Economics of Biodiversity Loss: The Research Agenda," *Ambio* 21 (1992): 201; Brian H. Walker, "Biodiversity and Ecological Redundancy," *Conservation Biology* 6 (1992): 19.

24. Gretchen C. Daily et al., "Ecosystem Services: Benefits Supplied to Human Societies by Natural Ecosystems," *Issues in Ecology* 2 (1997): 1–16.

25. Robert Costanza et al., "The Value of the World's Ecosystem Services and Natural Capital," *Nature* 387 (1997): 253–260.

26. David Tilman and John A. Downing, "Biodiversity and Stability in Grasslands," *Nature* 367 (1994): 363, 365.

27. David Tilman et al., "Productivity and Sustainability Influenced by Biodiversity in Grassland Ecosystems," *Nature* 379 (1996): 718.

28. Shahid Naeem and Shibin Li, "Biodiversity Enhances Ecosystem Reliability," *Nature* 390 (1997): 508.

29. F. Stuart Chapin III et al., "Biotic Control over the Functioning of Ecosystems," *Science* 277 (1997): 500–503; Ehrlich and Wilson, "Biodiversity Studies"; Carl Folke et al., "Biological Diversity, Ecosystems, and the Human Scale," *Ecological Applications* 6 (1996): 1018–1024.

30. Quoted in Steve Stuebner, "Jack Ward Thomas: Hail to the Chief," *High Country News*, December 13, 1993, p. 10; emphasis in the original.

31. Paul Ehrlich and Anne Ehrlich, *Extinction* (New York: Random House, 1981), pp. xi-xiii.

32. David Jablonski, "Extinctions: A Paleontological Perspective," *Science* 253 (1991): 755.

33. Folke et al., "Biological Diversity," p. 1020.

34. Jablonski, "Extinctions," p. 755.

35. R. Edward Grumbine, "Reflections on 'What Is Ecosystem Management?'" *Conservation Biology* 11 (1997): 43; Richard C. Bishop, "Endangered Species: An Economic Perspective," *Transactions of the 45th North American Wildlife and Natural Resources Conference* 45 (1980): 216.

36. Aldo Leopold, *Round River: From the Journals of Aldo Leopold*, ed. Luna B. Leopold (Oxford: Oxford University Press, 1993), p. 147.

37. Zygmunt J. B. Plater, "The Embattled Social Utilities of the Endangered Species Act," *Environmental Law* 27 (1997): 854.

38. Patrick Parenteau, "Rearranging the Deck Chairs: Endangered Species Act Reforms in an Era of Mass Extinction," *William and Mary Environmental Law and Policy Review* 22 (1998): 240–241.

39. Gary R. Lingle, "Economic Impact of Crane-Watching," *Braided River* 4 (1992): 8. *Braided River* is a newsletter produced by the Platte River Trust.

40. Holmes Rolston III, "Property Rights and Endangered Species," *University of Colorado Law Review* 61 (1990): 303–304.

41. The force of this point should be the same for most conservatives, whether they accept the scientific theory of evolution or believe in creationism. The forms of life are the handwork of God, no matter what process He used to create them.

42. Edward O. Wilson, "Resolutions for the 80s," *Harvard Magazine*, January–February 1980, p. 21.

43. Folke et al., "Biological Diversity," p. 1020; Walter V. Reid, "Creating Incentives for Conserving Biodiversity," in *Building Economic Incentives into the Endangered Species Act*, ed. Hank Fischer and Wendy E. Hudson (Washington, D.C.: Defenders of Wildlife, 1994), p. 45; Alan Randall, "What Mainstream Economists Have to Say About the Value of Biodiversity," in *Biodiversity*, ed. E. O. Wilson and Frances M. Peter (Washington, D.C.: National Academy Press, 1988), p. 219.

44. John B. Loomis and Douglas S. White, "Economic Benefits of Rare and Endangered Species: Summary and Meta-analysis," *Ecological Economics* 18 (1996): 197.

45. Matthew J. Kotchen and Stephen D. Reiling, "Estimating and Questioning Economic Values for Endangered Species: An Application and Discussion," *Endangered Species UPDATE* 15, no. 5 (1998): 77–83.

46. David Quammen, "Planet of Weeds," *Harper's Magazine,* October 1998, pp. 68–69. This article is based on an interview with David Jablonski.

47. Parenteau, "Rearranging the Deck Chairs," p. 280.

48. Oliver A. Houck, "Reflections on the Endangered Species Act," *Natural Resources and Environment* 10 (Summer 1995): 10.

49. These were federal projects or ones that required a federal permit, both covered by section 7 of the ESA; see below.

50. "The national media never got beyond caricature"; Plater, "Embattled Social Utilities," p. 859.

51. Plater, "Embattled Social Utilities," pp. 858–859.

52. Parenteau, "Rearranging the Deck Chairs," p. 257.

53. David S. Wilcove et al., "What Exactly Is an Endangered Species? An Analysis of the U.S. Endangered Species List: 1985–1991," *Conservation Biology* 7 (1993): 87.

54. Wilcove et al., "What Exactly Is an Endangered Species?" p. 92.

55. For an excellent summary of the provisions of the ESA and the ways in which it is applied, see the much cited article by Oliver A. Houck, "The Endangered Species Act and Its Implementation by the U.S. Departments of Interior and Commerce," *University of Colorado Law Review* 64 (1993): 277–370. For developments since then, see J. B. Ruhl, "Who Needs Congress? An Agenda for Administrative Reform of the Endangered Species Act," *New York University Environmental Law Journal* 6 (1998): 367–410.

56. Houck, "Endangered Species Act," p. 286.

57. Gardner M. Brown Jr. and Jason F. Shogren, "Economics of the Endangered Species Act," *Journal of Economic Perspectives* 12 (1998): 6.

58. Wilderness Society, *Endangered Species Act,* p. 19.

59. General Accounting Office, *Endangered Species Act: Types and Number of Implementing Actions,* GAO/RCED–92–131BR (Washington, D.C.: GPO, 1992), p. 2.

60. Parenteau, "Rearranging the Deck Chairs," p. 263.

61. Laura C. Hood, *Frayed Safety Nets* (Washington, D.C.: Defenders of Wildlife, 1998), p. 54. According to FWS, as of September 30, 1999, there were 525 approved recovery plans. But some cover more than one species and a few species have separate recovery plans for different parts of their ranges. The FWS does not say how many there are in either case; FWS, "Box Score."

62. Michael O'Connell, Response to "Six Biological Reasons Why the Endangered Species Act Doesn't Work and What to Do About It," *Conservation Biology* 6 (1992): 141.

63. Robert D. Thornton, "The Search for a Conservation Planning Paradigm: Section 10 of the ESA," *Natural Resources and Environment* 8 (Summer 1993): 23.

64. GAO, *ESA: Types and Numbers of Implementing Actions,* p. 19.

65. GAO, *ESA: Types and Numbers of Implementing Actions,* p. 19.

66. Houck, "Endangered Species Act," p. 368; see also pp. 321, 359–370.

67. GAO, *ESA: Types and Numbers of Implementing Actions,* p. 19. See also Donald Barry and David Hoskins, *For Conserving Listed Species, Talk Is Cheaper Than We Think,* 2d ed. (Washington, D.C.: World Wildlife Fund, 1994).

68. General Accounting Office, *Endangered Species Act: Information on Species Protection on Nonfederal Lands,* GAO/RCED-95-16 (Washington, D.C.: GAO, 1994).

69. That interpretation has been upheld by the Supreme Court: *Babbitt v. Sweet Home Chapter of Communities for a Great Oregon,* 115 S. Ct. 2407 (1995).

70. FWS, "Box Score."

71. Reid, "Creating Incentives for Conserving Biodiversity," pp. 43–44.

72. Jeffrey J. Rachlinski, "Noah by the Numbers: An Empirical Evaluation of the Endangered Species Act," *Cornell Law Review* 82 (1997): 381.

73. John Kostyack, "The Need for HCP Reform: Five Points of Consensus," *Endangered Species UPDATE* 16, no. 3 (1999): 48.

74. For acceptance by the business community, see Donald C. Baur and Karen L. Donovan, "The No Surprises Policy: Contracts 101 Meets the Endangered Species Act," *Environmental Law* 27 (1997): 768, 781; Eric Fisher, "Habitat Conservation Planning Under the Endangered Species Act: No Surprises and the Quest for Certainty," *University of Colorado Law Review* 67 (1996): 386; Fred P. Bosselman, "The Statutory and Constitutional Mandate for a No Surprises Policy," *Ecology Law Quarterly* 24 (1997): 710.

75. Baur and Donovan, "No Surprises Policy," p. 768; Ira Michael Heyman, "Property Rights and the Endangered Species Act: A Renascent Assault on Land Use Regulation," *Pacific Law Journal* 25 (1994): 166; Bosselman, "Statutory and Constitutional Mandate," p. 719. Bosselman indicates that establishing this certainty can increase the value of the land (p. 710).

76. See, for example, Michael J. Bean and David S. Wilcove, "The Private Land Problem," *Conservation Biology* 11 (1997): 1–2 [Environmental Defense Fund]; John Kostyack, "Reshaping Habitat Conservation Plans for Species Recovery," *Environmental Law* 27 (1997): 757 [National Wildlife Federation]; Hood, *Frayed Safety Nets*, p. vii [Defenders of Wildlife]. Some environmentalists are skeptical of the idea and many criticize the laxity with which some plans have been drafted; see below.

77. Quoted in Oliver A. Houck, "On the Law of Biodiversity and Ecosystem Management," *Minnesota Law Review* 81 (1997): 966.

78. Houck, "On the Law of Biodiversity," pp. 967, 971.

79. Houck, "On the Law of Biodiversity," pp. 969–970.

80. For the information on NCCP, see Dwight Holing, "The Coastal Sage Scrub Solution," *Nature Conservancy*, July–August 1997, pp. 16–24; Douglas P. Wheeler, "An Ecosystem Approach to Species Protection," *Natural Resources and Environment* 10 (Winter 1996): 7–9; Jon Welner, "Natural Communities Conservation Planning: An Ecosystem Approach to Protecting Endangered Species," *Stanford Law Review* 47 (1995): 319–361; Lynn E. Dwyer et al., "Avoiding the Trainwreck: Observations from the Frontlines of Natural Community Conservation Planning in Southern California," *Endangered Species UPDATE* 12, no. 12 (1995): 5–7.

81. Welner, "Natural Communities Conservation Planning," p. 346; John M. Gaffin, "Can We Conserve California's Threatened Fisheries Through Natural Community Conservation Planning?" *Environmental Law* 27 (1997): 793.

82. Wheeler, "Ecosystem Approach to Species Protection," p. 9.

83. John A. Baden and Tim O'Brien, "Toward a True ESA: An Ecological Stewardship Act," in *Building Economic Incentives into the Endangered Species Act*, p. 96.

84. Current policies have also solved some other earlier problems with the ESA. To ensure that the best scientific information is used, FWS and NMFS findings are now peer reviewed. To help landowners, when a species is listed the agency now provides as detailed a list as possible of actions that will harm the species and actions that will not. Previously, landowners did not have such information and were left in the dark. In addition, recovery planning teams today include experts in social and economic issues as well as biologists. See Ruhl, "Who Needs Congress?" pp. 390–392; Lynn E. Dwyer et al., "Property Rights Case Law and the Challenge to the Endangered Species Act," *Conservation Biology* 9 (1995): 730.

85. J. B. Ruhl, "While the Cat's Asleep: The Making of the 'New' ESA," *Natural Resources and Environment* 12 (Winter 1998): 187.

86. Ruhl, "Who Needs Congress?" pp. 398–400.

87. Hood, *Frayed Safety Nets*, p. 52.

88. Bruce B. Bingham and Barry R. Noon, "Mitigation of Habitat 'Take': Application to Habitat Conservation Planning," *Conservation Biology* 11 (1997): 127.

89. Daniel A. Hall, "Using Habitat Conservation Plans to Implement the Endangered Species Act in Pacific Coast Forests," *Environmental Law* 27 (1997): 803.

90. Doug Honnold et al., "Habitat Conservation Plans and the Protection of Habitat: Reply to Bean and Wilcove," *Conservation Biology* 11 (1997): 298.

91. Hood, *Frayed Safety Nets*.

92. Timothy H. Tear et al., "Recovery Plans and the Endangered Species Act: Are Criticisms Supported by Data?" *Conservation Biology* 9 (1995): 182.

93. Timothy H. Tear et al., "Status and Prospects for Success of the Endangered Species Act: A Look at Recovery Plans," *Science* 262 (1993): 976.

94. Bosselman, "Statutory and Constitutional Mandate," p. 713.

95. Thornton, "Search for a Conservation Planning Paradigm," p. 23.

96. Brown and Shogren, "Economics of the Endangered Species Act," p. 7.

97. Larry McKinney, "Reauthorizing the Endangered Species Act—Incentives for Rural Landowners," in *Building Economic Incentives into the Endangered Species Act*, p. 74.

98. Karin P. Sheldon, "Habitat Conservation Planning: Addressing the Achilles Heel of the Endangered Species Act," *New York University Environmental Law Journal* 6 (1998): 281.

99. Parenteau, "Rearranging the Deck Chairs," p. 281.

100. John F. Turner and Jason C. Rylander, "Conserving Endangered Species on Private Lands," *Land and Water Law Review* 32 (1997): 572.

101. Plater, "Embattled Social Utilities," p. 868.

102. *Pennsylvania Coal Co. v. Mahon,* 260 U.S. 393, 413 (1922). "The Supreme Court has uniformly rejected the proposition that diminution in property value, standing alone, can establish a taking"; Susan Shaheen, "The Endangered Species Act: Inadequate Species Protection in the Wake of the Destruction of Private Property Rights," *Ohio State Law Journal* 55 (1994): 466. Property rights are not now, and never have been, absolute. However, this is not the place for a detailed analysis of these rights. Rather, as argued below, the primary concern here is with fairness.

103. Blaine I. Green, "The Endangered Species Act and Fifth Amendment Takings: Constitutional Limits of Species Protection," *Yale Journal on Regulation* 15 (1998): 329–385; Robert Meltz, "Where the Wild Things Are: The Endangered Species Act and Private Property," *Environmental Law* 24 (1994): 369–417.

104. Parenteau, "Rearranging the Deck Chairs," p. 282.

105. Parenteau, "Rearranging the Deck Chairs," p. 284.

106. Oliver A. Houck, "Reflections on the Endangered Species Act," *Environmental Law* 25 (1995): 693–697.

107. Elizabeth Losos et al., "Taxpayer-Subsidized Resource Extraction Harms Species," *BioScience* 45 (1995): 448.

108. Jeff Opperman, *The Impacts of Subsidies on Endangered Species,* Research Paper no. 36 (Oak Grove, Ore.: Thoreau Institute, 1996), pp. 4–6.

109. Opperman, *Impacts of Subsidies on Endangered Species*, pp. 11–12.

110. David S. Wilcove, "The Promise and the Disappointment of the Endangered Species Act," *New York University Environmental Law Journal* 6 (1998): 276.

111. Welner, "Natural Communities Conservation Planning," p. 326.

112. Rachlinski, "Noah by the Numbers," pp. 376–377.

113. Rachlinski, "Noah by the Numbers," p. 378.

114. Rachlinski, "Noah by the Numbers," p. 383.

115. Rachlinski, "Noah by the Numbers," p. 384.

116. Rachlinaki, "Noah by the Numbers," p. 386.

117. Parenteau, "Rearranging the Deck Chairs," p. 275.

118. David S. Wilcove et al., *Rebuilding the Ark* (New York: Environmental Defense Fund, 1996), p. 3.

119. Hank Fischer, introduction to *Building Economic Incentives into the Endangered Species Act*, p. vii.

120. The Nature Conservancy, 4245 North Fairfax Drive, Suite 100, Arlington, VA 22203; <www.tnc.org>.

121. Reed F. Noss, "Some Principles of Conservation Biology, as They Apply to Environmental Law," *Chicago-Kent Law Review* 69 (1994): 899. Politicians who advocate reducing habitat protection under the law are effectively saying they do not want endangered species protected. Of course, they do not want to admit that outright.

122. Daniel B. Wood, "Backlash Against Urban Sprawl Broadens," *Christian Science Monitor,* December 16, 1999, pp. 1, 4.

123. Bruce Babbitt, "The Endangered Species Act and 'Takings': A Call for Innovation Within the Terms of the Act," *Environmental Law* 24 (1994): 362.

124. Opperman, *Impacts of Subsidies on Endangered Species*; Turner and Rylander, "Conserving Endangered Species on Private Lands," pp. 596–597.

125. Jim McKinney et al., "Economic Incentives to Preserve Endangered Species Habitat and Biodiversity on Private Land," in *Building Economic Incentives into the Endangered Species Act,* p. 12.

126. Opperman, *Impacts of Subsidies on Endangered Species,* p. 24.

127. Houck, "On the Law of Biodiversity." The thesis of this article is that the specific protection the ESA provides for individual species is critical because it provides a hard limit on agencies' discretion, below which they are not allowed to go.

128. Opperman, *Impacts of Subsidies on Endangered Species,* p. 9.

129. Wilcove et al., *Rebuilding the Ark,* p. 1.

130. See, for example, Fischer and Hudson, *Building Economic Incentives into the Endangered Species Act;* and *The Keystone Dialogue on Incentives for Private Landowners to Protect Endangered Species, Final Report* (Keystone, Colo.: Keystone Center, 1995).

131. Babbitt, "Endangered Species Act and 'Takings,'" p. 365.

132. Brett Schaerer, *Incentives for Wildlife,* Research Paper no. 35 (Oak Grove, Ore.: Thoreau Institute, 1996), p. 19.

133. McKinney, "Reauthorizing the Endangered Species Act," p. 75; Schaerer, *Incentives for Wildlife,* p. 10; Wilcove et al., *Rebuilding the Ark,* p. 12; *Keystone Dialogue,* pp. 26–33.

134. *Keystone Dialogue,* pp. 33–35.

135. For TDRs and their possible application, see James T. B. Tripp and Daniel J. Dudek, "Institutional Guidelines for Designing Successful Transferable Rights Programs," *Yale Journal on Regulation* 6 (1989): 369–391; Robert Bonnie and Michael Bean, "Habitat Trading for Red Cockaded Woodpeckers: Enhancing Recovery, Reducing Conflicts," *Endangered Species UPDATE* 13, no. 4–5 (1996): 7–9; Jon H. Goldstein and H. Theodore Heintz Jr., "Incentives for Private Conservation of Species and Habitat: An Economic Perspective," in *Building Economic Incentives into the Endangered Species Act,* pp. 55–58; Thornton, "Search for a Conservation Planning Paradigm," pp. 65–66; Elizabeth T. Kennedy et al., "Economic Incentives," *Journal of Forestry,* April 1996, pp. 22–26; Julian Conrad Juergensmeyer et al., "Transferable Development Rights and Alternatives After *Suitum,*" *Urban Lawyer* 30 (1998): 441–475.

136. Robert J. Lilieholm and Jeff Romm, "Pinelands National Reserve: An Intergovernmental Approach to Nature Preservation," *Environmental Management* 16 (1992): 335–343; see also Tripp and Dudek, "Institutional Guidelines," pp. 378–382; Juergensmeyer et al., "Transferable Development Rights," pp. 448–450.

137. For mitigation banking, see Royal C. Gardner, "Banking on Entrepreneurs: Wetlands, Mitigation Banking, and Takings," *Iowa Law Review* 81 (1996): 527–587; Goldstein and Heintz, "Incentives for Private Conservation," pp. 58–60; Lawrence R. Liebesman and David M. Platt, "The Emergence of Private Wetlands Mitigation Banking," *Natural Resources and Environment* 13 (1998): 341–344, 370–371.

138. For example, Randal O'Toole, "Proposal #1: Create a Biodiversity Trust Fund," *Different Drummer: The Endangered Endangered Species Act* 3 (Winter 1996): 52.

139. Goldstein and Heintz, "Incentives for Private Conservation," p. 61.

140. For mitigation fees, see Thomas W. Ledman, "Local Government Environmental Mitigation Fees: Development Exactions, the Next Generation," *Florida Law Review* 45 (1993): 835–871; Bosselman, "Satutory and Constitutional Mandate," pp. 726–728. Fees in HCPs vary widely, depending on the costs of mitigation. Some are only nominal; see Hood, *Frayed Safety Nets,* p. 48.

141. Thomas R. Bourland and Richard L. Stroup, "Rent Payments as Incentives," *Journal of Forestry,* April 1996, pp. 18–21; Michael Bean, "Incentive-Based Approaches to Conserving Red-Cockaded Woodpeckers in the Sandhills of North Carolina," in *Building Economic Incentives*

into the Endangered Species Act, pp. 23–25; Wilcove et al., *Rebuilding the Ark*, p. 13; Turner and Rylander, "Conserving Endangered Species on Private Land," p. 620.

142. Curtis H. Flather et al., "Threatened and Endangered Species Geography," *BioScience* 48 (1998): 365–376; Ruhl, "Who Needs Congress?" pp. 402–404.

143. Dobson et al., "Geographic Distribution of Endangered Species," p. 533.

144. Terry L. Anderson and Jody J. Olsen, "Positive Incentives for Saving Endangered Species," in *Building Economic Incentives into the Endangered Species Act*, pp. 110–111.

145. *Keystone Dialogue*, p. 11.

146. Meltz, "Where the Wild Things Are," p. 417. And we do not tax away windfall profits added to property by governmental actions, such as building roads and sewers and schools, awarding huge military procurement contracts in the area, etc.

147. Schaerer, *Incentives for Wildlife*, p. 22; Turner and Rylander, "Conserving Endangered Species on Private Lands," p. 604.

148. Russell Kirk, "Man Undoing Noah's Preservation Program," *New Orleans Times-Picayune*, June 1, 1965, sec. 1, p. 9.

149. Quoted in Parenteau, "Rearranging the Deck Chairs," p. 310.

150. Wilcove et al., *Rebuilding the Ark*, p. 16.

CHAPTER NINE

1. World Commission on Environment and Development, *Our Common Future* (Oxford: Oxford University Press, 1987), p. 43.

2. Russell Kirk, *Prospects for Conservatives* (Washington, D.C.: Regnery Gateway, 1989), p. 81.

3. *Rocky Mountain Institute Newsletter*, Spring 1994, p. 7; emphasis in the original.

4. Quoted in Albert Gore, *Earth in the Balance* (Boston: Houghton Mifflin, 1992), p. 191.

5. Kirk, *Prospects for Conservatives*, p. 173.

6. Garrett Hardin, "The Tragedy of the Commons," *Science* 162 (1968): 1243–1248.

7. Simon Fairlie et al., "The Politics of Overfishing," *Ecologist* 25, no. 2-3 (1995): 60.

8. Fairlie et al., "Politics of Overfishing," pp. 49–50, 60–61.

9. Terry L. Anderson and Donald R. Leal, *Free Market Environmentalism* (San Francisco: Pacific Research Institute for Public Policy, 1991), p. 124.

10. John Gray, *Beyond the New Right* (London: Routledge, 1993), p. 129.

11. Colin W. Clark, "Bioeconomics of the Ocean," *BioScience* 31 (1981): 231.

12. The editors, "Introduction," *Ecologist* 25, no. 2–3 (1995): 42.

13. Colin Woodard, "Troubles Bubble Under the Sea," *Christian Science Monitor*, September 10, 1997, p. 1.

14. Sam Walker, "Georges Bank Closes, Ending an Era," *Christian Science Monitor*, December 12, 1994, p. 1.

15. Todd Wilkinson, "Clinton's Green Agenda Goes Under the Sea," *Christian Science Monitor*, June 5, 2000, p. 3.

16. Walker, "Georges Bank Closes," p. 1.

17. Elisabetta Coletti, "In New England, No More Young Men and the Sea," *Christian Science Monitor*, April 15, 1999, p. 4.

18. Brad Knickerbocker, "World Dawdles as Its Fisheries Decline," *Christian Science Monitor*, September 23, 1998, p. 4.

19. Knickerbocker, "World Dawdles as Its Fisheries Decline," p. 4; Carl Safina, "The World's Imperiled Fish," *Scientific American*, November 1995, p. 50.

20. Edward Carr, "The Deep Green Sea (Survey)," *Economist*, May 23, 1998, p. 11.

21. Robert J. McManus, "America's Saltwater Fisheries: So Few Fish, So Many Fishermen," *Natural Resources and Environment* 9 (1995): 16.

22. Tatiana Brailovskaya, "Obstacles to Protecting Marine Biodiversity Through Marine Wilderness Preservation: Examples from the New England Region," *Conservation Biology* 12 (1998): 1239.

23. Carr, "Deep Green Sea," p. 12.

24. Peter Shelley et al., "The New England Fisheries Crisis: What Have We Learned?" *Tulane Environmental Law Journal* 9 (1996): 242.

25. Bob Holmes, "Biologists Sort the Lessons of Fisheries Collapse," *Science* 264 (1994): 1253.

26. Quoted in Holmes, "Biologists Sort the Lessons of Fisheries Collapse," p. 1252.

27. A. A. Rosenberg et al., "Achieving Sustainable Use of Renewable Resources," *Science* 262 (1993): 828.

28. Susan Pollack, "The Lobster Trap," *Sierra*, July–August 1998, p. 71.

29. Safina, "World's Imperiled Fish," p. 53.

30. Safina, "World's Imperiled Fish," p. 52.

31. Clark, "Bioeconomics of the Ocean," p. 234.

32. Council of Economic Advisers, *Economic Report of the President 1996* (Washington, D.C.: Government Printing Office, 1996), p. 151.

33. Peter H. Pearse, "Developing Property Rights as Instruments of Natural Resources Policy: The Case of Fisheries," in *OECD Documents: Climate Change: Designing a Tradeable Permit System* (Paris: OECD, 1992), p. 118.

34. "Financial Trawling," *Economist*, November 21, 1998, p. 59.

35. "Financial Trawling," p. 59.

36. David Malin Roodman, "Harnessing the Market for the Environment," in *State of the World 1996*, ed. Linda Starke (New York: Norton, 1996), p. 180.

37. Colin W. Clark and Gordon R. Munro, "Renewable Resources as Natural Capital: The Fishery," in *Investing in Natural Capital*, ed. AnnMari Jansson et al. (Washington, D.C.: Island, 1994), p. 358; Peter Weber, "Protecting Oceanic Fisheries and Jobs," in *State of the World 1995*, ed. Linda Starke (New York: Norton, 1995), pp. 32–33.

38. Fairlie et al., "Politics of Overfishing," p. 70.

39. Pollack, "Lobster Trap," p. 48.

40. On the destruction trawling can cause, see Peter J. Auster, "A Conceptual Model of the Impacts of Fishing Gear on the Integrity of Fish Habitats," *Conservation Biology* 12 (1998): 1198–1203; Paul K. Dayton, "Reversal of the Burden of Proof in Fisheries Management," *Science* 279 (1998): 821–822; Les Watling and Elliott A. Norse, "Disturbance of the Seabed by Mobile Fishing Gear: A Comparison to Forest Clearcutting," *Conservation Biology* 12 (1998): 1180–1197.

41. Louis W. Botsford et al., "The Management of Fisheries and Marine Ecosystems," *Science* 277 (1997): 514; Robert Costanza et al., "Principles for Sustainable Governance of the Oceans," *Science* 281 (1998): 198; Dayton, "Reversal of the Burden of Proof," p. 822.

42. Roodman, "Harnessing the Market," p. 180.

43. Gary E. Davis, "Seeking Sanctuaries," *National Parks*, November–December 1998, p. 42; see also Nigel Williams, "Overfishing Disrupts Entire Ecosystems," *Science* 279 (1998): 809; and Callum M. Roberts, "Rapid Build-up of Fish Biomass in a Caribbean Marine Reserve," *Conservation Biology* 9 (1995): 815–826.

44. "Helping Consumers Protect Ocean Resources," *Focus*, July–August 1999, pp. 1, 6. *Focus* is a newsletter published by the World Wildlife Fund.

45. "News in Brief," *Christian Science Monitor*, April 22, 1999, p. 24.

46. Robert Costanza, "Assuring Sustainability of Ecological Economic Systems," in *Ecological Economics* (New York: Columbia University Press, 1991), p. 333.

47. Costanza, "Assuring Sustainability," p. 333.

48. Kenneth Arrow et al., "Economic Growth, Carrying Capacity, and the Environment," *Science* 268 (1995): 520.

49. Dennis M. King, "Can We Justify Sustainability? New Challenges Facing Ecological Economics," in *Investing in Natural Capital*, pp. 331, 334.

50. Gretchen C. Daily et al., "Ecosystem Services: Benefits Supplied to Human Societies by Natural Ecosystems," *Issues in Ecology* 2 (Spring 1997): 2.

51. Robert Costanza et al., "The Value of the World's Ecosystem Services and Natural Capital," *Nature* 387 (1997): 253–260. This is the midrange figure; the range is from $16 to $54 trillion. Of course, "the economies of the Earth would grind to a halt without the services of

ecological life-support systems, so in one sense their total value to the economy is infinite"; p. 253. In the original article, they gave an incorrect figure for total global production of $18 trillion. At that time it was actually about $25 trillion. They corrected the figure later, in opening a symposium on the study and its methodology; Robert Costanza et al., "The Value of Ecosystem Services: Putting the Issues in Perspective," *Ecological Economics* 25 (1998): 69.

52. Robert Costanza et al., "Valuation and Management of Wetland Ecosystems," *Ecological Economics* 1 (1989): 339–340; emphasis in the original.

53. Janet N. Abramovitz, "Valuing Nature's Services," in *State of the World 1997,* ed. Linda Starke (New York: Norton, 1997), p. 95.

54. King, "Can We Justify Sustainability?" pp. 329–331.

55. Quoted in Donella H. Meadows et al., *Beyond the Limits* (Mills, Vt.: Chelsea Green, 1992), pp. 187–188.

56. Colin W. Clark, "Profit Maximization and the Extinction of Animal Species," *Journal of Political Economy* 81 (1973): 950–961; Colin W. Clark, "The Economics of Overexploitation," *Science* 181 (1973): 630–634.

57. Stephan Schmidheiny "Greener Greenbacks: Financiers Go Eco-efficient," *Christian Science Monitor,* March 14, 1996, p. 19; emphasis in the original.

58. Daily et al., "Ecosystem Services," p. 1.

59. Gretchen C. Daily, "Valuing and Safeguarding Earth's Life-Support Systems," in *Nature's Services* (Washington, D.C.: Island, 1997), p. 372.

60. See, for example, Michael A. Toman, "Economics and 'Sustainability': Balancing Trade-offs and Imperatives," *Land Economics* 70 (1994): 399–413.

61. Harvey Wasserman, "Cane Mutiny," *The Nation,* March 11, 1996, pp. 6–7; James Bovard, "Your Taxes Go to Sweetener but Sour the Everglades," *Christian Science Monitor,* February 14, 1994, p. 22.

62. Elizabeth Culotta, "Bringing Back the Everglades," *Science* 268 (1995): 1689.

63. Peter Katel, "Letting the Water Run into 'Big Sugar's' Bowl," *Newsweek,* March 4, 1996, p. 56.

64. Culotta, "Bringing Back the Everglades," p. 1688.

65. Warren Richey, "Farmers Curb Runoff, Help Everglades," *Christian Science Monitor,* November 10, 1997, p. 4.

66. Laura Helmuth, "Can This Swamp Be Saved?" *Science News* 155 (1999): 252–254; Culotta, "Bringing Back the Everglades," pp. 1688–1690.

67. Sheila Polson, "Fish May Be Winners Nationwide in Maine Dam Dispute," *Christian Science Monitor,* January 27, 1998, p. 12; "FERC Says Dam Must Go," *Outdoor News Bulletin,* December 12, 1997, pp. 3–4; Yvonne Zipp, "With a Dam's Demise, Hope for Reviving Rivers," *Christian Science Monitor,* July 2, 1999, pp. 1, 3. The *Outdoor News Bulletin* is a newsletter published by the Wildlife Management Institute.

68. "A Good Year for Alewives," *Economist,* July 29, 2000, p. 32.

69. Brad Knickerbocker, "U.S. Busts Up Dams in Bid to Help Fish," *Christian Science Monitor,* July 15, 1998, p. 3; Reed McManus, "Down Come the Dams," *Sierra,* May–June 1998, p. 16; *Dam Removal Success Stories* (American Rivers/Friends of the Earth/Trout Unlimited, 1999).

70. Knickerbocker, "U.S. Busts Up Dams," p. 3.

71. McManus, "Down Come the Dams," p. 17.

72. David James Duncan, "Salmon's Second Coming," *Sierra,* March–April 2000, p. 36.

73. Bob Schildgen, "Unnatural Disasters," *Sierra,* May–June 1999, p. 50.

74. Richard E. Sparks, "Need for Ecosystem Management of Large Rivers and Their Floodplains," *BioScience* 45 (1995): 173.

75. Peter B. Bayley, "Understanding Large River-Floodplain Ecosystems," *BioScience* 45 (1995): 153.

76. Schildgen, "Unnatural Disasters," p. 51.

77. Ken Bouc, "Missouri River Restoration," *Nebraskaland,* March 1998, p. 21.

78. Donald L. Hey and Nancy S. Philippi, "Flood Reduction Through Wetland Restoration: The Upper Mississippi River Basin as a Case History," *Restoration Ecology* 3 (1995): 5.

79. Zygmunt J. B. Plater, "Environmental Law as a Mirror of the Future: Civic Values Confronting Market Force Dynamics in a Time of Counter-Revolution," *Boston College Environmental Affairs Law Review* 23 (1996): 749–750.

80. Hey and Philippi, "Flood Reduction Through Wetland Restoration."

81. Hey and Philippi, "Flood Reduction Through Wetland Restoration," pp. 15–16.

82. Bouc, "Missouri River Restoration."

83. Laurent Belsie, "How a Flood Remade a Flood," *Christian Science Monitor,* December 3, 1998, p. 14.

84. Schildgen, "Unnatural Disasters," p. 56.

85. For a good analysis of the economic issues, see Michael Jacobs, *The Green Economy* (London: Pluto, 1991).

CHAPTER TEN

1. Nolan Clark, "The Environmental Protection Agency"; and Gordon S. Jones, "The Department of the Interior," in *Mandate for Leadership III: Policy Strategies for the 1990s,* ed. Charles L. Heatherly and Burton Yale Pines (Washington, D.C.: Heritage Foundation, 1989), pp. 218, 213, 322.

2. Robert Stavins and Thomas Grumbly, "The Greening of the Market: Making the Polluter Pay," in *Mandate for Change,* ed. Will Marshall and Martin Schram (New York: Berkley, 1993), p. 197.

3. David Malin Roodman, "Harnessing the Market for the Environment," in *State of the World 1996,* ed. Linda Starke (New York: Norton, 1996), pp. 168–169.

4. Terry L. Anderson, "Enviro-Capitalism vs. Enviro-Socialism," *Kansas Journal of Law and Public Policy* 4 (Winter 1995): 38.

5. John A. Baden and Richard L. Stroup, "The Environmental Costs of Government Action," *Policy Review* 4 (Spring 1978): 36; emphasis in the original. Stroup seems to have become more dogmatic in recent years.

6. See, for example, Terry L. Anderson and Donald R. Leal, *Enviro-Capitalists: Doing Good While Doing Well* (Lanham, Md.: Rowman & Littlefield, 1997); *PERC Reports,* December 1998, special edition devoted to stories of "enviro-capitalists"; and the regular section of *PERC Reports,* "Greener Pastures: Private Initiatives." PERC (Political Economy Research Center) is a libertarian organization in Bozeman, Montana, devoted to research in free market environmentalism.

7. Jane S. Shaw and Richard L. Stroup, "Not the Free Market," *PERC Reports,* June 1998, p. 6.

8. Terry L. Anderson and Donald R. Leal, "Free Market Versus Political Environmentalism," *Harvard Journal of Law and Public Policy* 15 (1992): 308–309. This issue contains a symposium on free market environmentalism.

9. This point also raises the issue of distribution of wealth which, to the best of my knowledge, the free market environmentalists never discuss. They simply assume that the current distribution of wealth is fine. This is an especially important problem in the preservation of nature because the corporate world has far more money than environmentalists ever will. Moreover, corporations use their money to buy resources for the purpose of making still more money from them. Environmentalists who want to preserve habitat have to buy it with only very limited opportunities for it to pay back the investments. As noted above, the habitat that land trusts acquire is normally removed from the profitmaking sector of our economy. Anderson and Leal contend that the major environmental organizations could do much more to preserve land because they have substantial budgets; their annual total was over $400 million, which could buy a lot of conservation easements; Terry L. Anderson and Donald R. Leal, *Free Market Environmentalism* (San Francisco: Pacific Research Institute for Public Policy, 1991), pp. 93–94. (The budget figures are from the late 1980s.) But that is like looking at the total gross income of, say, Exxon and remarking how much pollution control equipment that would buy. Just as Exxon has employees to pay and materials to buy, so do the major environmental organizations. Only The

Nature Conservancy has land preservation as its primary function, and it also spends a significant amount on basic scientific research. And $400 million per year is a mere drop in the bucket when compared to the resources of industry. At the same time, Exxon and Mobil together made more than that *every single day*; William Funk, "Free Market Environmentalism: Wonder Drug or Snake Oil?" *Harvard Journal of Law and Public Policy* 15 (1992): 514. A realistic assessment of their resources leads to the conclusion that environmental organizations could be easily outbid by any major corporation for virtually any given piece of land. Moreover, Anderson and Leal object to the common practice whereby The Nature Conservancy turns its acquisitions over to the government; *Free Market Environmentalism*, p. 8. A typical transaction is to purchase some important habitat adjoining a national wildlife refuge and then recover the costs by selling the land to the federal government, which buys it out of appropriations from the Land and Water Conservation Fund. That may be heretical in terms of libertarian dogma, but if The Nature Conservancy had not done that, it could never have preserved anywhere close to 12 million acres.

10. Anderson and Leal, *Free Market Environmentalism*, p. 3.

11. Anderson and Leal are not consistent; elsewhere they claim that depletion of a renewable resource can "make good economic sense"; *Free Market Environmentalism*, p. 114. But this violates the fundamental standards of stewardship and economic sustainability. Anderson and Leal, by the way, mistake sustainable development to mean governmental control and regulation of everything; *Free Market Environmentalism*, pp. 167–172. It most certainly does not. Sustainability is simply the goal and does not imply any particular policies to meet it.

12. As Thomas Michael Power and Paul Rauber point out, the belief that private ownership conserves natural resources "is hard to believe for anyone who has ever flown over the Pacific Northwest and seen the checkerboard of clearcut private lands next to still-intact bits of public forests, or peered beyond the beauty strips in Maine"; "The Price of Everything," *Sierra,* November–December 1993, p. 94.

13. Anderson and Leal, *Free Market Environmentalism*, chap. 9.

14. Anderson and Leal, *Free Market Environmentalism*, p. 3.

15. Anderson, "Enviro-Capitalism vs. Enviro-Socialism," p. 38.

16. Quoted in Andrew McFee Thompson, "Free Market Environmentalism and the Common Law: Confusion, Nostalgia, and Inconsistency," *Emory Law Journal* 45 (1996): 1331.

17. For example, Anderson, "Enviro-Capitalism vs. Enviro-Socialism," p. 39.

18. Thompson, "Free Market Environmentalism and the Common Law"; Ronald J. Rychlak, "Common-Law Remedies for Environmental Wrongs: The Role of Private Nuisance," *Mississippi Law Journal* 59 (1989): 657–698; David R. Hodas, "Private Actions for Public Nuisance: Common Law Citizen Suits for Relief from Environmental Harm," *Ecology Law Quarterly* 16 (1989): 883–908.

19. Michael C. Blumm, "The Fallacies of Free Market Environmentalism," *Harvard Journal of Law and Public Policy* 15 (1992): 377.

20. Edward Brunet, "Debunking Wholesale Private Enforcement of Environmental Rights," *Harvard Journal of Law and Public Policy* 15 (1992): 321.

21. Peter S. Menell, "Institutional Fantasylands: From Scientific Management to Free Market Environmentalism," *Harvard Journal of Law and Public Policy* 15 (1992): 504.

22. Blumm, "Fallacies of Free Market Environmentalism," p. 385.

23. Brunet, "Debunking Wholesale Private Enforcement," p. 313.

24. Brunet, "Debunking Wholesale Private Enforcement," p. 316.

25. Peter S. Menell, "The Limitations of Legal Institutions for Addressing Environmental Risks," *Journal of Economic Perspectives* 5 (1991): 100.

26. Brunet, "Debunking Wholesale Private Enforcement," p. 319.

27. Brunet, "Debunking Wholesale Private Enforcement," p. 314; Menell, "Limitations of Legal Institutions," pp. 95–100.

28. Thompson, "Free Market Environmentalism and the Common Law," p. 1360.

29. Menell, "Limitations of Legal Institutions," p. 103.

30. Quoted in Thompson, "Free Market Environmentalism and the Common Law," p. 1362; see also Menell, "Institutional Fantasylands," p. 505.

31. Peter Huber, "Panel II: Public Versus Private Environmental Regulation," *Ecology Law Quarterly* 21 (1994): 454.

32. Roger E. Meiners and Bruce Yandle, "Curbing Pollution—Case-By-Case," *PERC Reports*, June 1998, p. 9.

33. Thompson, "Free Market Environmentalism and the Common Law," p. 1358.

34. Brunet, "Debunking Wholesale Private Enforcement," p. 315.

35. Roger E. Meiners and Bruce Yandle, *The Common Law: How It Protects the Environment*, PERC Policy Series PS-13 (Bozeman, Mont.: PERC, 1998), pp. 18–19.

36. Meiners and Yandle, *Common Law*, p. 25.

37. For example, Meiners and Yandle, *Common Law*, p. 25.

38. In less ideological moods, some libertarians realize that there are problems that are not amenable to property rights solutions. Regulation, they concede, still has a place in environmental policy. See, for example, Anderson and Leal, *Free Market Environmentalism*, p. 22.

39. Tibor R. Machan, "Pollution and Political Theory," in *Earthbound*, ed. Tom Regan (New York: Random House, 1984), pp. 97–99.

40. For example, Anderson and Leal, *Free Market Environmentalism*, p. 147.

41. For example, Anderson, "Enviro-Capitalism vs. Enviro-Socialism," p. 39.

42. For example, Anderson and Leal, *Free Market Environmentalism*, pp. 16, 22, 40, 42; Richard L. Stroup and John A. Baden, *Natural Resources* (San Francisco: Pacific Institute for Public Policy Research, 1983), p. 87.

43. Anderson and Leal, *Free Market Environmentalism*, p. 22.

44. Carl Pope, "What's Wrong with FME?" *PERC Reports*, March 1998, p. 18.

45. Anderson and Leal, *Free Market Environmentalism*, pp. 4–5.

46. Anderson and Leal, *Free Market Environmentalism*, p. 11.

47. Anderson and Leal, *Free Market Environmentalism*, p. 47.

48. Russell Kirk, *The Conservative Mind*, 7th ed. (Chicago: Regnery Books, 1986), pp. 44–45.

49. Anderson and Leal, *Free Market Environmentalism*, p. 4.

50. Anderson and Leal, *Free Market Environmentalism*, p. 16.

51. See Chapter 5.

52. Anderson and Leal, *Free Market Environmentalism*, p. 22.

53. Anderson and Leal, *Free Market Environmentalism*, p. 15.

54. Anderson and Leal, *Free Market Environmentalism*, p. 14.

55. Anderson and Leal, *Free Market Environmentalism*, chap. 6.

56. At one point, Anderson and Leal seem to realize that there may be a trade-off between recreational use and wildlife habitat; *Free Market Environmentalism*, p. 11.

57. Anderson and Leal, *Free Market Environmentalism*, p. 22.

58. Peter H. Morrison et al., *Ancient Forests in the Pacific Northwest* (Washington, D.C.: Wilderness Society, 1991).

59. John Gray, *Beyond the New Right* (London: Routledge, 1993), p. xi.

60. Gray, *Beyond the New Right*, p. 63.

61. Gray, *Beyond the New Right*, p. 62.

62. Gray, *Beyond the New Right*, p. 126.

63. Gray, *Beyond the New Right*, p. 63.

CONCLUDING THOUGHTS

1. Erich Pica, ed., *Green Scissors 2000* (Washington, D.C.: Friends of the Earth, 2000).

2. Donald Worster, *The Wealth of Nature* (New York: Oxford University Press, 1993), pp. 218–219.

3. Richard M. Weaver, *The Southern Tradition at Bay*, ed. George Core and M. E. Bradford (New Rochelle, N.Y.: Arlington House, 1968), p. 391.

4. Russell Kirk, *Prospects for Conservatives* (Washington, D.C.: Regnery Gateway, 1989), pp. 260–261.

INDEX